From Temples to Garden Estates and Academies

How did the temple-scape of Suzhou transform into garden-scape during the 13th to 16th centuries? This book investigates the landscape evolution of China during the Yuan-Ming socio-political transformation and provides a previously unrepresented Buddhist and hydraulic history of Chinese gardens.

Utilising historical GIS mapping and 3D architectural modelling, the research uncovers spatial strategies employed by monks, the gentry, and the government. It argues that by seizing hydraulic estates from Buddhists, the gentry and government rose as a landed class and cultural elites, establishing the literati garden. Yuan temples featured a mound-field-river topography which reclaimed urban wetlands. In temples such as Shizi Lin, scenes were crafted as Chan gong'ans to support Buddhist practice towards enlightenment, underpinning a monastic architectural transformation in the 14th century. In the Ming dynasty, the gentry and government exploited the hydro-topography while erasing Buddhist traces by converting temples into gardens. The book forms a theoretical model of triangulating the garden practice with socio-hydro spaces and advocates a new garden history enabled by historical mapping and modelling.

This book is aimed at students and scholars of East Asian history, temples and gardens, and landscape architecture and design.

Pania Yanjie Mu is a postdoctoral research associate at the Institute of East Asian Art History, Heidelberg University, Germany, and co-founder of P. M. Architects. She received her PhD from the Faculty of Architecture, the University of Hong Kong.

Routledge Research on Gardens in History
Series Editors:
Karen Jones
University of Kent, UK
James Beattie
Victoria University of Wellington, New Zealand

Planting a City in the Tropical Andes
Plants and People in Bogotá, 1880 to 1920
Diego Molina

From Temples to Garden Estates and Academies
Landscape Transformation of Suzhou During the 13th–16th Centuries and Beyond
Pania Yanjie Mu

From Temples to Garden Estates and Academies

Landscape Transformation of Suzhou During the 13th–16th Centuries and Beyond

Pania Yanjie Mu

NEW YORK AND LONDON

First published 2025
by Routledge
605 Third Avenue, New York, NY 10158

and by Routledge
4 Park Square, Milton Park, Abingdon, Oxon, OX14 4RN

Routledge is an imprint of the Taylor & Francis Group, an informa business

© 2025 Pania Yanjie Mu

The right of Pania Yanjie Mu to be identified as author of this work has been asserted in accordance with sections 77 and 78 of the Copyright, Designs and Patents Act 1988.

All rights reserved. No part of this book may be reprinted or reproduced or utilised in any form or by any electronic, mechanical, or other means, now known or hereafter invented, including photocopying and recording, or in any information storage or retrieval system, without permission in writing from the publishers.

Trademark notice: Product or corporate names may be trademarks or registered trademarks, and are used only for identification and explanation without intent to infringe.

ISBN: 978-1-032-48057-2 (hbk)
ISBN: 978-1-032-48058-9 (pbk)
ISBN: 978-1-003-38716-9 (ebk)

DOI: 10.4324/9781003387169

Typeset in Sabon
by Deanta Global Publishing Services, Chennai, India

To my soulmate Maolin, who loves research and believes in me.

Contents

x   *Contents*

Figures

Tables

Acknowledgements

Many people I am grateful to upon the publication of my first book. I would like to first express my heartfelt thanks to Duncan Campbell for his recommendation of my research and consistent support during the manuscript's revision. When reviewing my PhD thesis as a draft for the book, he suggested a critical reflection on the key terms used, such as 'gentry' and 'literati,' and working towards a garden history which could enrich sinology. He also emphasised a meticulous translation of classical texts, leading to my full translation of all cited classical texts in the endnotes. Though these translations were finally largely tailored away due to the word cap, the process made my interpretation of the history more solid and made me more aware of the author's perspective and layers.

My series editors, James Beattie and Karen Jones, encouraged me to adopt a comparative view between Chinese and European Landscape studies, to contextualise Chinese garden history within the global context. Their guidance and feedback were instrumental in transforming my thesis into a book appealing to a broad audience and further inspired my thinking of Chinese architecture as a global history. As an author, I developed chapters from case studies towards the book introduction and conclusion; my editors took a reversed focus to ensure that the theoretical framework and methodologies in the introduction and conclusion are transferable and hold broad implications for landscape history. The press editors, Max Novick and Louise Ingham, patiently resolved issues I encountered, making the submission process both exciting and seamless.

My dissertation committee member, Zhou Ying, frequently engaged with me in discussions about urbanisation and wrote extensive comments and suggestions in detail. Her enthusiasm for architectural history and broad research scope was genuinely encouraging. The art historian Flora Li-tsui Fu enriched my understanding through detailed discussions about Yuan-Ming period paintings and artists in Suzhou. Our collaboration on research projects, such as mapping the painting scroll by Wu Zhen and Wang Meng, prompted me to critically reflect on the artistic agendas of paintings as referenced in this book. The positive review from the Chinese garden expert Lu Andong offered me significant confidence. I am particularly drawn to his application of deictic theory in garden studies and have read his seminal paper many times. When he participated in my thesis defence via Zoom in 2020 (since it was not feasible for him to travel to HKU due to Covid-19), I was exhilarated to see the author.

While revising this book during 2021–2023, I embarked on the postdoc programme in East Asian Art History at Heidelberg University under the supervision of Sarah E. Fraser. I am grateful to the institute for providing me with a lovely office under a giant London plane tree, a serene space where I could concentrate on writing while being refreshed by the rustling leaves, chirping birds, and soothing rain outside the window. Transitioning from architecture to an art faculty, from Hong Kong to Germany, introduced me to a fresh environment with new friends and teachers. This relocation expanded my research focuses and corpora, challenging me to reframe knowledge structure and values. The art history training in the postdoc programme deepened my understanding of pictorial space. When teaching literati paintings in the course 'Visual Analysis,' I adopted an art historical perspective, which differed from the architectural focus used in the book. This new perspective allows me to distinguish the dual references in Chinese paintings: both to the pictorial iconography and to the actual landscape on site.

At Heidelberg University, courses on Buddhism and Dunhuang enriched my research on the Chan philosophy and Buddhist visuality at Shizi Lin Temple. Eugene Wang's lecture and seminar deepened my understanding of Buddhist cosmology and the culture of Chinese topography, aiding my reflections on the hydro-topography of the Yuan dynasty temple-scape. The use of 3D modelling methods in my postdoc research and teaching on Dunhuang inspired me to clarify how I used 3D modelling to investigate historical landscapes.

During the revision period, I presented related topics and received valuable feedback from the audience which helped to diversify the research angles. I am grateful to these conference and panel organisers who provided me with such opportunities. In the 2019 Conference on Gardens in Nagoya, I got to connect with scholars studying Mediterranean landscape Roman and Italian gardens, such as Yukiko Kawamoto, Nicholas Purcell, and Fabian Jonietz. The invited talk on hydraulic mappings of Suzhou at the University of Venice, organised by Pietro Daniel Omodeo and Heiner Krellig, informed me of the potential of hydraulic perspective in global studies. The water city series organised by them also inspired me to draw comparisons between the hydraulics of Suzhou gardens and the reclaim scheme in European cities such as Venice and the Netherlands.

The book evolved from my thesis, completed during my four years of PhD at the University of Hong Kong, Faculty of Architecture, from 2016 to 2020. I am grateful for the scholarship received from the Hong Kong government, which supported me in building my own research in Chinese architecture—a field where funded PhD programmes are relatively less globally. I express deep gratitude to my PhD supervisor, Weijen Wang, whose discussions with me, often over coffee, were both inspirational and transformative. I have a very cherished memory with him on the fieldtrips to Suzhou and Daegu, and the lessons I learned from him extend beyond academics. The vibrant and debating environment in the Faculty of Architecture is crucial to my development. Eunice Seng's methodology courses deepened my understanding of the social history of architecture, Cole Roskam's rigorous review during my PhD candidature exam has been particularly formative, and Zhu Tao's critical discussions with me are inspirational. My experience as a teaching assistant for surveying heritage buildings in Macao enhanced my approach to 3D modelling as a historical inquiry and refined the aesthetics of my drawings. Teachers from the landscape faculty, such as Gavin Coates, provided me with guidance in analysing the hydraulic flow of urban landscapes.

Craig Clunas's books have been inspirational in my research trajectory. I remembered that years ago when I read his *Fruitful Sites*, pondering: 'How could I write a book like this?' I then visited him in Oxford in 2017. He welcomed me warmly and encouraged my project. When I expressed admiration for his work, he smiled and humbly replied, 'I wish my book to be like a brick on the ground, upon which new scholars can step and move on.' I also would like to thank Johannes Küchler, who welcomed my visits and took the time to carefully read my thesis. I fondly recall a winter's day when Maolin and I visited his beautiful garden, where we had Chinese tea and in-depth discussion.

My thanks also extend to scholars who have discussed with me research topics and writings related to this book, including Sahar Khan, Huang Xiao, Zhou Hongjun, Wang Jin, Zhao Yanzhe, Mu Xiaodong, and Liu Yan. I feel especially thankful to Sahra Khan for her instruction on my thesis writing during my second and third years in the PhD programme. I also appreciated the assistance of Cheng Shi, Pan Wang, Qian Yu, Li Yingchun, Li Luke, Ni Jie, and Xu Ruqing in the copyright obtainment of paintings and photos. I am indebted to the monks who accommodated me during my field trips in Suzhou, to whom I extended detailed thanks in relevant chapters. My appreciation also extends to locals in Suzhou, such as Zhu Haijun, Yan Wenyun from the Suzhou Garden Monitoring Bureau, Fan Yezhou from the Suzhou Planning Bureau, and Wang Zhicheng from the Suzhou Gazetteer Office, who received my visits facilitated access to resources. Additionally, I value the scholars who actively responded to my emails, whose contributions I have noted in the relevant sections of this book.

My heartfelt appreciation goes to my husband, scientist Peng Maolin, for his unwavering support and confidence throughout my academic career. Our 'grand expedition' in 2017, cycling through cities in the eastern Lake Tai region to explore 49 gardens, as listed in Tong Jun's book, is my precious memory. Each we cycled 30–50 kilometres, 'measuring the landscape by feet,' as Maolin advocated. Our excitement in discovering suburb landscapes was mixed with the scorching summer heat, sweat, thirst, and the sweetness of watermelons from roadside vendors. Maolin also accompanied me on many field trips, guiding me in streamlining data organisation, optimising workflow, and providing insights into historical research from a scientist's view. I genuinely appreciate his harsh criticisms, which push me to transcend. As my saying for him goes: 'He kicks me up when I fail, criticises me down to the bone.'

My thanks also extend to Maolin's supervisor, Philip Wigge, who is open to transdisciplinary cooperation and provided me with working spaces in his lab. We shared many enjoyable moments discussing architecture. Maolin's parents generously backed me with financial support so I could focus on research without financial strains. I hope this book would be encouraging for young scholars who struggle to establish themselves. We could break new ground, and we will.

Chronology

Sui 581–618
Tang 619–907
    Early Tang 618–704
    High Tang 705–780
    Mid-Tang 781–847
    Late Tang 848–906
Five Dynasties 907–960
    Wu-yue Kingdom 907–978
Song Dynasties 960–1279
    Northern Song 960–1127
    Southern Song 1127–1279
        r. Jianyan 1127–1130
Yuan Dynasty 1206–1368
    Yuan began in Suzhou: 1275
Zhang Shicheng Regime 1356–1367
Ming 1368–1644
    Early Ming 1368–1465
    Middle Ming 1488–1521

    r. Hongzhi (1488–1505)
    r. Zhengde (1506–1521)
High Ming 1521–1566
    r. Jiajing (1522–1566)
    r. Longqing (1567–1572)
Late Ming 1573–1643
    r. Wanli (1573–1619)
Qing 1644–1911
    Early Qing 1644–1661
    r. Shunzhi
High Qing 1662–1795
    r. Kangxi 1662–1722
    r. Yongzheng 1723–1736
    r. Qianlong 1736–1795
Middle Qing
    r. Jiaqing 1796–1820
    r. Daoguang 1821–1850
Late Qing 1851–1911

Notes: The reign periods mentioned in the book are specified. The High Ming period is periodised by anti-Buddhism and the conversion of temples into family estates or Confucian institutions, particularly in Suzhou. Some local histories of Suzhou, such as the Wu-yue Kingdom, the Yuan Dynasty in Suzhou, and the Zhang Shicheng Regime, deviate from the general Chinese dynastic periods.

In the manuscript, Chinese and Japanese names follow the East Asian tradition, where the family name precedes the given name. Monks' names typically consist of a *zi* (courtesy name) and a given name, and they are traditionally addressed by their courtesy name. When referencing Chinese sources in the manuscript, the English translation of the title is used. Abbreviations for referencing figures and historical details including: cf.: Latin *confer*, meaning 'compare'; b.: the birth date; d.: died; fl.: Latin *floruit*, denoting when a person was active. ca.: Latin *circa*, indicating an approximate date.

Introduction

This book investigates the transformation of temples into gardens in Suzhou during the 13th to 16th centuries and beyond, a Buddhist and hydraulic history of gardens which has been long overlooked and unrepresented by the scholarship. The classical gardens of Suzhou, since being listed as UNESCO World Heritage in 1997, have been established and exported worldwide as exemplars of Chinese literati style. Three of these heritage gardens—Surging Wave Pavilion (*Canglang ting*), Lion Grove Garden (*Shizi lin*), and Unsuccessful Politician's Garden (*Zhuozheng yuan*)—have long been recognised as epitomes of the highest craft and literati culture of the Chinese garden.[1] Scarce Buddhist traces can be found in them and other Suzhou gardens today. A closer historical examination, however, reveals that the three gardens used to be temples during the 13th to 14th centuries of the Yuan dynasty. The garden of Surging Wave Pavilion was part of Southern Chan Temple (*Nanchan si*), Lion Grove Garden was formerly the renowned Shizi Lin Temple, and Unsuccessful Politician's Garden used to be Great Propagation Temple (*Dahong si*). Why were Buddhist traces erased?

During the Yuan dynasty, 372 temple estates in Suzhou (Fig. 0.1, Fig. 0.3, Table A.1, Table A.6) far outweighed the few recorded garden estates and academies (Fig. 0.2, Table A.2, Table A.3). Approximately half percent of these temples were built in former dynasties and continued their prosperity during the Yuan, while another half were newly founded. Two hundred and fifty-three of them were in the suburb areas, spreading amongst rural villages and mountains, and 119 were in the intra-mural area. Regarding that there were 65 residential quarters inside the city of 14.2 square kilometres, it could be deduced that each residential block owned two temples on average. Temples were readily accessible to citizens.

Temples' widespread presence and grandeur during the Yuan dynasty were vividly captured in the writings of poet Gao Qi (1336–1373), he wrote:

DOI: 10.4324/9781003387169-1

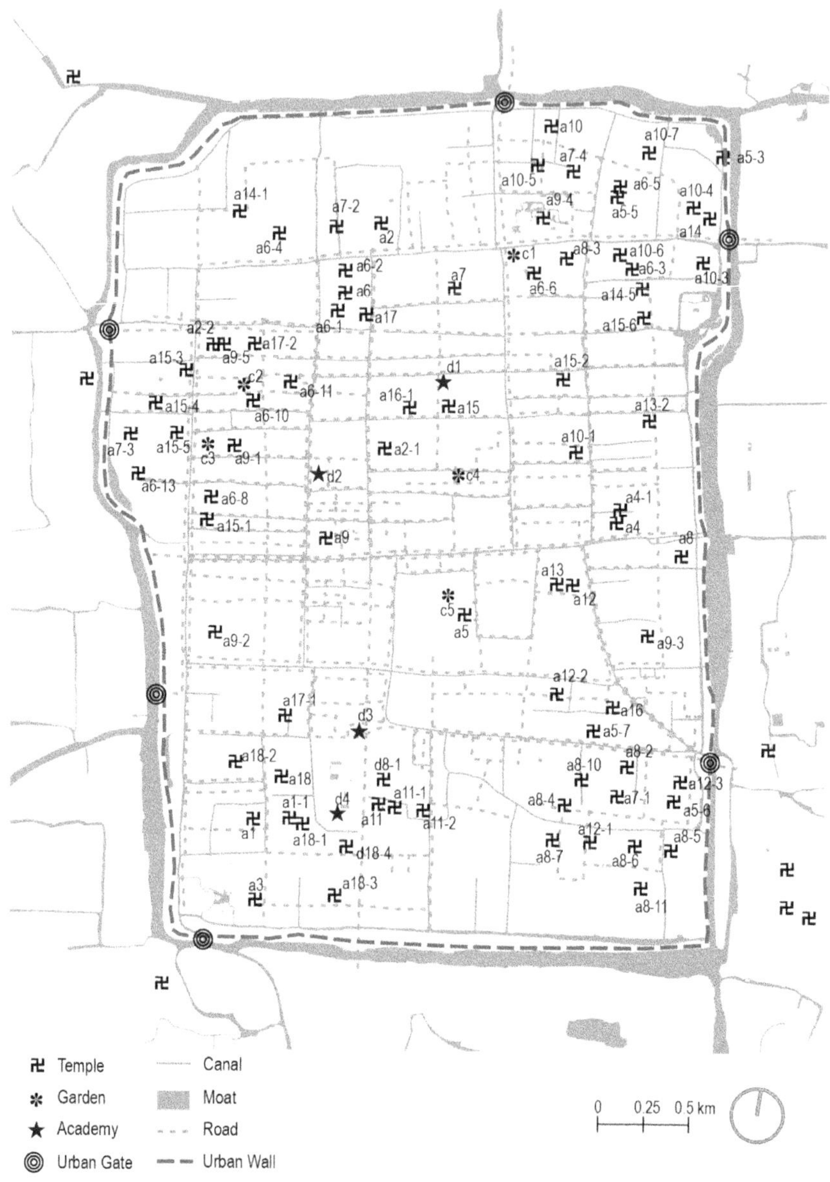

Figure 0.1 GIS map of temples, gardens, and academies in Suzhou's intramural area in the Yuan dynasty. The canals and streets in the GIS map are based on *Pingjiang Map*. Among the 119 intramural temples identified by Wang Ao in *GSZ*, 81 are feasible to be identified on the map. For the names of temples, gardens, and academies, see Table A.1, A.2, and A.3, respectively. Source: Author's mapping and drawing.

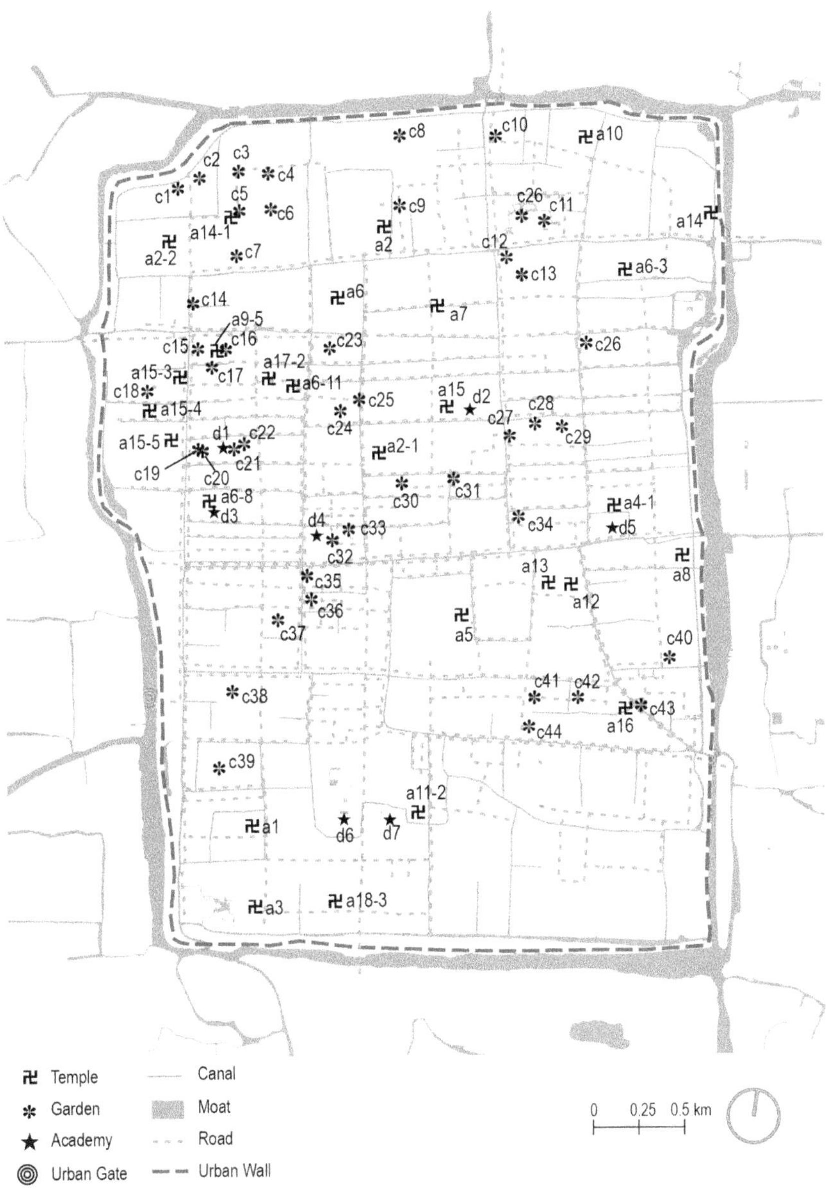

Figure 0.2 GIS map of temples, gardens, and academies in Suzhou's intramural area in the Ming dynasty. The Canal Network is based on a Map of the Waterways Management within the Suzhou Prefecture City by Zhang Guowei. For the names of these temples, gardens, and academies, see Tables A.1, A.4, and A.5, respectively. Source: Author's mapping and drawing.

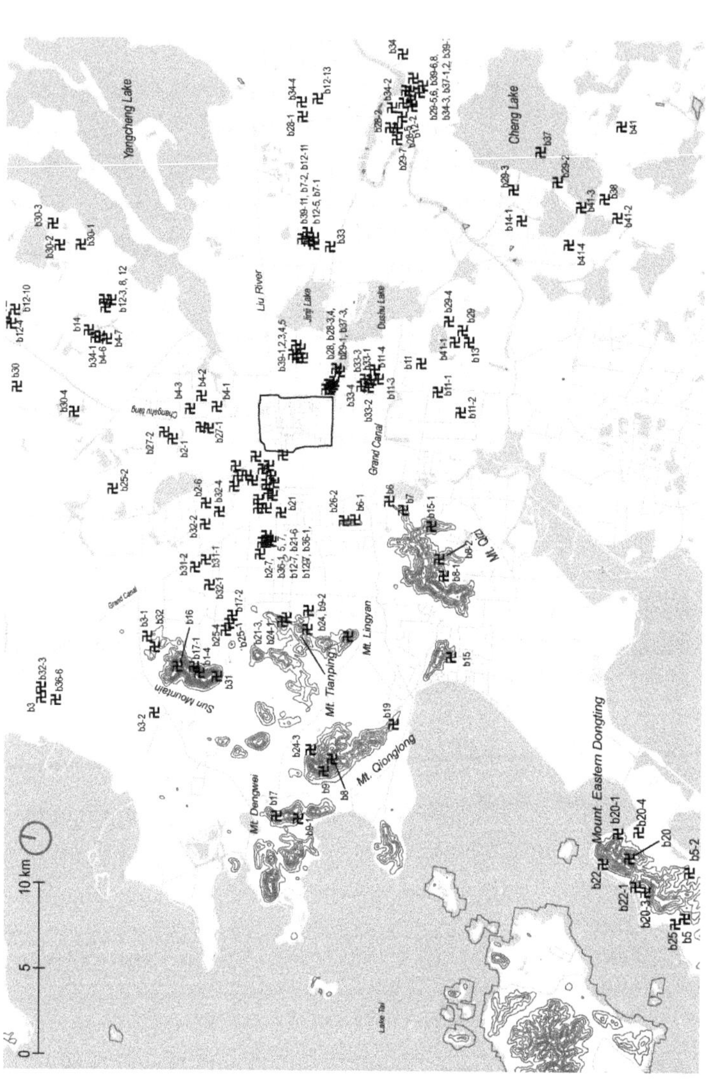

Figure 0.3 GIS map of temples in Suzhou's extra-mural area during the Yuan Dynasty. For temple names, see Table A.6. For names of temples in the densely concentrated area to the northwest of the urban wall, which are not identified in the map, see Fig. 2.16. Source: Author's mapping and drawing. Among the 253 intra-mural temples identified by Wang Ao, 161 are identifiable on the GIS map.

The temples and monasteries in Suzhou are the most prosperous in comparison to other cities. They are located in essential wards of the city, in spiritual mountains, and along rivers in hundreds. Their marvellous plinths, pavilions, and buildings stand next to each other, and the sound of bells and reading sutras resonates with each other.[2]

Shizi Lin Temple, which was portrayed by Gao Qi's friend Xu Ben (1335–1380) (Fig. 0.4a), provided a visual glimpse of temples in the Yuan, though remarked by Gao Qi as incomparable to the grandiose complexes of many other temples. The traveller Macro Polo (1254–1324) recorded that a majority of citizens in Suzhou were Buddhists.[3] Further examination of historical records reveals the diverse range of Buddhist practices taking place within the city. Eminent monk masters migrated to Suzhou, and local Buddhist leaders emerged, fostering a vibrant religious community. Citizens worshipped Buddha, discussed and read sutras, took tonsure, and built or devoted to temples. Local officials and the imperial court enthusiastically supported temples' construction and promoted Buddhism.[4] Given the density of temples and the dynamic religious urban life, the urban-scape of Suzhou at that time could be termed as 'temple-scape.'

In contrast, by the end of the Ming dynasty, the number of temples was decreased by three-fourths, with only 27 temples remaining within the city walls and around 120 temples outside (Fig. 0.2). Conversely, the number of garden estates increased to 62, with 13 constructed in the early Ming period and 49 in the Middle and Late Ming.[5] Confucian institutions, including academies, schools, and shrines, increased to about ten. The rate of temple construction fell down to 0.029 per year, far behind the rate during the Yuan. In contrast, the construction rate of the latter two estates increased significantly.[6]

The garden-scape was observed by the Ming magistrate Yuan Hongdao (1568–1610). After visiting most of them, he praised the flourishing gardens in the city.[7] The prospered garden estates during this period gave rise to the painting genre known as 'by-name' paintings. Typical works include the album *Thirty-One Scenes of Unsuccessful Politician's Garden* by Wen Zhengming (1470–1559) (Fig. 0.4b) and the painting leaf of Surging Wave Pavilion by Wen Boren (1502–1575) (Fig. 0.4c). The literature genre of 'garden record (*yuan ji*)' also established itself during this time, numerous pieces of which were anthologised in *Supplementary Anthology of Literary Essence from the Capital of Wu.*[8]

Joanna Handlin Smith has pointed out this garden mania in the Ming: 'the building of private gardens originated in the Han Dynasty and was somewhat prevalent during the Song, but during the late Ming, construction gained real momentum, both in the quantity of gardens and in the

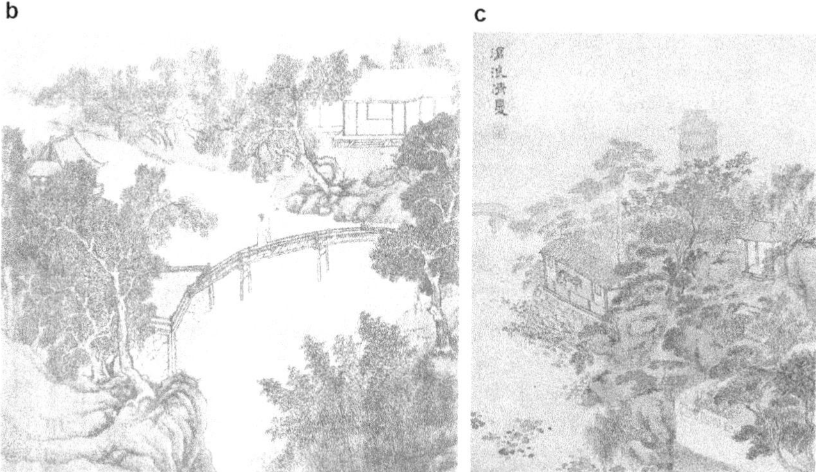

Figure 0.4 Paintings depicting temples in the Yuan dynasty and gardens in the Ming. (a) Attributed to Xu Ben (1335–1380), 'Lion Peak,' leaf no.1, in the album*Painting of Shizi Lin (Shizi lin tu)*, ink on paper, 22.5×27.1 cm. (b) Wen Zhengming (1470–1559), 'Little Flying Rainbow.' Leaf no. 3, in the album *Thirty-One Scenes of Unsuccessful Politician's Garden*, 1533, ink on paper. (c) Wen Boren (1502–1575). 'Brisk Summer in Surging Wave (*Canglang qingxia*),' leaf no. 3, in the album *Ten Scenes of Gusu* (Gusu shijing tu), 1554, colour ink on paper, 32×25.6 cm. Source: (a) (c) Taipei: Collection of the National Palace Museum; (b) Copy from Kate Kerby, *An Old Chinese Garden*.

amount of documentation they generated. Indeed, a veritable mania for garden-building swept through centres of prosperity.'[9] The bloom of garden constructions and representations in the Ming could be termed as 'garden-scape.'

How did the transition from temple-scape to garden-scape interact with estate transfers from Buddhists to the gentry and the local government? Tracing the conversion processes of 11 temple estates in gazetteer records reveals three main features of the transference. First, there was an immediate decline in Buddhist estates following the overthrow of the Yuan dynasty. Temples' dilapidated status set the stage for subsequent encroachment. While the encroachment of temples began during Zhang Shicheng's reign and the Early Ming period, significant transfers of Buddhist estates occurred during Jiajing's reign in the High Ming period. Most Buddhist estates were converted into gentry's gardens or Confucian institutions, with a few transformed into ancestral halls or government offices. Despite the competition for estate acquisition, some temples managed to survive. Overall, the number of temple institutions decreased while gentry estates expanded.

Temples Converted into Family Estates

Four cases being investigated are Shizi Lin Temple, Great Propagation Temple, Illuminous Celebration Temple (*Zhaoqing si*), and Return to the Origin Temple (*Guiyuan si*). Great Propagation Temple was constructed by Panquan Youlan (fl. 1290s) in the early Yuan dynasty and was then converted to Wang Xianchen's (fl. 1490s–1540s) Unsuccessful Politician's Garden. Illuminous Celebration Temple was established by the Yuan court and converted into another Wang family's estate. Return to the Origin Temple was established by the family of Cao Ruli (fl. 1281) and converted into Xu Taishi's (1540–1598) family estate called Western Garden (*Xi yuan*). Shizi Lin Temple was founded by Tianru Weize (1286–1354) and was converted into Lu family's agricultural field.

Temple Estates Converted into Confucian Institutions

Five temples were recorded as converted into academies, shrines, and schools. Four of these conversions were made by the magistrate Hu Zuanzong (1480–1560) during his three years' post in Suzhou. In 1523, he converted Forever Pacification Temple (*Yongding si*) into Golden Country Academy (*Jinxiang shuyuan*) and Dragon Prosperity Temple into Hejing Academy (*Hejing shuyuan*). In 1525, he turned Bright *Virtue* Temple into Learn the Way Academy (*Xuedao shuyuan*). Two years later, he converted Witty Retreat Cloister (*Miaoyin an*) into Shrine of the King of Qi (*Qi*

Wang miao). The most violent attack on temples happened in the 1540s, when the magistrate Shu Ting (fl. 16th c.) transformed Ten Thousand Longevity Chan Temple (*Wanshou chan si*) into Changzhou County School (*Changzhou xianxue*).

What were the socio-political and hydrological dimensions influencing and triggering the dissolution of monastic estates? How did garden practice transfer from temples in the Yuan to literati estates in the Ming? Why did Buddhism decline after its prosperity in the Yuan dynasty, fading away in the city, while the literati culture and Confucianism rose? This book delves into the evolved vision and ideology in landscape during the estate transformation and investigates the changing social structure during the contestation over hydraulic gardens.

Garden Studies, Spatial, and Social Inquiries

Spatial Approaches

From the 1920s to 1980s, researchers introduced perspectival spatial concepts to investigate Chinese gardens, facilitated by photography and architectural drawings. Florence Lee Powell's 1926 work *In the Chinese Garden* was one of the earliest to use photographs alongside garden plans to represent the garden.[10] In the 1930s, architect Tong Jun surveyed about 49 gardens around the eastern Lake Tai region. He documented these gardens in photos and detailed plans, and published *Gazetteer of Jiangnan Gardens* in 1937.[11] Tong Jun also contributed papers on Chinese gardens to the *T'ien Hsia Monthly Journal*, edited by Lin Yutang, fostering Western debates of Chinese gardens as integral to the Chinese way of living.[12] Tong Jun developed an iconography of Chinese gardens as embodied by the Chinese character '園': '囗' symbolises the enclosing walls, '土' the pavilion and gazebos, '口' the central pond, and '衣' signifies rocks and trees.[13]

Tong Jun's publications on Chinese gardens sparked interest from the West, leading to a heightened appreciation of gardens as a national essence of Chinese culture.[14] The book *Chinese Houses and Gardens* edited by Lee Shao-Chang brought together Tong Jun and other three authors to explore the interplay between Chinese gardens and houses.[15] The second half of the book showcased garden photography by Henry Inn, whose framing of garden scenes through architectural openings such as doors and lattice windows became a canonical composition in the field.[16] Osvald Siren's *Gardens of China* (1949), published in 1949, expanded the scope of study to include imperial gardens in the north. Siren adopted a multi-faceted approach, integrating classical literature, paintings, and site analysis to interpret gardens.[17]

During the Maoist era of the 1950s to 1970s, as foreign access to China was limited, the spatial studies of Suzhou garden were mainly conducted by Chinese architectural historians represented by Liu Dunzhen from Nanjing Institute of Technology and Chen Congzhou of Tongji University in Shanghai. They were responsible both for restoring and documenting the gardens. Liu's team conducted comprehensive surveyed drawings of 15 Suzhou gardens and published *Classical Gardens of Suzhou*.[18] At the same time, Chen Congzhou published *Suzhou Gardens*.[19] Their meticulous photography and detailed drawings, facilitated by their involvement in garden restoration, captured the gardens in a tranquil state prior to their opening to the tourists. Based on these pictorial representations, spatial concepts such as 'depth of scenes (*jing shen*),' 'layers of scenes (*ceng ci*),' and 'frame the view (*kuang jing*)' were developed. Their studies argued that Chinese gardens exemplified mastery in spatial manipulations such as configuring and dividing spaces, plotting vista lines, and using architectural elements to frame sceneries.

While Liu and Chen's works largely construed an impression that gardens in their photos were historical remains from the imperial era, it is essential to recognise that the gardens they photographed were, in fact, reconstructions after the 1950s. Their photos and drawings were not in the archaeological nature as the photos and survey drawings of Dunhuang made by Sergei Oldenburg's team in the 1910s, which was an as-found documentation of historical sites.[20] Many of the garden design techniques argued by Liu and Chen were not historical but the outcome of their modern renovations.

The reopening of access to mainland China in 1978 saw a surge in Western publications of Chinese gardens. Maggie Keswick's *The Chinese Garden: History, Art, and Architecture*, first published in 1978, was richly illustrated with Chinese paintings, colour photographs, redrawn plans following Liu's earlier survey drawings, and the author's on-site observations.[21] A similar work which received less attention was Edwin T. Morris's *The Gardens of China: History, Art, and Meanings*.[22] Keswick advanced the concept of 'literati garden (*wenren yuan*),' drawing on literati paintings, a genre extensively studied in Chinese art history.[23] This theme was further focused in R. Stewart Johnston's book *Scholar Gardens of China* in 1992.[24] The 1980s also witnessed an international export of Suzhou gardens, exemplified by the construction of a replica of Master of Fish-nets Garden (*Wangshi yuan*) as Ming Xuan in the Astor Chinese Garden Court in the Metropolitan Museum of Art in New York.[25]

The spatial studies were continued yet criticised in the late 1990s, represented by scholars such as Feng Jin, Stanislaus Fung, and Lu Andong. Feng Jin traced the concept of 'scene (*jing*)' in traditional Chinese literature, distinguishing its modern scholarly interpretations from historical applications.[26] Lu Andong compared the visuality in Chinese paintings and visibility in modern scholarship, pointing out that scenes in Chinese gardens should not be understood in a perspectival and picturesque way— 'stroll' differs from 'movement,' and 'realm' in Chinese gardens does not equal the modern conception of 'space.'[27] Stanislaus Fung further challenged the perspectival approach in studying Chinese gardens, aiming to capture the non-perspectival spatial effects in the existing Suzhou gardens through narrative and photography.[28]

The perspectival methods employed for Chinese gardens during the 1920s to 1980s were rooted in the development and study of Renaissance and picturesque gardens since the 15th century. Denis Cosgrove highlighted the linear perspective as central to understanding European landscape visuality.[29] Palladio manipulated distances between buildings and the scenery to create views commanded by interior spaces and further represented scenes in perspectival drawings.[30] The study of Palladio's design of San Giorgio Maggiore revealed that the architect oriented the church façade to oppose the palace Palazzo Flangini Fini across the canal.[31] The Renaissance period also saw the development of trompe-l'oeil technique and the concept of landscape as a theatrical stage for viewers' gaze.[32]

The earliest picturesque architect, Willian Kent (1685–1748), used multiple landscape drawings to represent garden sceneries composed of 'fore-, middle, and background' layers, predating Liu Dunzhen's 1960s concept of layers.[33] Kent, inspired by Palladio, constructed theatrical scenes within the landscape. His concept of 'prospect' involved framing views and arranging vistas within gardens, using light, shadows, and artefacts to create focal points.[34] John Rocque (1704–1762) assisted Kent by providing plans for these designed gardens.[35] William Chambers (1723–1796) applied these emerging picturesque concepts to study gardens in China, presenting the first perspective drawing of a Chinese garden in 1757.[36]

Lancelot Brown (1716–1783) transformed the walled and geometric garden style and established the picturesque style as irregular landscapes in the 17th century with gentle turfs, meandering paths, lakes, scattered veterans, and clumps.[37] This picturesque style was particularly configured in plans drawn by Brown, such as in his renowned design for Bowood estate in Wiltshire.[38]

Lancelot Brown's picturesque technique was further advanced by his successor, Humphry Repton (1752–1818), renowned for his Red Books. These volumes effectively communicated spatial layouts and redesigned views to his clients, typically beginning with a site map followed by renovation proposals. Repton introduced a unique overlaying technique to demonstrate the enhancement of the viewing experience. Beneath is the drawing of the designed site view, with the renovated part being covered by the overlay recording the current view. As readers flip the overlayed segment, they compare the existing site with Repton's vision, clearly conveying which aspects were deemed inadequate and how they were refined.[39]

The method of correlating plans with perceived views gained a wide practice as the picturesque garden style spread across Europe. J. F. Karl's 1792 album, *Twenty-Four Scenes of the Garden at Freundenhain*, showcased 24 different scenes from various vantage points in the garden, with each viewpoint plotted on the plan.[40] Frederick Piper's 1811–1812 study of British picturesque gardens similarly employed this approach, which identified the vista lines arranged in the gardens.[41]

The picturesque method in which plans were coordinated with perspectival drawings persisted after the invention and application of the camera. Photographic images gradually superseded landscape designers' perspectival drawings for site documentation.[42] It is noted that photographs adhere to the perspective principle that underpins earlier drawings. With the onset of the colonial expansion, this method was transferred to the study of Chinese gardens, exemplified by Powell's photography and plan of gardens in Suzhou, as discussed at the beginning of this section.

Social Approaches

In the past two decades, the social approach towards Chinese garden studies was spearheaded by art history and sociology scholars such as Craig Clunas, Joanna Smith, Antonia Finnane, and David Sensabaugh.[43] Unlike the post-modernist critiques of the validity of picturesque methods in studying Chinese gardens, historians criticised the existence of a singular 'Chinese garden.' The opening statement in Craig Clunas's book *Fruitful Site: Garden Culture in Ming Dynasty China* is thought-provoking: 'This book does not present a history of "the Chinese garden". It is written out of a distrust that such a thing exits.'[44] Historians criticised earlier studies for overlooking the social and economic dimensions of garden practice, often portraying gardens as timeless artwork exclusive to the literati.

Scholars adopting the social approach are more attentive than spatial-focused scholars to the fact that the physical form of current Chinese gardens largely resulted from renovations in the 1950s. They re-examined historical evidence, contextualising gardens in specific spatial and temporal settings, and contended that gardens' meanings and functions evolved over time. Concerning the food crisis and assessing gardens' agricultural outputs, Clunas argued that Wu Kuan's (1435–1504) Eastern Estate (*Dong Zhuang*) in the Early Ming and Wang Xianchen's Unsuccessful Politician's Garden in the High-Ming were actively engaged in diverse farming. Drawing on Pierre Bourdieu's theory of social distinction, Clunas highlighted how the upper gentry's penchant for luxurious artificial rockery gardens in the late Ming served to differentiate them from other social strata.[45]

Antonia Finnane's research on gardens in Yangzhou has examined the merchants' social network facilitated by their gardens.[46] She posits that these gardens enticed Emperor Qianlong's (1711–1799) visits thereby securing privileges for their owners. Their writings showcase how Chinese garden studies could contribute to broader social and cultural debates, expanding beyond artistic connoisseurship.

Other scholars' research on individual garden cases have demonstrated the changing nature of garden ownership and settings, thus a garden's history cannot be reduced into a single account. David Sensabaugh has traced the history of Lion Grove Garden, which was initially a temple, transformed into a garden through the influence of Xu Ben and Ni Zan's (1306–1374) paintings, as well as the promotion by Emperor Qianlong. The emperor's replication of the Lion Grove Garden in his palaces brought it national acclaim.[47] Zou Hui's research further reveals that Emperor Qianlong's interpretation of the artificial mountain in Lion Grove Garden diverged from its original Daoist connotations.[48] Examining the evolution of the Surging Wave Pavilion, from its construction by Su Shunqin (1008–1048) to its later renovation by Song Luo (1634–1714), Xu Yinong argues that the pavilion's symbolic meaning shifted from representing Su Shunqin's frustration to symbolising Song Luo's political success.

Since the 1980s, the study of European gardens has increasingly embraced social approaches, exploring the interplay between societal changes and landscape design. In *Property and Landscape*, Williamson offers an in-depth analysis of how the enclosure movements in England reshaped patterns of landholding, leading to the development of picturesque gardens.[49] A decade later, in *Polite Landscape*, he advocates for a more inclusive garden history that considers not only prominent designers but also key sites and landowners.[50]

Williamson places the rise of the Brownian landscape within the broader socio-economic context during the 1750s to 1770s. He posits that the rise of the gentry class, along with their cultivation of refined tastes and polite society within their expansive estates in the countryside, was instrumental in elevating the picturesque garden style above the previous geometric, which had been favoured by monarchs and aristocrats.[51] This interpretive lens, acknowledging the impact of landownership and societal structures on landscape style, invites the comparison with the evolution of the literati garden in Ming dynasty China.

Denis Cosgrove's *Social Formation and Symbolic Landscape* (1984) highlights the social transition from feudalism to capitalism and investigates how the Italian cityscape, such as the composition of Piazza San Marco, evolved to express Renaissance humanism, giving symbolic readings of the social dynamics between aristocracy, popular republicanism, and the monarchy.[52] His book, *Palladian Landscape*, further carried out this approach, exploring how Palladio's architectural designs were prescribed by nobilities' land use patterns. For instance, the mansions in Godi Estate and Villa Barbaro were designed to oversee the agricultural productions on site.[53]

Within the social analysis framework, gardens are often viewed as estates. The close intersection of these two has been investigated by Keirstead Thomas.[54] Similarly attentive to the land possession before the landscaping, his work *The Geography of Power in Medieval Japan* (1992) grounded gardens in the estate system of medieval Japan, illuminating how gardens manifested social power.[55] The divergence between social and spatial studies of gardens prompts an intriguing question: how could nuanced spatial analysis of historical gardens effectively engage with historical contexts?

Garden and Religion

A notable aspect in the study of Chinese gardens is their rare analysis in conjunction with religious contexts, particularly Buddhism and Confucianism. Gardens and temples of China have been distinguished as two separate discourses. For instance, the well-received book on Chinese gardens by Maggie Keswick does not include any temples, whereas books on Chinese Buddhist Architecture, such as Wang Guixiang study on the evolution and transmission of Buddhist temples and Zhang Shiqing's works on monasteries in southern China, do not mention garden practices.[56]

This dichotomy between garden culture and Buddhism, however, is hard to justify, considering that several dynasties, such as the Sui, Tang, and Yuan, were predominant Buddhist states. The pervasive influence of Buddhism in these empires makes it less likely that gardens of the time were unaffected by Buddhist visions. Buddhist murals in Dunhuang, for instance, illustrated religious activities, such as the contemplation of the pure land, as unfolded in idyllic garden landscapes.[57] The Buddhist origin in India fostered a profound connection with the concept of nature as being tamed. Gregory Schopen highlights that in early Indian Buddhism, the ideal monastery was envisioned as a garden.[58] This notion resonates with Christian traditions as well, where the ideal monastic setting is frequently depicted as a garden.[59] These parallels underscore the role of gardens as conducive to spiritual reflection and contemplative practices.

Studies of gardens in the medieval period frequently highlighted their connection to religious rituals. Following the collapse of the Western Roman Empire and the spread of Christian monasticism, monastic gardens gained prominence from the 6th century onward. Gardens served multiple roles in churches and monasteries, integrating the practical, spiritual, and aesthetic. In *Medieval English Gardens* (2014), Teresa McLean explores the multifaceted roles of church gardens, from cultivating medicinal herbs and providing floriculture education to representing the imagined paradise.[60] Roberta Gilchrist's 2005 study on Norwich Cathedral utilised archaeological findings for an in-depth analysis of clergies' garden plots.[61] Tom Williamson further investigates how land concentration within religious institutions conditioned their spatial privilege and incentivised gardening practices.[62]

In Mirei Shigemori's extensive 26 volumes of *The History of Japanese Gardens*, most garden cases selected are temples.[63] Shigemori identified ten temples as epitomising garden achievements in Japan's Kamakura period (1185–1333), which corresponds to China's Song to Yuan dynasties. These ten temples include the renowned Saiho-ji and Tenryu-ji. Musō Soseki (1275–1351), the Linji monk in close relation to Ashikaga Shogunate government, was acclaimed as the best garden designer in the Kamakura period. For gardens in the Muromachi era (1336–1573), a period corresponding to China's late Yuan and Ming dynasties, the author selected another ten temples as the representative gardens.[64]

These scenarios of religious gardens worldwide propel the investigation of the garden as monastic practice realising Buddhist vision, transcending the dichotomised discourses between Chinese gardens and temples. Pertinent questions arise in the current study: how did Buddhists craft garden estates when they ascended to the landowning class during the Yuan dynasty? Moreover, how has religion been marginalised in the discourse on gardens in Chinese historiography? The scant consideration of the religious dimension in Chinese garden studies can be partly attributed to the dominance of historical evidence produced by the gentry class. Furthermore, the Buddhist reformation carried out in China since the 1950s has significantly suppressed Buddhism and Confucianism as public belief—the Chinese gardens were modernised.

Theoretical Framework

The theoretical framework of the current study is structured around the triangulation of estate transference, landscape transformation, and hydraulic evolution. It provides a new theoretical model to tackle the social transformation from Buddhist society to the gentry society, from the Buddhist medieval towards the literati and Confucian pre-modern.

Estate Transference

The estate studies worldwide generally share three common grounds. First, an estate is a bounded piece of land property owned by individuals, a family, or an institution. The landed ownership was validated by registers and maps. As the land market developed to facilitate estate transference, estates became both economic and social capital. Owners' site renovation and culture representation could thus increase the value of an estate.

Second, estates, at their inception since the medieval era, served for both dwelling and self-sustained agricultural production. In Japan, estates are closely tied to Buddhist societies and the Shoen system.[65] While in England, estates were often referred to as villas, with a focus on mansions and their immediate surroundings.[66] After the medieval, the production role of estates gradually diminished, as observed in the Late Ming and Qing dynasty Suzhou, where aesthetic viewing and touring were emphasised. The habitability and productivity in estates in Suzhou were highly associated with land reclamation.

Strategies of estate investment reveal a local response to state control. Different types of estates were subjected to varied taxation and corvee labour obligations. The privileges that the state granted to certain estates influenced the contentions between estate owners and further drove the landscape transformation from monastic to literati and Confucian configurations.

Transformation of Landscape Visuality

The concept of landscape is multifaceted, embodying both the physical and perceptual.[67] The physical changes of a garden include the successive owners' inheritance, erasure, adaptation, and development of the former owners' landscape. The physical change is also multi-layered, including the topography, plantation, hydraulics, and architecture. The current study analysed these on-site transformations using 3D modelling.

On the other hand, the idea of landscape is visionary, implied by its suffix '-scape,' which refers to ways of seeing and ideology.[68] People manifest and forge ideologies through crafting landscapes, which in turn shapes their vision and thinking. In this regard, landscapes serve as spatial arenas for competing cultures.

Landscape also held agency of social power, as estate owners showcase their social standing by gathering in gardens and disseminating estate representations.[69] The diverse pictorial and textual representations of the temple-to-garden cases in this research support an interrogative reading of their authors' perspectives and ideologies. Comparing these historical representations with 3D models offers further insights into the exaggeration, highlights, and omissions in the representation, pointing to the cultural agendas they embody.

Hydraulic Evolution

Distinct from other forms of architecture, gardening inherently integrates plants and water management, making water a crucial element. Water features are prevalent across garden traditions in arid and wetland geographies, from Persian and Roman gardens to Italian Renaissance and French formal gardens, and the English picturesque garden.[70] Traditionally, water elements in gardens were examined primarily for their decorative and symbolic values. Recent studies have shifted focus to explore the mechanism of creating water features, their ecological roles in gardens, and how rivers and lakes influenced garden topography and layout.

Tom Williamson's edited volume *What Did Capability Brown Do for Ecology?*, published in 2017, marked an early attempt to apply an ecological perspective to the study of Brownian landscapes. John Barnatt noticed how Capability Brown, in creating the lake at Chatsworth, widened the river and added an ornamental canal, which involves tremendous earthwork.[71] Wendy Bishop's 2012 work *Ornamental Lakes* delved into the engineering methods used to construct picturesque garden lakes, tracing their evolution from functional predecessors like fish ponds and castle moats to a focus on aesthetics.[72]

The current research diverges from these hydraulic approaches in several ways. First, it positions gardens as intermediary hydraulic infrastructure within urban, environmental, and cultural contexts. This aspect highlights the role of garden estates in land reclamation and urbanisation. It further examines the deterioration of urban canals in the late 16th century which shifted the function of water features in gardens from productive to ornamental, indicating the loss of a hydro-ecosystem.

Second, the hydrological process inside garden estates was the mechanism that shaped the garden topography of Suzhou during the 13th to 16th centuries. In contrast to Mediterranean gardens, such as those developed in Roman and Renaissance periods, where water was precious and dependent on aqueducts to transport, water was abundant and even excessive in Suzhou before the 15th century: the primary concern being the conversion of wetlands into habitable dry land. Third, in contrast to the radical shift from the Buddhist landscape towards the literati and Confucian ones, the hydraulic infrastructure inherited by the gentry from the previous Buddhist estates indicates a topographical continuity.

Technical Framework

The technical methods applied in the current research, GIS mapping and 3D modelling, assist to integrate raw historical data into spatial inquiries at both urban and estate scales. The two methods assist to differentiate the physical constructions of a historical estate in history from its cultural representations, which were usually confounded in landscape studies. The historical modelling further provides subversive historical reading.

GIS Mapping

GIS mapping has a long history in landscape studies since the 1960s and witnessed an increased transference into historical research and conservation over the past two decades. The GIS mapping operates within global projection systems, assigning coordinates to objects on the earth's surface. A GIS map layer comprises points, lines, or polygons. Points are used to mark locations, lines to trace linear features such as roads and rivers, and polygons to represent land areas. Each layer is associated with a table, inventorying attributes of the contained geographical shapes such as identity number, name, location, construction time, etc. (see Appendix, Table A.1 to A.6).

The attribute table facilitates sorting and filtering geo-features based on customised criteria. Researchers can use spatial tools within GIS for various analyses, including the concentration of geo-features, their inter-relationships, and viewshed and visibility analysis. The overlaying function allows for examining relations of features across different layers.

The current research uses GIS to map out the fabric of canals, rivers, and the distribution of temples, gardens, and academies in Suzhou during the Yuan and Ming periods. The first step is charting the canals and streets in *Pingjiang Map* (Fig. 0.5a, Fig. 0.5b), assuming that canal systems in the Yuan dynasty primarily inherited from the Southern Song dynasty. Objects depicted in *Pingjiang Map* were projected onto different layers in the GIS map (Fig. 0.5c), and further indexed in attribute tables with object identity numbers, names, coordinates, etc. (Fig. 0.5d). Other related features such as temples and residential wards (*fang*) in *Pingjiang Map* were also mapped out (Fig. 1.4).

Since *Pingjiang Map* does not conform to a unified metric measurements, projecting it onto a GIS map requires identifying anchor points co-shared by *Pingjiang Map* and the current city. Many bridges depicted in *Pingjiang Map* have archaeological remains, allowing them to be accurately placed on the GIS map.[73] In historical records, bridges served as spatial nodes at the ends of or intersections between streets and rivers. Therefore, bridges are used as anchor points for projecting *Pingjiang Map* onto the GIS map (Fig. 0.5e), and streets and rivers were identified accordingly. By correlating their positions with bridges and rivers, the locations of temples and residential wards in the GIS map could be deduced. A similar procedure is applied to map out rivers in *Map of the Waterways Management within the Suzhou Prefecture City* (Suzhou fucheng nei fenzhi shuidao tu) which provides the network of urban canals in the Late Ming, as illustrated in the GIS map in Figure 0.2.[74]

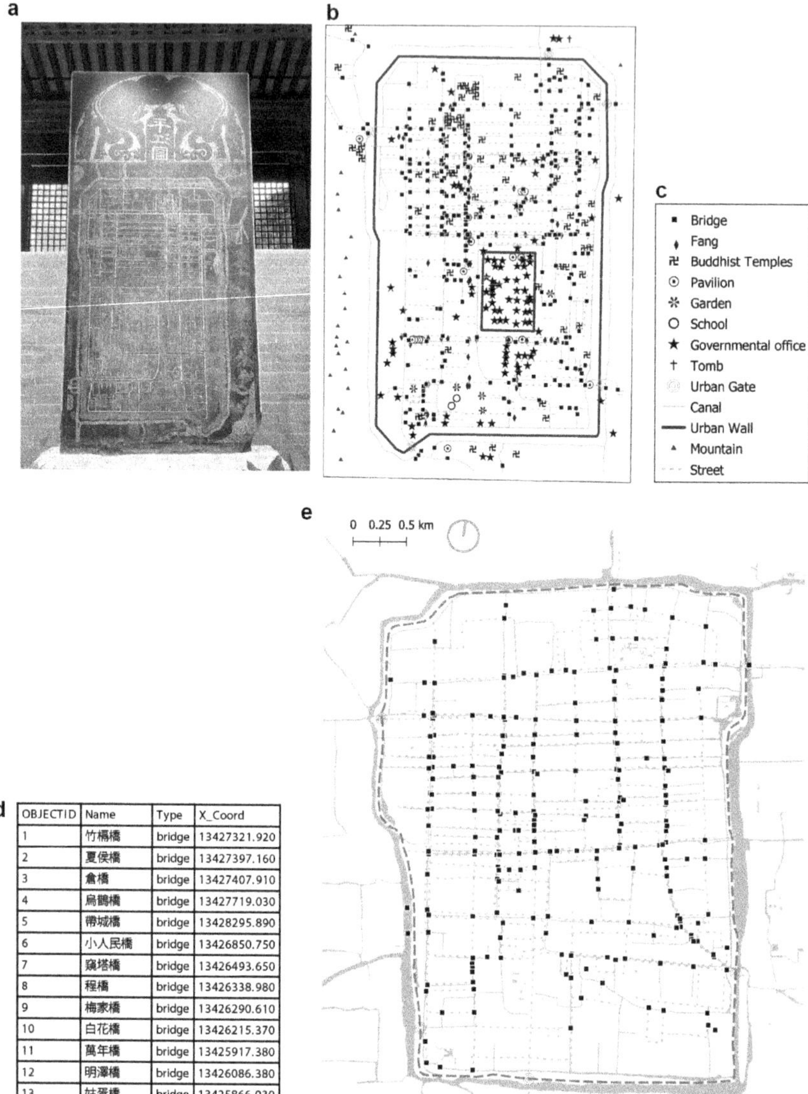

OBJECTID	Name	Type	X_Coord
1	竹柵橋	bridge	13427321.920
2	夏侯橋	bridge	13427397.160
3	倉橋	bridge	13427407.910
4	烏鵲橋	bridge	13427719.030
5	帶城橋	bridge	13428295.890
6	小人民橋	bridge	13426850.750
7	窺塔橋	bridge	13426493.650
8	程橋	bridge	13426338.980
9	梅家橋	bridge	13426290.610
10	白花橋	bridge	13426215.370
11	萬年橋	bridge	13425917.380
12	明澤橋	bridge	13426086.380
13	姑胥橋	bridge	13425866.030

Figure 0.5 GIS mapping process diagram of *Pingjiang Map*. (a) *Pingjiang Map* of 1229. Courtesy of Xu Ruqing. (b) Projecting canals, streets, bridges, temples, etc. into GIS Map. (c) Layers of objects in GIS Map. (d) Attribute table in GIS database. (e) The resulting GIS map comprises bridges, canals, and streets. The left-facing swastika (卍) is a traditional symbol used in East Asian cultures to represent Buddhism. In China, this symbol has been widely associated with Buddhism since the 3rd century. The symbol was often embossed on the chest, palms, or feet of Buddha statues. It is commonly used in maps or tourist signs to denote temples in countries such as China, Japan, and Korea.

Records of temples, gardens, and academies during the Yuan to Ming dynasties were extracted to be inventoried in GIS tables from gazetteers such as *Gazetteer of Gusu (Gusu zhi)* and *Gazetteer of Wu County in the Reign of Chongzhen (Chongzhen wuxian zhi)* (see Table A.1 to A.6).[75] Other resources, such as later historical maps and archaeological reports, are also consulted for supplementary information. Each estate is represented by a point symbol on the GIS map, with attributes such as constructed year, locations and owners charted in the table. Overlaying the layers representing temples, gardens, and academies in the Yuan and Ming respectively, the GIS map of Yuan and Ming estates are mapped out in Figures 0.1 to 0.3.

Translating Paintings into 3D Modelling

Digital models assemble discrete historical data, translate them into architectural drawings, aid analysis, and convince argumentation. The body of evidence, composed of ancient paintings, pictures, and texts, are fragmentary historical representations of artefacts and landscapes. They were framed with agendas from their authors' point of view. Historical evidence includes and omits information, aggrandises or underplays physical reality, and even contradicts itself. In contrast, a site's historical model requires a comprehensive index of various aspects, such as artefacts, layout, topography, hydrology, vegetation, measurements, as well as its urban context. Many of these practical aspects were downplayed especially in East Asian historical representations. Historical models thus form an arena where historians are required to integrate historical data, verify its authenticity and consistency, and take construction aspects into analysis.

The model could generate isometric, perspective, and sectional drawings to facilitate historical enquiries. The isometric view presents an estate layout in relation to the urban hydraulic context. Overlaying the isometric drawings of the different phases an estate going through reveals changes in its boundary, configuration, and topography. The current research particularly used successive isometric drawings to trace the contention of boundaries in estate transference, the competing ideologies, and owners' modifications of the hydro-topography.

Perspectival drawings simulate historical figures' viewing experience to assist historical analysis of their scenic interpretation. Sectional drawings reveal the hydrological process of the garden. They reveal the underground structure of an estate, such as the foundation of architecture, the rammed earth layers of an embankment, and the extension of vegetation roots. The sectional drawing challenges the existing garden analysis of only above-ground artefacts and vegetations by bringing the subterranean into the front of public purview.

Book Structure

The book is composed of five chapters. The first and second chapters investigate monks' landscaping in the Yuan. Chapter 1 identifies the canal-and-polder reclamation scheme developed in the eastern Lake Tai region in medieval urbanisation. Using sectional drawings to investigate the hydrologic process, it finds that monks in the Yuan invented a hydro-topographical pattern of 'mound-field-river' to transform the wetland in their temples into habitable and productive gardens. Chapter 2, 'Constructing the Temple-scape,' first examines monk Tianru Weize's translation of scenes from Mount Heaven Eye (*Tianmu shan*) to Suzhou in his construction of Shizi Lin Temple. The gong'an scenes crafted by Tianru showcase the Buddhist ways of seeing and assist Chan Buddhism inquiries. Then the chapter zooms out to outline different Buddhist sects' and patrons' temple constructions and patrons in the Yuan. It finds that spiritual pursuits, economic investment, and political propagation integrated and contributed to the prosperous temple-scape in the Yuan.

The third and fourth chapters examine the transformation of temples into literati and Confucian gardens in the Ming respectively. Shizi Lin Temple's transformation into Lion Grove Garden through five stages demonstrates how the gentry established an iconic literati garden on a temple by demarcating. Great Propagation Temple's conversion into Unsuccessful Politician's Garden showcases that the gentry contended with Buddhists on the estates with established hydro-topography. They erased Buddhist traces on site to create tranquil and reclusive scenes suitable for poetic literati tours. The government's transformation of Southern Chan Temple's transformation into Surging Wave Pavilion shows local officials' successive encroachment on monastic lands and contention with Buddhists on estates facing the public canal. The waterfront of former temples was choreographed to promote Confucianism. The anti-Buddhism and conversion of temples peaked in the High Ming, with three hydro-spatial patterns established and embedded in the literati and Confucian landscape.

Chapter 5 delves into the political and economic factors that affected the decline of temple-scape and the rise of the gentry garden. It argues that by seizing hydraulic estates from Buddhists, the gentry and government rose as prominent landholders and cultural elites. The book concludes with advocating a new garden history which applies 3D modelling and GIS mapping to investigate the hydro-social transformation.

Notes

1 The four famous Suzhou gardens (*sida mingyuan*) are identified in a dynastic order: Surging Wave Pavilion represents the Song dynasty, Lion Grove Garden

for the Yuan, Unsuccessful Politician's Garden for the Ming, and Lingering Garden for the Qing. Chen,*Shuoyuan Zhongguo Mingyuan*, 16.

2 Gao, 'Preface to the Twelve Odes of Shizi Lin 師子林十二詠序 (Shizi Lin shi'er yong xu),' in *JSJ*, juan xia, 2–3.
3 See Feng, trans., *Travels of Marco Polo*, 316–318.
4 The daily experience of religious life will be elaborated in Chapter 2.
5 See Mei, 'Study on the Changes in the Scale of Ming and Qing Dynasty Suzhou Gardens and Their Relationship with Urban Transformation,' 153–169.
6 The construction rate of Yuan temples in Suzhou was calculated by Marmé, *Suzhou: Where the Goods of All the Provinces Converge*, 25.
7 Yuan, 'Yuan Ting Jilüe 園亭紀略 (A Brief Account of Gardens and Pavilions),' in *Collated Anthology of Yuan Hongdao*, 180–182. Translated by Clunas, *Fruitful Sites*, 74–75.
8 *WDWCXJ*.
9 Smith, 'Gardens in Ch'i Piao-Chia's Social World: Wealth and Values in Late-Ming Kiangnan.'
10 Powell, *In the Chinese Garden.*
11 Tong, *Gazetteer of Jiangnan Gardens*, 1–2.
12 Tong, 'Chinese Gardens, Especially in Kiangsu and Chekiang.'
13 Tong, *Gazetteer of Jiangnan Gardens*, 7.
14 Tong, 'Chinese Gardens, Especially in Kiangsu and Chekiang.'
15 Inn, *Chinese Houses and Gardens.*
16 Ibid.
17 Sirén, *Gardens of China.*
18 Liu's team's work was first published as *Suzhou de Yuanlin* (Gardens of Suzhou) in 1957. Then the drawings and writings were extensively revised to the well-known work, *Suzhou Gudian Yuanlin* (Classical Gardens of Suzhou).
19 Chen, *Suzhou Gardens.*
20 See the Russian team's archaeological documentation of Dunhuang in State Hermitage Museum, *Dunhuang Artistic Relics in the Collection of the State Hermitage Museum*, vol. three, 5. For an overview of the archaeological survey of Mogao caves, see Wu, *Spatial Dunhuang*, 53–65.
21 Keswick, *The Chinese Garden: History, Art & Architecture.*
22 Morris, *The Gardens of China: History, Art, and Meanings.*
23 See the historical formation of scholar's painting and literati painting in Bush, *The Chinese Literati on Painting*, 1–12; see also Cahill's discussion literati painting tradition in Suzhou in *Parting at the Shore*, 57–96.
24 Johnston, *Scholar Gardens of China.*
25 Murck and Fong, *A Chinese Garden Court.*
26 Feng, 'Jing, the Concept of Scenery in Texts on the Traditional Chinese Garden.'
27 Lu, 'Lost in Translation.'
28 Fung, 'Non-Perspectival Effects in the Liu Yuan.'
29 Cosgrove, 'Prospect, Perspective and the Evolution of the Landscape Idea.'
30 Cosgrove, *The Palladian Landscape*, 225.
31 Savoy, 'Palladio and the Water-Oriented Scenography of Venice.'
32 Cosgrove, *The Palladian Landscape*, 246–247.
33 See the four contributions of Kent as categorised by Hunt, *The Picturesque Garden in Europe*, 26–35.
34 Hunt, *The Picturesque Garden in Europe*, 26–35.
35 See the 1735 Plans and views of Wanstead House and Park in the Borough of Redbridge in Weber, *William Kent: Designing Georgian Britain*, 32.

36 See the perspectival drawing of a Cantonese garden from Chambers, *Designs of Chinese Buildings, Furniture, Dresses, Machines and Utensils*, 1757, pl.9. Also see Rinaldi, ed., *Ideas of Chinese Gardens*, 303–342.
37 Brown and Willamson, *Lancelot Brown and the Capability Men*, 7–9.
38 Ibid., 92–114.
39 See Repton's mapping overlay drawings of Ouston in Yorkshire in Eyres and Lynch, *On the Spot*, 31–48.
40 See J. F. Karl's twenty-four scenes in Hunt, *The Picturesque Garden in Europe*, 169.
41 See Fredrik's drawings in Piper, *Ängelsk Lustpark, English Park*.
42 Ackerman, 'The Photographic Picturesque.'
43 Clunas, 'Nature and Ideology in Western Descriptions of the Chinese Gardens'; Clunas, *Fruitful Sites*; Finnane, *Speaking of Yangzhou*; Sensabaugh, 'The Lion Grove in Space and Time.'
44 Clunas, *Fruitful Sites*, 8. The social approach in art historical writing is discussed in Clunas, 'Social History of Art.'
45 See Clunas, *Superfluous Things*, 2. Clunas extensively transferred Pierre Bourdieu's theory of cultural capital and social distinction to the study of material culture in Suzhou. For a detailed examination of class tastes and lifestyles, see *Distinction*, 257–317.
46 See Finnane, *Speaking of Yangzhou*, 90–116.
47 Sensabaugh, 'The Lion Grove in Space and Time.'
48 Zou, 'The "True Wonder" in Emperor Qianlong's Garden Labyrinths.'
49 Williamson and Bellamy, *Property and Landscape*. Also see Williamson, 'Understanding Enclosure.'
50 Williamson, *Polite Landscapes*, 4–9.
51 Ibid., 109–113.
52 Cosgrove, *Social Formation and Symbolic Landscape*, 113–117.
53 Cosgrove, *The Palladian Landscape*, 131.
54 Keirstead, 'Gardens and Estates: Medievality and Space.' I would like to thank Thomas Keistead for sending this paper via email.
55 Ibid.
56 Keswick, *The Chinese Garden*; Zhang, *The Painting of Five Mountains and Ten Monasteries and Chan Temples in Northern Song Dynasty*; Zhang, *Buddhist Temples of Jiangnan in China*; Wang, *History of Buddhist Architecture in China*. The typical seven-hall layout of Chan temples, as identified by Zhang, contained no garden spaces. Other researches on Chinese temples in the context of Chinese architecture include Steinhardt, *Chinese Architecture*; Prip-Møller, *Chinese Buddhist Monasteries*; Liang, *A Pictorial History of Chinese Architecture*.
57 Wu, '*Reborn in Paradise*.' Many portrayals of religious activities in garden settings could be found in other Dunhuang murals, such as on the southern wall of cave 220, where the Buddha is depicted preaching above a lotus emerging from the water under the Bodhi tree.
58 Schopen, 'The Buddhist "Monastery" and the Indian Garden.'
59 Paul Meyvaert, 'The Medieval Monastic Garden,' in *Medieval Gardens*, ed. Macdougall, 23–54.
60 Mclean, *Medieval English Gardens*. Also see Amherst, *A History of Gardening in England*.
61 Gilchrist, *Norwich Cathedral Close*.
62 Williamson, *Environment, Society and Landscape in Early Medieval England*, 6–35.

63 Shigemori, *Japanese Garden History with Drawings*.

64 Ibid.

65 Goodwin, 'Claiming the Land: Ch ō gen and the Development of Ōbe Estate,' 231–52; Piggott, 'Estates: Their History and Historiography,' 3–36.

66 Damian, 'As Estates Faded'; Williamson, 'Estate Landscapes in England.'

67 Stackelberg, *The Roman Garden*, 9. Stackelberg categorises the garden space as the physical space and the represented space.

68 Cosgrove, *Social Formation and Symbolic Landscape*, 1–14.

69 See Mitchell's theorisation of landscape as symbolising, instrumenting, and exerting social power in *Landscape and Power*, 1–4.

70 See the water features in Roman gardens in Stackelberg, *The Roman Garden*, 35–41.

71 Rotherham and Handley, eds., *What Did Capability Brown Do for Ecology?* 49–62.

72 Bishop, *Ornamental Lakes: Their Origins and Evolution in English Landscapes*, 17–61.

73 Bridge records and their archaeological remains are inventoried in Suzhou difangzhi bangongshi, *Gazetteer of Old Bridges in Suzhou*; Qu, *Gazetteer of Rivers and Canals of Suzhou*.

74 See the map in Zhang, *Complete Record of Hydraulic Management in the Region of Wu*, vol. one, 7–10.

75 Key gazetteers consulted include *GSZ*, Niu and Wang, *Gazetteer of Wu County in the Reign of Chongzhen*; Canglang qu zhi bianzuan weiyuan hui, *Gazetteer of Pingjiang District*.

1 Garden Estates
Transforming Urban Hydrology

To transform the wetlands below the sea level, the 'canal and polder' reclamation scheme was developed in the eastern Lake Tai region during the 7th to 11th centuries. This scheme reclaimed marshy areas into agricultural polders and habitable settlements connected by canals. The current research proposes to understand the city of Suzhou during the medieval period as a large, enclosed polder. Within its encircling moat and urban walls, the grid of urban canals divided the city into small polder units. These plots gradually evolved into estates integrating residence, agriculture, and gardening.

The marshy condition of estates necessitated owners' landscaping to establish habitable dwellings and fertile fields. During the 13th to 14th centuries, temples in Suzhou developed a 'mound-field-river' topography to transform swampy land plots into sustainable estates. Mounds were placed at the highest elevation, fields as the intermediary, and rivers at the lowest. The mound, river, and dispersed architecture showcased a landscape with Chinese natural aesthetics. The hydro-topographical landscape facilitated rainwater runoff, regulated subsurface infiltration, and fostered habitats with optimal conditions for diverse crop cultivation. The rammed earth technique, common in regional land reclamation, was adapted to craft topographies on an estate scale.

Monastic estates during the 13th and 14th centuries Suzhou widely employed this landscaping technique, particularly in the city's low-lying areas. Monks interpreted sceneries and managed the topography. It can be inferred that temples, through their garden practices, accelerated Suzhou's urbanisation during the Yuan. The current study offers a micro-hydraulic perspective on the formation of cities.[1]

Canal-and-Polder Scheme and the City as a Polder

Suzhou city is nestled on the low-lying plains of the Lake Tai basin, flanked by mountain ranges and Lake Tai to the west, and a higher plain to its

DOI: 10.4324/9781003387169-2

east, as shown in Figure 1.1a.[2] Water originating from Lake Tai slows as it approaches Suzhou on its journey towards the sea (Fig. 1.1b,c). This slow-moving water characteristic led to Suzhou being named in the Song dynasty as 'Prefecture of Levelled Rivers (*Pingjiang fu*).'[3] While the low elevation ensures a steady supply of water, Suzhou also faces the challenge of waterlogging and flooding.

The wetland soil condition of Suzhou resembles that of the Netherlands and Venice during the medieval period.[4] The terrain features an abundance of wetlands, including swampy bogs, numerous ponds, rivers, and ephemeral streams.[5] The topsoil in this region primarily contains sand deposits from the Yangtze river, making it fertile and nutrient-rich, perfect for rice cultivation. Rice polders thus have been extensively constructed across this area from ancient to the present, as shown in Figure 1.1a.

Three major canals were excavated early on to regulate water flow from Lai Tai to the sea. Wusong and Lou rivers, east of Suzhou, are regarded to have taken shape as early as the Xia period (2070–1600 B.C.), with scholars attributing their creation to the great Yu.[6] Xu river, linking Lake Tai to the city, was excavated by Wu Zixu (559–484 B.C.) in 506 B.C. (Fig. 1.1c).[7] These canals not only directed waters from the lake to the sea but also drained excess water from wetlands and converged ephemeral streams. They facilitated transportation between settlements, facilitating their development into towns and cities. Initially, polders were discreetly constructed among these regional canals.

The walled city of Suzhou was first established by the King of Wu and Wu Zixu as Helü city in 514 B.C. When locating the city, Wu Zixu particularly considered the land topography to find a high land less prone to flooding. It was recorded that 'he investigated the land and water, imitated the heavens and earth, built a large city with a circumference of forty *li*, and a smaller city of ten *li*. He opened eight gates to represent eight winds' (Fig. 1.2a).[8]

The current urban moat spans around 18 kilometres in length with average width varying from 30 to 100 metres. The urban wall, initially constructed by rammed earth, was later reinforced by bricks around the 13th century. Despite being recognised by scholars primarily for its role in military defence, the urban wall also functioned as a dam for flood control, with water gates acting as sluices. During heavy rainfall, when the water level rose above safe limits, the water gates would be put down to prevent further water inflow, thus protecting the inner urban areas from inundation.[9]

Predating the collective land reclamation in Venice and the Netherlands, which began in the 12th and 16th centuries, respectively, massive land reclamation in the Lake Tai region was underway during the 9th to 10th

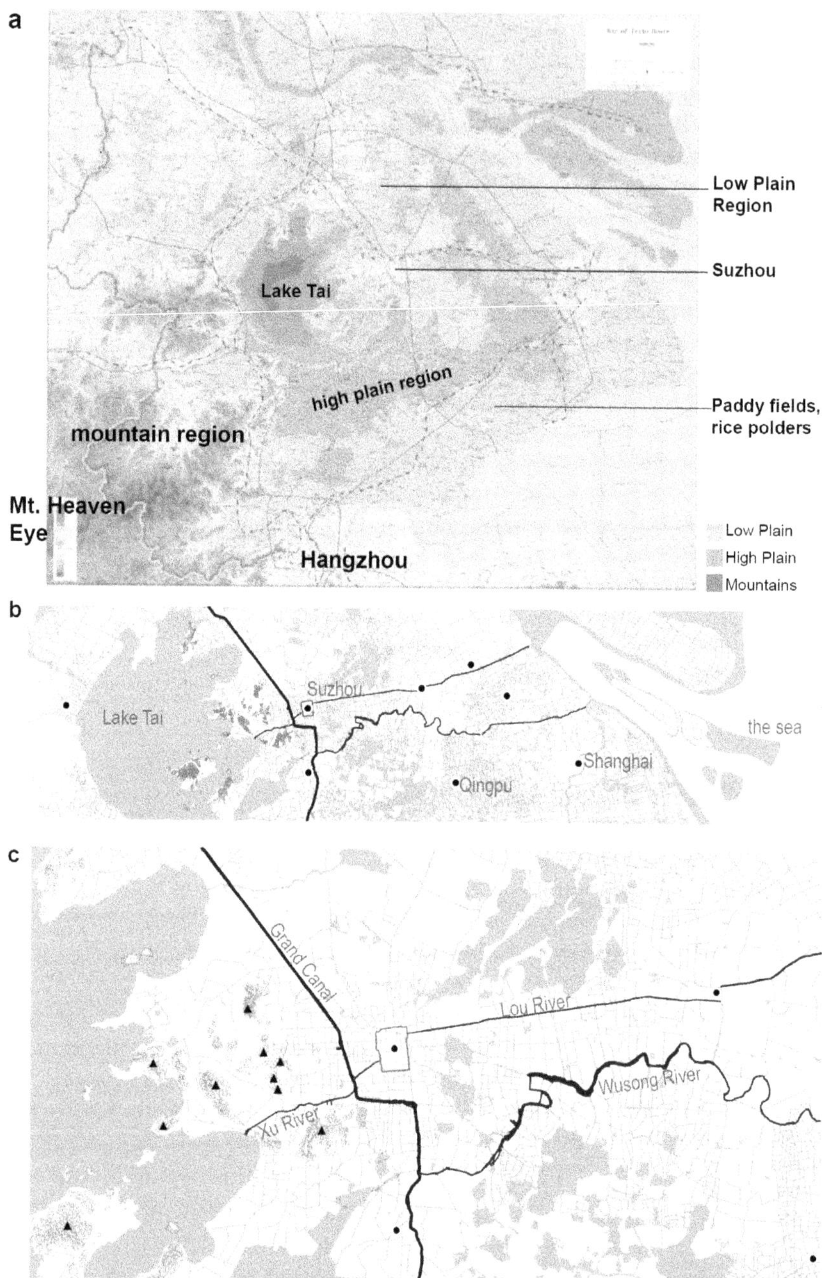

Figure 1.1 Topography and primary canals of Lake Tai region. (a) Suzhou region on the low plain region (below 5 metres in elevation) of Lake Tai estuary. Source: Courtesy of Qin Boqiang. (b), (c) Regional canals linking Lake Tai to the sea. Source: Author's mapping and drawing.

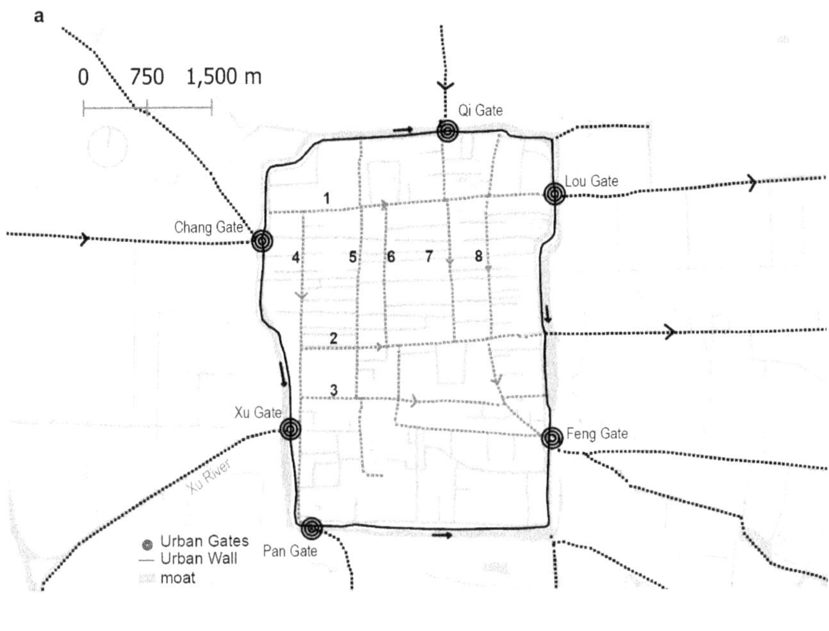

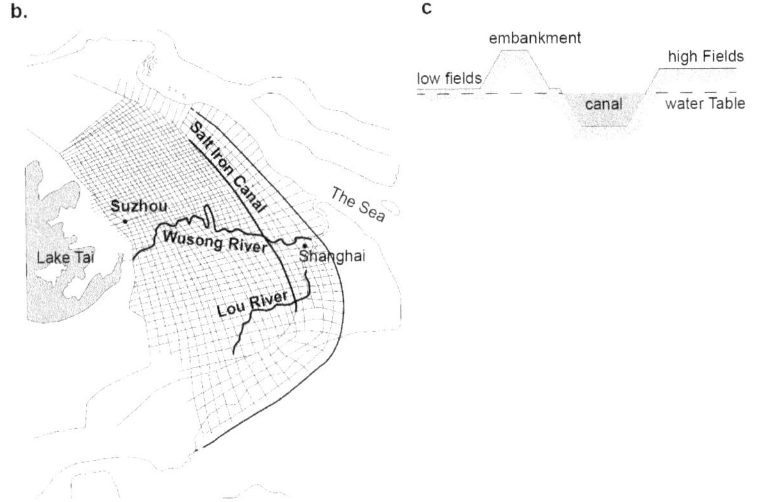

Figure 1.2 Regional and urban water management of Suzhou in history. (a) Urban canals during the Song dynasty, arrows indicating water flow direction. (b) The 'canal and polder' infrastructure in the eastern Lake Tai region. Source: Author's drawing based on Wang, *History of Hydraulic Technologies in Lake Tai Region*, 85. (c) Conceptual section of polders. Source: Author's drawing based on Wang, ibid., 122.

centuries.[10] Starting from the Middle Tang period (781–847), the government implemented a systematic hydro-agricultural strategy known as 'canal and polder' planning, or '*tang pu wei tian*' in Chinese. This approach configured a network of canals in a chessboard pattern to divide the region into polders. The latitudinal canals running in west-east direction were referred to as '*tang*,' while the longitudinal canals stretching north-south were termed '*pu*.'[11] Figure 1.2b provides a conceptual illustration of this canal-and-polder infrastructure.[12]

The government also dredged Salt Iron canal (*Yantie tang*), running longitudinally, separated the lower western plain from the higher eastern plain.[13] An embankment was likely built to protect the lower fields, as depicted in the sectional view shown in Figure 1.2c. Commissioners of Agriculture (*Yingtian shi*) were appointed to supervise labour and oversee the construction of these polders. This strategic reclamation significantly advanced agricultural development in the south of China. With the widespread implementation of labour-intensive rice agriculture and multiple cropping, Suzhou outpaced the north of China in agricultural productivity.[14]

The Wu-yue Kingdom (907–978) further enhanced this hydro-agriculture strategy, emphasising the concurrent construction of canals and polders, a practice which was lauded by hydrologists in history.[15] The official overseeing these projects was titled the 'Official of Managing Water and Constructing Fields (*Dushui yingtian shi*).' This coordinated approach ensured that newly constructed polders could be immediately irrigated from the canals, facilitating prompt agricultural use and settlement establishment. Another benefit of this co-construction was that the soil excavated from dredging canals could be used to build embankments and pile up dry foundations of architecture. In terms of flood control, the combination of canals and embanked polders effectively managed water levels.

Polders delineated by intersecting canals often underwent further landscaping to optimise agricultural productivity. *Wang Zhen's Agricultural Manual* records two types of polders which were prevalent in the Yangtze and Huai River regions. The first type, known as 'enclosed field (*wei tian*),' featured embankments stacked around the periphery to protect against flooding.[16] This surrounding embankment was reinforced with willows (Fig. 1.3a). A meandering river ran through the polder's centre, connecting it to external water sources. From this central river, channels and ditches branched out to irrigate rice and mulberry fields. Several settlements were strategically located on the higher ground.

The second type, 'sandy field (*sha tian*)' shared similarities with the first but without enclosed embankments (Fig. 1.3b). Wang Zhen (1271–1368) illustrated this type as more naturalistic, with several mounds dispersed within. Buildings were constructed on the mounds' embankments. Internal

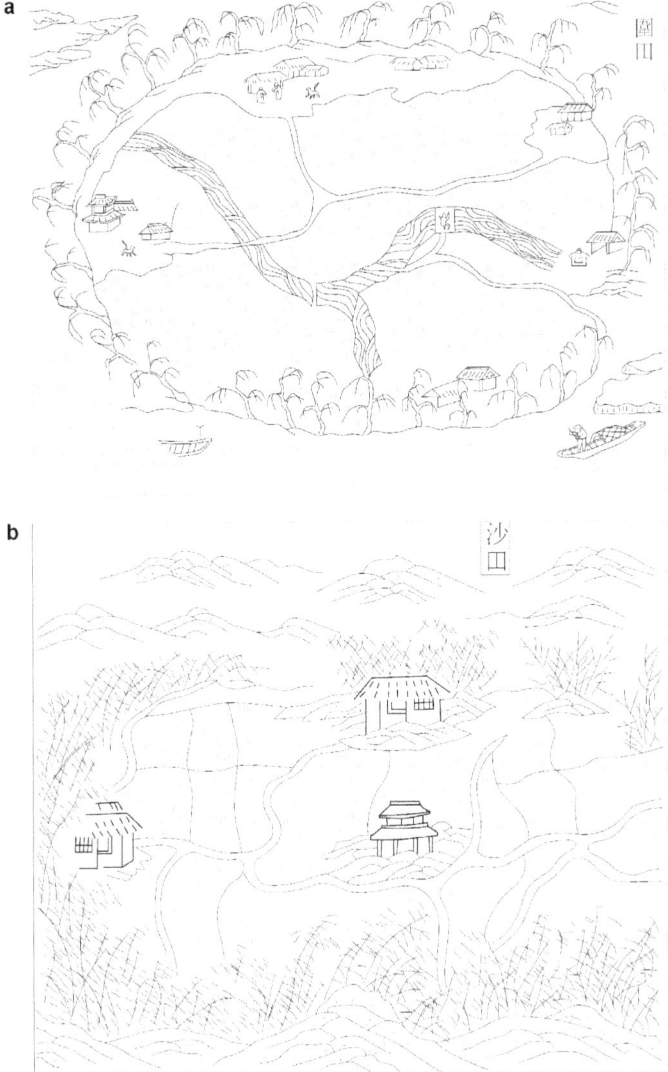

Figure 1.3 Types of polder fields in *Wang Zhen's Agricultural Manual*. (a) The enclosed field. (b) The sandy field. Source: Author's traced drawing following Wang Zhen, 185–186, 199–200.

rivers further segmented the paddy fields, and water bamboos were grown along the banks.[17]

Within this regional planning framework, I propose to understand the city of Suzhou as a large polder encircled by urban walls functioning as a dam. This perspective is validated by Fan Zhongyan's (989–1052) observation about polders in the 9–10th centuries. He noted: 'In the Jiangnan region there used to be polders, one polder is as large as around 5 kilometres (10 *li*), as a large city.'[18] The dimension of Suzhou city is approximately 4.5 by 5 kilometres, which closely matches Fan's identification of the length of one large polder. My second evidence came from Geng Ju's (fl. 17th c.) hydraulic study of historical polders in Lake Tai region, which metaphorically compared large polders to walled cities and small polders to courtyards.[19]

Wang Zhen's poem of the sandy field polder also alludes that a polder could be a city: 'Beneath, they embrace and surround the city's land; above, the sky hanging over like a canopy of heaven.'[20] If we consider the embankments in Wang Zhen's depiction to be as high as the urban wall, this imagery closely resembles that of a walled city. Supporting this, Liu Jianguo's archaeological research on pre-historical settlements and towns in China's Jianghan plain has revealed that many prehistorical cities were reclaimed polders, comprising extensive agricultural fields within and urban walls serving for flood control.[21] Furthermore, Wang Zhen's illustration of the second type of polder resembles the topography observed in the monastic estates in the Yuan.

Estates Demarcated by Canals

Pingjiang Map, dated by scholars to 1229, depicted an intricate network of urban canals within the city, arranged in a chessboard pattern.[22] The feasibility to locate its bridges to the current city allows a projection of *Pingjiang Map* onto a GIS map for further analysis (Fig. 0.5). The Mid-Tang poet Bai Juyi's (772–846) mention of 390 bridges has led scholars to surmise that the urban canal system depicted in *Pingjiang Map* may exist around the early 9th century.[23] The 17th-century hydrologist Zhang Guowei (1595–1646) later identified the major urban canals as 'three horizontals and four verticals.' The major canals would be 'three horizontals and five verticals' during the 9th to 14th centuries, considering the third vertical identified by Zhang, Jinfan canal (*Jinfan jing*), gradually diminished during the 14th to 16th centuries. According to the Suzhou government's survey of rivers, I identified the flowing direction of each major canal in Figure 1.2a.[24]

Similar to the intersecting regional canals which divided the eastern Lake Tai region into large polders, the canal fabric inside the city segmented the land into smaller plots, potential for future polders and estates. *Pingjiang Map* recorded that many large estates occupied a single land plot (Fig. 0.5a). These include notable temples such as Repay Kindness Temple (*Bao'en si*) and Heavenly Bestowed Benevolence Temple (*Chengtian Nengren si*). The inner citadel, initially an imperial estate as the palace of

Wu-yue Kingdom, later became the seat of local government and encompassed the largest land unit. Noble families' garden estates, such as Han Shizhong's (1089–1151) Surging Wave Pavilion and the Yang general's garden, each owned one large land plot. Meanwhile, most of the land plots were divided into contain two or more owners' properties.

Considering scholars' research on the urban transition from residential ward planning into a market-oriented economy during the Tang-Song period, it could be deduced that in the Tang dynasty, the 60 residential wards were walled. After the demolition of the city in Song-Jin war, the walls of these residential wards were not rebuilt during urban reconstruction, leaving only the gates as historical markers. Translating the location of residential wards and temples in *Pingjiang Map* into GIS map and incorporating the elevation contours (Fig. 1.4), it becomes evident that during the 9th to early 13th centuries of the Tang-Song period, residential wards, temples, and official buildings were primarily situated on the city's higher plains above 10 metres, especially the central area where the inner citadel is situated, which occupied a plateau reaching 15 metres in elevation. In contrast, during the Yuan dynasty, temples were often built in the marshy areas at the city's four quadrants, at an elevation around 6 metres, as exemplified by Grand Clouds Cloister (*Dayun an*) and Great Propagation Temple. This preference for higher lands is also observed in the urban planning of Chang'an, the establishment of settlements in Venice, and prehistorical settlements in the Jianghan plain.[25]

Land was typically not considered as private property during the Tang dynasty, nor was it commoditised due to the implementation of the Equal Field System (*Juntian zhi*).[26] In the Song dynasty, as landed properties became purchasable, individuals and local communities exerted increased control in landholdings and their estates developed. One earliest known land deals in Suzhou was made by Fan Zhongyan, who purchased approximately 0.68 square kilometres of land near the city wall to establish inheritable lands for his clan.[27] Another notable acquisition was by Su Shunqin, who bought the site of Surging Wave Pavilion for 40,000 taels. These examples from local gazetteers indicate that estate transactions were not only feasible but also recorded.

In the 13th century of the Yuan dynasty, the development of temple estates gained momentum.[28] Two monastic transaction records provide valuable insights into the layout and configuration of these estates. One stele record details the strategic expansion of Capable Kindness Cloister (*Nengren an*, Fig. 1.4(1)), a modest cloister situated on the city's low-lying plain. This cloister was juxtaposed by public canals to its north and south:

Zhen, who viewed Suzhou as a base for the travel and visitation of Chan monks, lacked a permanent residence for them. He then bought the land

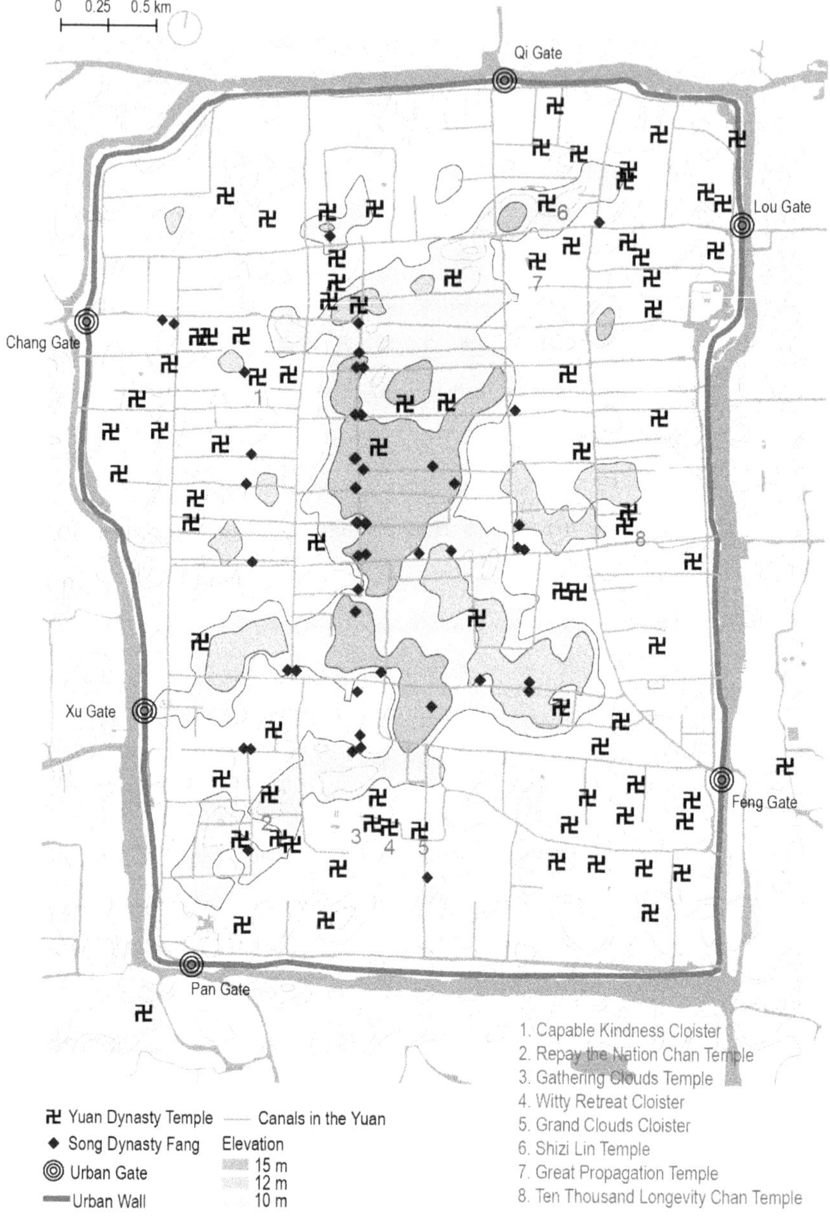

Figure 1.4 GIS map showing the Tang and Song residential wards on high ground, and Yuan dynasty temples on low-lying ground in Suzhou city. Source: Author.

from the Yan family near Yangjia bridge in 1322. The land bordered the river to the north and south, and it extended approximately 72.7 meters. The cloister was named 'Capable Kindness (*Nengren*)', to reside the itinerant monks.

Lay Buddhist Guo'an supported this by donating funds for the creation of Avalokitesvara and Arhat statues from the entrance towards to the residential quarters. Additionally, land to the right of the cloister, belonging to the Yu family of Lake Tai, was purchased to create a garden, stretching approximately 25.9 meters in length east to west.

In 1331, [Zuzhen further extended the area] by purchasing land from the Dai family to the left of the cloister to create another garden. This garden reached rivers to the north and south, and stretching about 26.52 meters east to west. [Zuzhen's purchase of lands] extended to Songlin [district], where he bought fields about 0.04 square kilometres to support the community. What his heart intended, the donors fully responded.[29]

The record shows the components of a Yuan dynasty monastic estate. The architectural part, 'the basis of the cloister (*an ji*),' accounts for 12%. It was possibly in a traditional Suzhou residential layout, with courtyards arranged in sequence.[30] Complementing this were two gardens located to the east and west, making up 9% of the estate. The majority, 80%, is composed of agricultural fields outside the city for food supply. Their expanse was comparable to Unsuccessful Politician's Garden in the Ming period. According to Wang Zhen's agricultural writings, which suggest 10 multiples of *mu* is sufficient for a small group, the estate's 62 *mu* fields could feasibly support a community of around 30 monks.[31] This layout reflects a harmonious blend of the practicality in architecture, aesthetics in garden, and productivity in the fields.

The second stele record is about Repay the Nation Chan Temple (*Baoguo chansi*, Fig. 1.4(2)) written by Zhu Yunming (1460–1526):

[He] brought the pictorial register of the foundation of the temple, which was established by Master Tong, to me for a record. According to the record of that time, the temple stretched about 214 meters (700 *chi*) to the south of the road and north of the moat. Its east was near Huoshao pond, and west towards the Red and White Spider Ditch, spanning 122 meters (400 *chi*). It faced the official thoroughfare in the front, surrounded by water on three sides. The temple also managed vast mountains and fields. It also had seven sub-temples under its name, truly magnificent.[32]

Similar to Capable Kindness Cloister's adjacency to the urban canal, Repay the Nation Chan Temple was flanked on three sides by water bodies: an official canal, a ditch, and a pond. In addition to this, it possessed agricultural fields outside the city, crucial for sustaining the community's daily food consumption. In contrast to the brief records of Song dynasty estate transactions, the Yuan dynasty temples' detailed documentation of estate boundaries and purchases through registers, maps, and steles reveals a sophisticated approach to estate management. This comprehensive documentation marked a significant shift in the way property rights were defined, recorded, and transferred, indicating advancements in land administration. It points to potential changes in the legal and cultural frameworks concerning land ownership and transaction.

Landscaping Inside Monastic Estates

When Yuan monks purchased land for establishing temples, particularly in the low-lying areas, what site conditions did they encounter? Insights into this are provided by Su Shunqin's account of the Surging Wave Pavilion. He highlighted the significant waterlogging in these areas, covering an expanse as large as 0.67 square kilometres. There were also ephemeral streams where he could navigate by boat. The land was densely covered with trees, grasses, and bamboo, giving the impression of a secluded suburban landscape rather than an urban setting.[33] These details suggest a typical wetland environment, highlighting the undeveloped and water-abundant land plots inside the city.

Modelling Shizi Lin Temple in the Yuan Dynasty

The current research reconstructs a model of Shizi Lin temple in the Yuan to investigate the landscaping strategies employed by monks. Modelling the 14th-century temple involves a synthesis of surveyed drawings, archaeological remains, historical maps, visual depictions, and written records, with the consideration of the site's constructions across different phases (Table 3.1). Given the often unmeasured and subjective nature of Chinese painting, textual records serve to verify pictorial evidence and distinguish construction facts from artistic imagination. Historical maps also offer more precise geographical information.

Though historical writings often follow a chronological structure—subsuming to this convention this book presents Yuan temples in Chapters 1 and 2, followed by the Ming practices in Chapters 3 and 4—the modelling process for a Yuan dynasty temple such as Shizi Lin adopts a reversed order. It begins with an analysis of the current site, situating it in the chronology of the site's construction and restoration. From there, the process works backwards. It constructs a model of the site in an imperial period when historical resources were most comprehensive. By tracing the construction, alteration, and removal of artefacts, vegetation, and landforms from each historical period, the temple's appearance in its earlier stages could be comprehensively deduced.

(1) Establishing historical boundaries in GIS map.

The biography of Shizi Lin (Table 3.1) indicates that in the 1640s of the Early Qing period, the site of Shizi Lin was divided into the northern Lion Grove Garden and the southern temple part. *Map of Gusu City* (1745, Gusu chengtu) shows the site was bordered by Panru valley to its north and Lindun road to its west. Its southern street corresponds to Qingli Bridge, and the eastern street corresponds to the current Yuanlin road.[34] Thus the boundary of the site before division could be identified in the GIS map (Fig. 1.5a), with the surveyed plan of Lion Grove Garden made by Liu Dunzhen's team in the 1950s being merged into it.

(2) Translating Shizi Lin painting from *Grand Ceremony of the Southern Inspections (Nanxun Shengdian)*into the site model of the 18th century.

Gao Jin's (1706–1779) painting 'Shizi Lin' in *Grand Ceremony of the Southern Inspection*, created in 1771 (Fig. 1.5b), offers a comprehensive portrayal of Shizi Lin site during the High Qing period. In all the images of Figure 1.5, the pond and mound are indexed as 1 and 2, respectively. The central mound and the pond at the current site roughly correspond to those in Gao Jin's painting, though the painted mound area was larger. Other artefacts in Gao Jin's painting are numbered from 3 to 16. Further correspondence could be established between the archaeological remains of the current site (Fig. 1.5a) and artefacts in Gao Jin's painting (Fig. 1.5c), including the northern hall facing the pond (3), the northern pavilion facing the mound (4), and the bridge (5). The current platform to the east of the bridge (a6) corresponds to the imperial stele pavilion (b6), and the location of the current courtyard to the south of the mound (a7) corresponds to that in the 18th century (b7). The current residential hall (a8) corresponds to the imperial book tower (*yushu lou*, b10).

The wall between Lion Grove Garden and the southern temple is deduced from the current south wall of the garden (a9). The current location of the five pines in Gao Jin's painting could also be relatively deduced. Thus the plan of Lion Grove Garden and the southern temple in the 18th century is deduced as in Figure 1.5c, and a model is further constructed in Figure 3.10.

(3) Deducing Shizi Lin Temple in the Yuan dynasty (14th c.).

The area of Shizi Lin Temple in the 14th century consisted of both the northern and southern parts. Wang Yi's (d.1374) *Record of Touring Shizi Lin* identified the temple as featuring a central mound encircled by a river that connected to a public canal (see Appendix, Translation 1).[35] Assuming

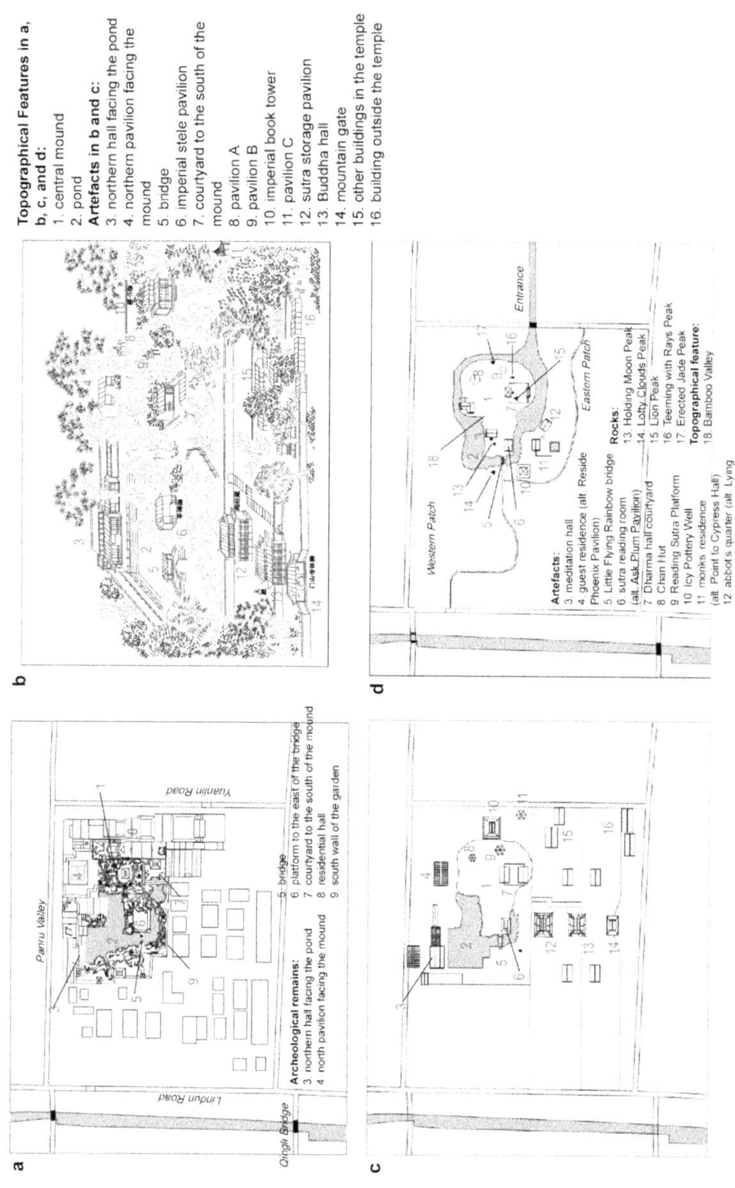

Figure 1.5 Deducing layout of Shizi Lin Temple in the Yuan. (a) The current plan of Shizi Lin in the GIS map. (b) Gao Jin, Shizi Lin, in *Grand Ceremony of the Southern Inspections*, 1771. Author's traced drawing based on Gao Jin. (c) Plan of Shizi Lin during the Qing (18th c.). (d) Plan of Shizi Lin in the late Yuan and Early Ming (14th c.). Source: Author.

the mound retained its shape during the 14th to 18th centuries, despite its ground subsidence due to a drop in urban groundwater level, the location and form of the mound during the 14th century are deduced in Figure 1.5d. The course of the surrounding river was inferred by connecting the pond of the 18th century to the urban canal located east of the site in the 14th century.[36] The varying width of the riverbed was deduced by referring to Xu Ben's painting album *Twelve Scenes of Shizi Lin*.

An inventory of the temple's artefacts, rocks, trees, and agricultural plantations could be created by referring to Xu Ben's *Twelve Scenes of Shizi Lin*; Wang Yi's, *Record of Touring Shizi Lin*, and *Anthology of the Recorded Scenery of Shizi Lin* (Shizi Lin jisheng ji). The major ten artefacts in the temple, as indexed in Figure 1.5d, include: the meditation hall (d3), the guest residence (d4, also named 'Reside Phoenix Pavilion (*Qifeng ting*)'), Little Flying Rainbow bridge (d5, *Xiao feihong*), the sutra reading room (d6, also named as 'Ask Plum Pavilion (*Wenmei ge*)'), the Dharma hall courtyard (d7), Chan Hut (d8, *Chan wo*), Reading Sutra Platform (d9, *Fanjing tai*), Icy Pottery Well (d10, *Binghu jing*), the monks' residence (d11, also named 'Point to Cypress Hall (*Zhibai xuan*)'), and the abbot's quarter (d12, also named 'Lying Clouds Chamber (*Woyun shi*)').

There were six rocks recorded, including Holding Moon Peak (d13, *Tuyue feng*), Lofty Clouds Peak (d14, *Angxiao feng*), Lion Peak (d15, *Shizi feng*), Teeming with Rays Peak (d16, *Hanhui feng*), Erected Jade Peak (d17, *Liyu feng*), and Residing Clouds Peak (*Qiyun feng*).[37] Trees recorded include lots of pines and bamboos, as seen in Xu Ben's leaf no.5 *Chan Hut* and leaf no. 6 *Bamboo Valley* (*Zhu gu*) (Fig. 1.6a), respectively.[38] Monk Tianru's record mentioned rows of cypress to the walls and the eastern and western vegetable patches.[39]

Several artefacts' locations in the 14th century could be traced to those in the 18th century and further to the current site, including the Dharma hall courtyard, Ask Plum Pavilion, and Little Flying Rainbow bridge. Lion Peak rock inside the Dharma hall courtyard as depicted by Xu Ben's leaf no. 1 (Fig. 0.4a) thus could be located. Giving their comparable size and scale, the location of Reading Sutra Platform and Chan Hut could be deduced. The corresponding relationships between artefacts are identified in Table 1.1.

Referring to Wang Yi's detailed record (see Appendix, Translation 1), the other seven artefacts' locations could be deduced. Wang Yi used Lion Peak (d15) to locate other rocks and identified Teeming with Rays Peak (d16) to the east of Lion Peak, Holding Moon Peak (d13) to the west of Lion Peak, and Chan hut (d8) to the north of Lion Peak. He also mentioned that Teeming with Rays Peak (d16) was to the west of Reading Sutra Platform (d9), and Erected Jade Peak (d17) was also next to this platform. The location of Bamboo valley (d18) was to the north of the mound

Table 1.1 Correspondence of Artefact Locations in the 14th Century, 18th Century, and the Current

The 14th c. (Fig. 1.5d)	The 18th c. (Fig. 1.5b,c))	The current (Fig. 1.5a)
Little Flying Rainbow bridge (d5)	The bridge (b5, c5)	The bridge (a5)
Ask Plum Pavilion (d6)	Imperial stele pavilion (b6, c6)	Platform to the east of the bridge (a6)
Dharma hall courtyard (d7)	Courtyard to the south of the mound (b7, c7)	Courtyard to the south of the mound (a7)
Chan Hut (d8)	Pavilion A (b8, c8)	None-exist
Reading Sutra Platform (d9)	Pavilion B (b9, c9)	None-exist

behind Ask Plum Pavilion (d6), thus the guest residence (d4) in this valley could be located (Fig. 1.5d). The meditation hall (d3) was facing Holding Moon Peak (d13), as depicted in Xu Ben's leaf no. 3 (Fig. 2.13b). The locations of Lying Clouds Chamber (d12) and Point to Cypress Hall (d11) were to the south of the Dharma hall courtyard. Wang Yi also recorded that 'the location of Point to Cypress Hall (d11) was aligned with Ask Plum Pavilion (d6).' The pavilion of Icy Pottery Well (d10) was identified by Wang Yi as to the west of Ask Plum Pavilion (d6). Xu Ben's *Twelve Scenes of Shizi Lin* (Fig. 1.6) plays a crucial role in modelling the appearance of buildings and rocks. By rotating the 3D model to simulate views in Xu Ben's painting, which depicted spatial relations between the mound, artefact, and the river, it is feasible to further justify and verify the configuration of artefacts in the model.

Hydro-analysis: Shizi Lin Temple

The historical model of Shizi Lin is instrumental in analysing the hydraulic process that occurred in the topography, and it aids in further deducing the topo-hydro patterns applied in the Yuan dynasty Suzhou temples. The isometric model in Figure 1.7a shows that the temple featured a topographical pattern consisting of mound, fields, and river. The sectional drawing in Figure 1.7b further clarifies these areas as located on three distinctive elevation levels from high to low. At the highest level is the mound, where rainwater flows down along the slope. The middle tier is occupied by agricultural fields. In Shizi Lin these fields were planted with vegetables. The lower level was the riverbed which channelled water to the fields. The embankment at the mound's base provides a sturdy foundation for architectural structures, including the Dharma hall courtyard, monks' residence, and the abbot's quarter.

Figure 1.6 Leaves in *Painting of Shizi Lin*. (a) 'Bamboo Valley,' leaf no. 6. (b) 'Ask Plum Pavilion,' leaf no. 10. Source: Taipei: Collection of the National Palace Museum.

It could be deduced that the 'mound-field-river' topography leverages gravity to maintain water equilibrium. In dry seasons, the river within the estate acts as a reservoir, ensuring a steady water supply essential for irrigation and the landscape's aesthetic appeal. During rainy seasons, the topography facilitates rainwater run-off, as analysed in Figure 1.7a. The

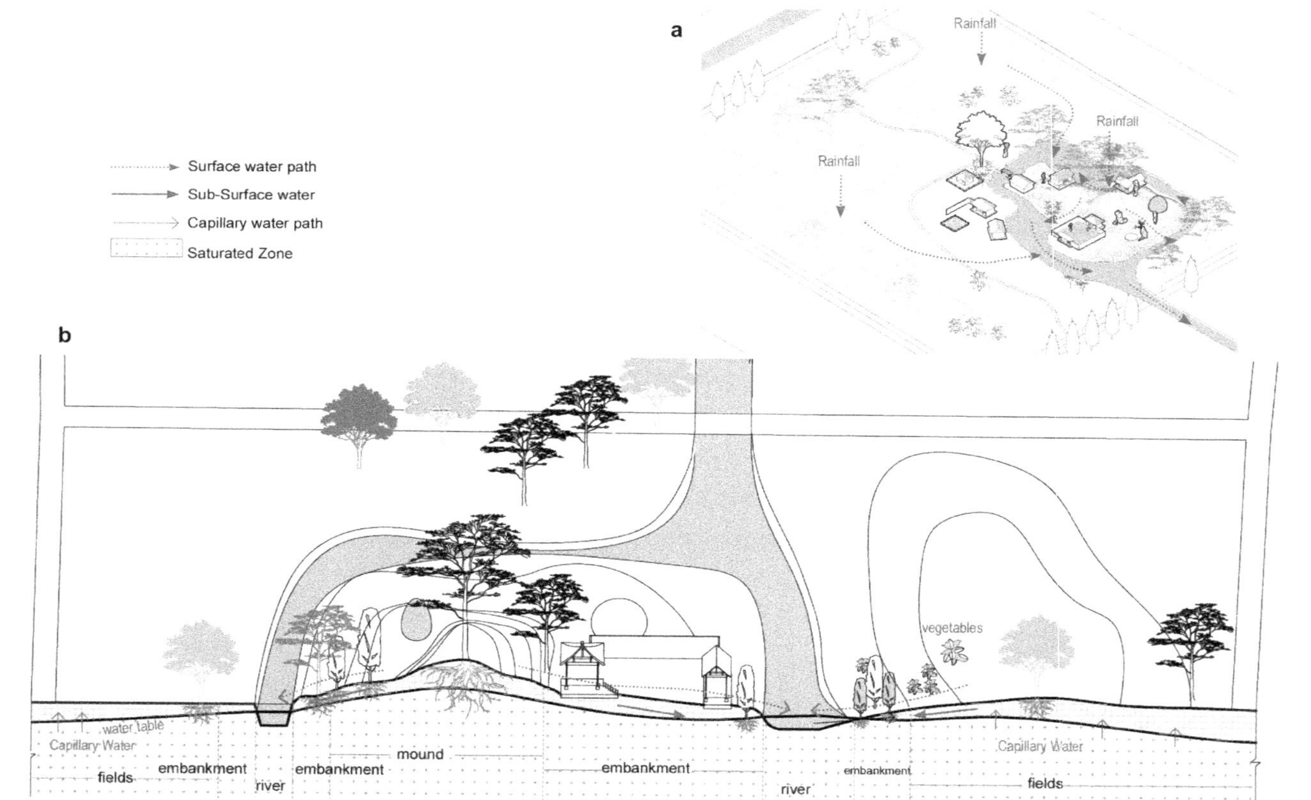

Figure 1.7 Sectional drawing analysing the hydrology of Shizi Lin Temple in 1342, Yuan Dynasty. (a) Surface water run-off. (b) Sub-surface water flow. Source: Author.

slope of the mound ensures the direction of excess rainwater to the river by gravity. The valley's concave shape also collects and channels the excess rainwater runoff. The river at the lowest level then discharged the water into the public canal.

A detailed reading of Xu Ben's leaves *Bamboo Valley*, *Ask Plum Pavilion*, and *Little Flying Rainbow* (Fig. 1.6a, Fig. 1.6b, Fig. 2.12a) reveals two other intricate topographical manipulations of the riverbed. First is the altitude difference of the riverbed. The northwest part of the riverbed was higher than its southeast part, thereby utilising gravity to enhance water flow. The cascades formed by the water running across different heights, as depicted in Figure 1.6, not only add visual and auditory appeal but also aerate the water, improving its quality by increasing oxygenation.

A second topographical characteristic of the riverbed is its varying width, alternating between broad and narrow. This serves to regulate the speed of water flow. In the narrow channel, as shown in leaf no. *6 Bamboo Valley* in Figure 1.6a, the water velocity increased. This would prevent stagnation and sedimentation, keeping the water fresh and reducing the likelihood of algae bloom and mosquito breeding. In wider channels, as depicted in leaf no. *9 Ask Plum Pavilion* in Figure. 1.6b, the slower movement of water would encourage the deposit of sediments, further cleansing the water and creating habitats for aquatic life.

Applying Rosemary Charlton's water pathway model in the section of Shizi Lin temple illustrates how the 'mound-field-river' topography creates a mosaic of habitats suitable for artefacts and various types of cultivation (Fig. 1.7b).[40] This topography not only promotes surface water flow but also regulates subterranean water flow. Consequently, there is less water retention in the mound and fields, leading to drier conditions that are suitable for vegetation thriving in dry soil. This also explains why there were abundance of pines on the mould, since pines are well-adapted to grow in less fertile and drier soils compared to other tree species.

The embankment at the base of the mound was situated above the water table level and distanced from saturated zones. This location offers a dry foundation for architecture by preventing water percolated from the saturated zone. The placement of the architecture at the sloping end of the mound also ensures that the surface water was efficiently discharged into the river during rainy seasons.

The placement of vegetable patches at lower elevations near the river takes advantage of the naturally high moisture content of the soil. Due to their adjacency to the saturated zone, the vegetable fields absorb water from the underground, reducing the frequency of artificial irrigation. The closeness to the river also allows for easily fetching water for irrigation. Tianru mentioned that there were eastern and western patches in the temple.[41] In a poem, he depicted monks carrying water to irrigate vegetables, suggesting that the temple was cultivating its vegetables, while grains were sourced externally.[42]

By transforming the wetland into a garden with manageable hydrology and shaping the topography on both macro and micro levels, Tianru established his temple as religious, habitable, and productive. Although this landscape appears natural in Xu Ben's painting, it is actually an exquisite example of hydraulic adaptation. The Yuan Hanlin academician Zhang Zhu (1287–1368) appraised Tianru's conversion of the wild land into a garden:

> The old Chan master of the temple retreated to dwell and obtained this famous garden. Secluded among the extraordinary cliffs and ravines, it is distanced from the hustle of dusty market.
>
> [He] Cleared paths by cutting off the ancient dense grass, dredged a pond to reveal its true origin. Reside Phoenix Pavilion faces the bridge across the ravine, and Ask Plum Pavilion is adjacent to Point to Cypress Hall.[43]

Hydro-analysis: Temples Around Surging Wave Pavilion

The isometric models and sectional drawing of Grand Clouds Cloister reveal a similar hydro-topographical pattern in water management. The temple was founded by monk Shanqing (ca. 14th c.) in 1340 of the late Yuan. It similarly followed the configuration of mound, field, and river, arranged on three altitudes from high to low (Fig. 1.8). The mound formed the highest level, the fields occupied the second elevation, meanwhile the pond was situated at the lowest. A large pond was dredged in front of the temple, its name 'Release Life Pond (*Fangsheng chi*)' indicates the fishes raised in, and it was likely dredged by the monks.

The surface and subterranean water flow facilitated by the topography similarly contributed to a mosaic of habitats, suitable for plants' various moisture needs from the soil. Shen Zhou's painting of this temple identifies a planting scheme adapted to the topography (Fig. 1.9). The mound located at the highest elevation was covered with evergreen trees such as juniper and pines on the upper slope to prevent soil erosion. It can be further deduced that the roots of these trees, by holding the soil of the artificial mound together, reduced the impact of water runoff which could wash soil away. Moreover, the canopies of these trees can lessen the force of rain hitting the ground, further mitigating soil erosion.

Water bamboos were planted on the ridge, in the valley, and along the embankment. Their extensive roots can grasp and integrate soil to prevent erosion. There were also willows and sophora trees planted on the embankment.[44] The flat and large area to the north of the temple, adjacent

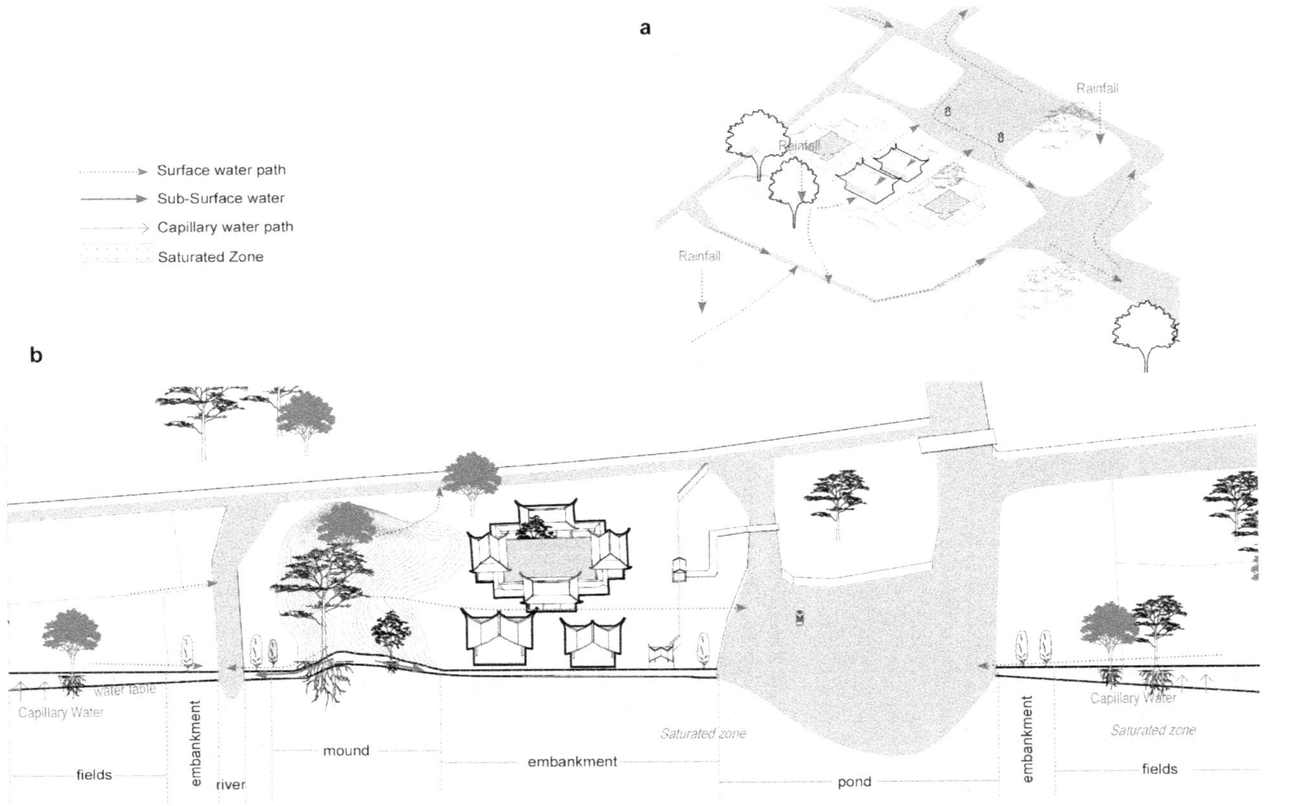

Figure 1.8 Sectional Drawing analysing the hydrology of Grand Clouds Cloister in the Yuan Dynasty. (a) Analysis of surface water run-off. (b) Analysis of sub-surface water flow. Source: Author.

Figure 1.9 Shen Zhou, *Painting of Thatched Cloister* (Cao an tu juan), 1497, handscroll, ink, and colour on paper, 29.5×155 cm. Shanghai: Shanghai Museum.

to water, encouraged the monks to lay out paddy fields for growing rice. Shen Zhou depicted the bunds which further divided the field into rice paddy plots, helping to retain water within each plot. It could be deduced that placing rice paddy fields on the lower altitude next to water would ensure consistent irrigation.

Another detail from Shen Zhou's painting is the slight tilt of the embankment towards the water (Fig. 1.9). The slope of the ground within the architectural complex ensures that water flowing from the mound is quickly drained into the river, as illustrated in Figure 1.8a. This prevents waterlogging in the courtyard and maintains a comfortable living. This topographical configuration was praised by Shen Zhou as exemplifying the best geomantic feature: 'Geographers say it possesses all four mythical creatures. The wind and spirit are rich and vibrant. From this perspective, observing all the monasteries in the city of Wu, none can compare.'[45]

Hydro-analysis: Great Propagation Temple

The third case is Great Propagation Temple, situated in the less developed northeast area proximate to a major canal to its south. The 3D modelling process of this temple will be elaborated in Chapter 3. As shown in Figure 1.10a, the temple featured a river at its centre, which connected to the major canal outside. Some part of the river was enlarged to form a pond. The temple featured ten mounds along the shore.

The rainwater run-off analysis in Figure 1.10 shows that during rainy days, rainwater would follow diverse pathways to converge in the pond and the river. The surface water runoff scheme, as analysed in Figure 1.10, could generate soil with varying degrees of moisture and solidity. The slope of mounds would be particularly suitable for fruit trees and vegetables, though without textual evidence of this temple during the Yuan era we could hardly conclude its plantation.

As deduced in Chapter 3, the temple similarly had trees such as pines, elms, and junipers planted on tops and slopes of the mounds. Bamboos were planted along the valley, and willows along the shore. They later became characters in the scenes created by Ming artist Wen Zhengming, such as *Xiang Bamboo Valley* (*Xiangyun wu*, Fig. 3.12a), *Listening to Windblown Pines* (*Tingsongfeng chu*, Fig. 3.19a), and Thoughts Afar Terrace (*Yiyuan tai*, Fig. 3.19b).

Crafting Topography

The topo-hydro patterns and the architectural and planting schemes in these three temples allow a further deduction of the process of crafting topography in Yuan monastic estates. The first was to dredge an inner river which bordered the planned agricultural fields meanwhile connected with the outside urban canal. Employing rammed earth technique, the excavated soil could be used to enhance embankments or pile up the mounds. When moulding the topography, stones and timber stilts were used to anchor and fortify the shore. This rammed earth technique was a traditional method widely used in reclaiming wetlands, building dikes, and constructing the foundations of artifacts since ancient times.

After establishing the landform, strategic planting was used to reinforce the terrain. Timber trees, such as pines and cypress, were usually planted around the summit of the mound. Their deep and robust roots prevent soil erosion and maintain the mound's structural integrity. The lush branches and foliage of these trees resist the top-down pressure of wind and rainfall, further securing the mound. Along the shore, species with sprawling root systems, such as bamboo and willow, were planted. Their extensive roots bound together the discrete particles of earth, clay, and gravel from dredging, forming a cohesive, organic barrier against the forces of water and aiding in sediment accumulation.

Slopes of mounds are well-suited for fruit trees because they ensure good drainage, preventing water from pooling at the roots and causing rot. The slope aids in directing excess water away from the trees while maintaining sufficient moisture. Vegetable patches could be placed on the higher fields,

Figure 1.10 Sectional drawing analysing the hydrology of Great Propagation Temple in the Yuan Dynasty. (a) Surface water run-off. (b) Sub-surface water flow. Source: Author.

which are less prone to flooding, and the soil is likely to be well-drained, ideal for vegetables that require well-aerated soil. Rice paddies are being placed on the lower fields.

The initial phase of constructing the estate's hydraulic infrastructure, which involved extensive dredging and piling up earth to create mounds and embankments, was a labour-intensive and costly endeavour. This foundational work is more feasible for owners with an enduring land ownership, such as temples. These religious institutions benefited from an organised community of monks and the capability to ensure consistent, multi-generational site management. Their long-term commitment to the land was critical for transforming wetlands into productive and sustainable estates.

Conclusion

The three temples, Shizi Lin, Grand Clouds Cloister, and Great Propagation Temple, demonstrate the innovation of monks in the Yuan dynasty in creating hydraulic infrastructure within their estates, particularly in low-lying urban areas where wetland conditions required further landscaping. The hydro-topographical model of 'mound-field-river' leveraged natural topography to control rainwater runoff, facilitating a self-sustaining ecosystem. The mounds provided elevated ground safe from floods; the fields mediated between the dry and saturated zone for agriculture; and the rivers ensured water supply and irrigation. Their placement at different elevations with sloping surfaces ensures efficient surface water flow, regulates sub-surface infiltration, and further creates diverse habitats for plant growth.

This topo-hydrological scheme transformed marshy wetlands into habitable monastic estates capable of sustaining temple communities through agricultural production, which usually took up at least half of a temple's monastic land. The hill-like topography also facilitated garden scenes supporting Buddhist rituals, as observed at in Shizi Lin Temple, which will be discussed in the next chapter. These monastic estates were predecessors to the productive gardens in the Middle Ming and aesthetic gardens in the subsequent Late Ming and Qing periods.[46] The collective hydraulic reclaiming strategy in medieval China, the canal-and-polder scheme, was transformed into the mound-field-river topography practiced in monastic estates. The study of the hydraulic pattern in these temples provides referential models for understanding the 40 new temples constructed inside the city and the 69 new temples constructed outside the city, for which pictorial and textual evidence is inadequate.

Notes

1 My gratitude to Gavin Coates for his insightful discussions with me on the hydrological process in the urban landscape and his revision for initial diagrams.
2 This elevation of the western Lake Tai region is described by Qin, *Lake Taihu, China*, 2–5.
3 Zhang Guowei provided a detailed hydrological description of Suzhou: 'Suzhou Prefecture was originally wetland. It is high in the east, west, and north, but only the southwest is low and sunken. Lake Tai (region of Ju) is an expansive and boundless area of water. From there, the water flows into ports and channels of the northwest border. Additionally, various waters south of Liangxi converged at Suzhou, thus making Suzhou a true marsh country. However, here the water momentum gradually becomes gentle, which is why it is called "Pingjiang" (meaning "level river").' Zhang Guowei, 'Pictorial Explanation of Water Hydraulics of the Entire Region of Suzhou Prefecture (Suzhou fu quanjing shuili tushuo),' in *Complete Record of Hydraulic Management in the Region of Wu*, ed. Zhang and Cai, vol. 1, 5. The name 'Pingjiang' dates back to 974 in the Northern Song dynasty, as noted in Wang et al., eds., *Suzhou Shigang*, 129.
4 Christian Nolf et al. conducted a comparative study of the western Lake Tai region and the Rhine-Meuse-Scheldt Delta in the Netherlands, identifying both as low-lying, nutrient-rich wetlands. The highlighted differences are: the sediment load of the Yangtze river delta (the western Lake Tai estuary) is a hundredfold greater than that of the Rhine, and the Rhine-Meuse-Scheldt Delta's wetlands are more sanitised. See Nolf et al., 'Delta Management in Evolution.'
5 The challenge of wetland feature of the region was consulted by King Helü with Wuzi Xu on establishing the city. The King expressed his concerns about the military and environmental difficulties: '吾国在东南僻远之地, 险阻润湿, 有江海之害. 内无守御, 外无所依. 仓库不设, 田畴不垦, 为之奈何? Our country is located in a remote and secluded area in the southeast, characterised by treacherous terrain and damp conditions, plagued by the perils of rivers and seas. Lacking defences internally and without support externally, with no granaries established and fields left uncultivated, what can be done about this situation?' See Zhao, *Wu and Yue in Spring and Autumn*, 30–31. Yang et al.'s mapping shows that the eastern Lake Tai region has lots of wetlands, which offer referencing soil samples for studying historical wetland conditions. See Yang et al., 'Knowledge-Based Raster Mapping Approach to Wetland Assessment.'
6 See Zhen, *History of Hydraulic Technologies in Lake Tai Region*, 21–22, 268–269.
7 See Qu, ed., *Gazetteer of Rivers and Canals of Suzhou*, 82.
8 '阖庐乃委计于子胥, 使之相土尝水, 象天法地. 筑大城周四十里, 小城周十里. 开八门以象八风.' In Zhu, *Supplementary Records to the 'Illustrated Guide to Wu Commandery*,' 5.
9 The archaeological discovery of a sluice gate at the Qi Gate in Suzhou is detailed in the report by Archaeological Team of Suzhou Museum, 'Reports on Excavation of the Foundation of Ancient Water Gates in Suzhou.' Additionally, consult Jia Dan's 郏亶 (1038–1103) 'Shang Shuili Shu 上水利書 (On Water Hydraulics)' written in 1070, where he discussed the role of the Qi Gate as a dam: 'The old city of Suzhou used to have dams at all five gates. Nowadays, the area below the city walls is commonly referred to as "below the dam", and the Qi Gate still retains the name "old dam". This implies that once the dikes

were complete, there was no place for water to accumulate. The dams were constructed out of the fear that a sudden surge of water might flood into the city. 故蘇州五門舊皆有堰. 今俗呼城下為堰下, 而齊門猶有舊堰之稱. 是則堤防既完, 則水無所瀦容. 設堰者, 恐其暴而流人於城也.' In Zhang, ed., *Complete Record of Hydraulic Management in the Region of Wu*, 492–508.

10 See Ammerman, 'Venice before the Grand Canal'; Adams, 'Competing Communities in the "Great Bog of Europe",' 35–76.

11 Jia Dan described this hydraulic network as 'the traces of ancient water management include longitudinal channels known as "pu", and horizontal embankments known as "tang". There were also gates, weirs, streams, and drains, all strategically laid out like a chessboard. To my recollection, there were more than two hundred and sixty such places in total. Now, wishing to follow the methods of the ancients, let every seven li be a longitudinal "pu", and every ten li be a transverse "tang". 故古人治水之跡, 縱則有浦, 橫則有塘. 又有門堰涇瀝而棋布之. 亶所能記者, 總二百六十餘所. 今欲略循古人之法, 七里為一縱浦, 十里為一橫塘.' In ibid.

12 This conceptual map of *tang* and *pu* was proposed by Zhen, *History of Hydraulic Technologies in Lake Tai Region*, 85. The mapping of the ninth to tenth centuries' canal and polder system can be largely enhanced by applying GIS techniques and integrating archaeological findings of early polders and dikes.

13 See Zhen, *History of Hydraulic Technologies in Lake Tai Region*, 14.

14 See the agricultural and economic development of Suzhou during the Tang period in Marmé, *Suzhou: Where the Goods of All the Provinces Converge*, 43–45.

15 See Zhen, *History of Hydraulic Technologies in Lake Tai Region*, 14–15.

16 Refer to Wang, *Wang Zhen's Agricultural Manual*, 184–188.

17 Ibid., 199–202.

18 Fan Zhongyan wrote in his *Answering the Imperial Edict and Proposing Ten Measures* 答手诏条陈十事 (Da Shouzhao tiao chen shi shi) that '江南旧有圩田, 每一圩方数十里, 如大城. 中有河渠, 外有门闸. 旱则开闸引江水之利, 涝则闭闸拒江水之害. 旱涝不及, 为农美利. Inside them were rivers and channels, and outside were gated sluices. In times of drought, the gates would be opened to draw in the benefits of river water, and in times of flood, they would be closed to prevent the harm of river water. Thus, neither drought nor flood would affect them, benefitting agriculture greatly.' Quoted in Geng Ju, 'Discuss Dredging Baimiao and Other Rivers and Banks 議浚白茆等河浦申 (Yi jun Baimao deng hepu shen),' in *Complete Record of Hydraulic Management in the Region of Wu*, ed. Zhang and Cai, 772–789.

19 '盖大围如城垣, 小䧇如院落, 二者不可缺一. Thus, a big polder is as large as a walled city, while small polders resemble courtyards. The two are completely reliant to each other.' In ibid.

20 '俯納環城地, 穿懸覆幕天' in Wang, *Wang Zhen's Agricultural Manual*, 184–188.

21 Consult various drone images in in Liu et al., *Prehistoric Water Management Civilization in the Jianghan Plain*, 2023. I express thanks to prof. Liu Jianguo's discussion with me on the hydraulic context of prehistoric urban formation during his visit to Heidelberg University in October 2023.

22 See *Pingjiang Map* in Zhang, *The Atlas of Ancient Suzhou*.

23 Bai's poem 'Vermilion railings [frame] 390 bridges 紅欄三百九十橋' was excerpted in Fan, *Gazetteer of Wu Prefecture*, 236.

24 The water flow direction could also be deduced from records in Zhang, *Complete Record of Hydraulic Management in the Region of Wu*.

25 Lu and Xiao, 'Taking the High Ground'; Ammerman, 'Venice before the Grand Canal'; Liu, *Prehistoric Water Management Civilization in the Jianghan Plain.*
26 von Glahn, *The Economic History of China*, 208–254.
27 '范文正公, 蘇人也. 平生好施與, 擇其親而貧疎而賢者, 咸施之. 方貴顯時, 於其里中買負郭常稔之田千畝, 號曰義田, 以養濟群族. Fan Wenzheng was a local man of Suzhou. All his life he loved giving charity. He chose of his close relatives who were poor, and those of his distant relatives who were worth and made charitable gifts to all of them. At the time when he was famous and held high positions, he bought a thousand *mu* of land near the city walls of his native place which bore regular harvests. He called these lands "yi tian" (charity land).' In Qian Gongfu 錢公輔 (1023–1074), '范文正公義田記 The Record of the Charitable Lands,' in Fan, *Gazetteer of Wu Prefecture*, 201. Translated by Denis Twitchett, 'The Fan Clan Charitable Estate, 1050–1760,' with author's modification. I am grateful to Michael Marmé for bringing this paper to my attention and his email discussion about purchasable estates during the Tang and Yuan dynasties.
28 See von Glahn's study of temple estates in the Tang dynasty in *The Economic History of China*, 200–204.
29 '震以姑蘇爲禪衲遊訪之所, 而無所歸. 及至治二年, 買仰家橋嚴氏地. 南北至河, 東西延二百三十三尺. 卓庵曰 "能仁", 以駐飛錫. 有果庵居士吳志英者佐之, 施財作觀音羅漢之像, 自門而室. 及買庵基之右太湖余氏地爲園, 南北至河, 東西延八十三尺. 至順二年, 復買庵基之左戴氏民地爲園. 南北至河, 東西延八十五尺. 爰及松陵, 贍衆之田六十二畝. 心之所, 施者畢應.' In Shishi Zuying, 'Nengren An Ji 能仁菴記 (Record of Capable Kindness Cloister),' in *WDFC*, vol. 10, shang, 88–89. Songlin is in the current Wujiang county to the south suburb of Suzhou. This record was inscribed onto a stele with Zhao Mengfu's calligraphy.
30 For traditional Suzhou residential layouts, refer to the 1958 survey conducted by Tongji Daxue Jianzhu Gongcheng Xi Jianzhu Yanjiu Shi, *Suzhou Old Residences Reference Drawing Catalogue*, 4.
31 Refer to Wang, *Wang Zhen's Agricultural Manual*, 184–188. Also see Clunas' analysis of the agricultural yields of garden estates in Clunas, *Fruitful Sites*, 38–59.
32 '及泰定丁卯通师所立寺基圖簿, 示予求记. 按当时所载寺南距路, 北距堑, 延七百尺. 东距火烧池, 西距红白二蜘蛛沟, 袤四百尺. 前出官衢, 三面阻水. 所辖山场, 阡陌甚广. 别有下院七区. 可谓盛矣.' from 'Chici Suzhou Baoguo Chansi Ji 勅賜蘇州报国禅寺记 (Record of Imperial Endowed Repay the Nation Chan Temple in Suzhou)' in Zhu, *Anthology of Zhu Yunming*, ed. Xue Weiyuan, 506.
33 Refer to Su Shunqin, 'Canglang Ting ji 滄浪亭記 (Record of Surging Wave Pavilion),' in *CLXZ*, shang juan, 13–14.
34 See *Map of Gusu City* in Zhang, *The Atlas of Ancient Suzhou*.
35 See a full translation of Wang Yi's, *Record of Touring Shizi Lin* in Appendix, Translation 1.
36 See urban canals of the 14th century in Figure 0.1 and also Figure 1.4.
37 The location of Reside Couds Peak is not specified in historical records.
38 For leaf no. 5, refer to the open digital source of Taipei Palace Museum in https://painting.npm.gov.tw/Painting_Page.aspx?PaintingId=3826.
39 '門外檜行之側各作矮博墙, 開二小門入東西圃. Outside the gate, on either side of the row of juniper trees, low brick walls were constructed, with two small gates opening into the east and west gardens.' in *YL*, vol. 8.
40 Charlton, *Fundamentals of Fluvial Geomorphology*, 22.

41 Wang Zhen mentioned several vegetables as suitable to be planted, and their soil should be kept in moisture in *Wang Zhen's Agricultural Manual*, vol. 7, 12–14.

42 '門外檜行之側各作矮塼墻, 開二小門入東西圃. Short brick walls were built outside the temple, next to rows of juniper trees. Two small gates were opened to provide access to the western and eastern patches'; '道人肩水灌畦蔬, 托鉢船歸粟有餘. People of the way shoulder water to irrigate vegetable plots; returning by boat with their begging bowls filled with surplus grain,' in *YL*, 0802a06.

43 *JXJ*, shang, 8–9.

44 Thank Wang Ruijun for his discussion with me in identifying trees in Shen Zhou's painting.

45 '地理家謂其四獸俱全, 風氣藏鬱. 以是觀之吳城諸蘭若, 莫之及矣.' In the post-inscription of Shen Zhou's painting.

46 I am grateful to Lu Andong for reminding me to pay attention to the transition from productive to aesthetic in Middle Ming and Late Ming gardens during my thesis defence.

2 Constructing the Temple-scape

This chapter is structured into three parts to present a comprehensive picture of the temple-scape of Suzhou during the Yuan dynasty. The first part is a detailed case study of monk Tianru Weize's (1286–1354) construction of Shizi Lin Temple. Translating scenes from Mount Heaven Eye and establishing scenic gong'ans on site, the religious activities in Shizi Lin temple integrated scenes with Buddhist philosophies. The second part, 'Monks,' broadens the scope to investigate the constructions of temples by three Buddhist sects, Chan, Tiantai, and Huayan Buddhism. The flourishing of diverse sectorial practices in Suzhou went in tandem with the expansion of monastic communities, architecture, and garden estates. Through constructing temples, monks established a network of mutual support, which in turn incentivised further temple construction.

The third part, 'Patrons,' delves into temples sponsored by the central court, Uyghur officials, and local citizens. The central court founded temple institutions to promote a social order aligned with the imperial ideal. Uyghur officials supported temples to practice Buddhism and forge a sense of connection as migrants. Local citizens constructed temples as a strategy to obtain monkhood, which privileged them to secure the property and enhance their economic status. The diverse array of motivations shaped the temple-scape and reflected a complex interplay between religion, economy, and politics in Suzhou during the early 14th century.

Translating Scenes from Mount Heaven Eye to Shizi Lin in Suzhou

Shizi Lin, also known as 'Lion Grove Garden,' located in the north-eastern quadrant of Suzhou city, has been acclaimed as one of the representative Chinese gardens (Fig. 2.1a).[1] Its iconic layout centres on an artificial rockery hill surrounded by a pond.[2] The name 'Lion Grove' derives from the resemblance of the Lake Tai rocks on its artificial hill to lions. The current site's abundant inscriptions and refined interior setting allude to its literati past. This poetic image, however, was both ascertained by the

DOI: 10.4324/9781003387169-3

Figure 2.1 Lake Tai rocks in Lion Grove Garden and Lion Cliff in Mount Heaven Eye. (a) Lion Grove Garden, present-day photo. Source: Courtesy of 699pic.com. (b) Lion Cliff in Mount Heaven Eye. Source: Author.

modern scholarship and the site's renovation in the 1950s. In contrast to this singular perception, the garden's history was multifaceted: it was initially established as a temple and subsequently changed hands several times.[3]

This section delves into the site's overlooked Buddhist history in the late 14th century, when it was established by Tianru Weize, a Chan monk who migrated from Western Mount Heaven Eye in Hangzhou to Suzhou.[4] Tianru named it 'Shizi Lin' temple (*Shizi lin*), which literally means 'Masters' Disciple's Temple.' 'Masters' refers to his prominent teachers in Mount Heaven Eye's Lion Grove Temple (*Shizi lin si*), a renowned Chan centre symbolised by Lion Cliff (*Shizi yan*) during the early 14th century (Fig. 2.1b).[5] Tianru modelled his new temple in Suzhou simulating the mountain's scenic features and artefacts.

Prior to his migration to Suzhou in the mid-14th century, Tianru was trained in the lineage of Mount Heaven Eye in Hangzhou for 13 years. The spectacular rocky features in Mount Heaven Eye inspired this sect to develop a 'scenic gong'an' pedagogy, critiquing the literary gong'an pedagogy pervasively used inside monasteries during the Song dynasty, the 10th–13th centuries. In the new pedagogy, practitioners reached their enlightenment by interpreting and crafting scenes in the landscape, through which philosophical ideas were derived and demonstrated. Tianru inherited this pedagogy in his establishment of Shizi Lin temple in Suzhou and imitated several scenes from Mount Heaven Eye to implement. He further developed the scenic gong'an method by reproducing scenes from historical textual gong'ans and renovating interior spaces to command outside views for monastic activities.

This pedagogical change pinpointed a philosophical transition from 'Buddha-centred' to 'self-centred.' The new pedagogy also entailed several architectural changes in Shizi Lin, distinguishing it from the former Song dynasty temples: Shizi Lin temple did not have the iconic Buddha hall; its monastic interior opened towards the scenic views; oriented towards sceneries, the monastic buildings scattered amongst the landscape and possessed an extroversive feature, contrasting with the axial layout of enclosed buildings in former Song dynasty temples.

The research employs cultural translation theory as a framework to interpret the nuances of Tianru's translation of scenes across regions, and from the literary to the spatial. It triangulates the interplay between the Buddhist philosophical shift, pedagogical change, and the evolution of religious space and visuality. 3D models demonstrate the assistive roles of scenes in monastic rituals. By unravelling the Buddhist history of a typical Chinese garden, the research challenges the dichotomised discourse of Chinese Garden and Chinese Temple.

Scenes and Architecture as Gong'ans

The lineage in which Tianru was trained originated from Mount Heaven Eye during the early Yuan dynasty of the late 13th century. Mount Heaven Eye, located in the western suburb of the capital of the former northern Song dynasty, Lin'an, has an elevation of 1500 metres (see Fig. 1.1a). It features a spectacular arrangement of rocks that seem to grow out of the terrain, as well as fantastic water features like ravines and ponds (Fig. 2.2).[6] The highest peak is Lofty Clouds Peak (*Angxiao feng*), with a spectacular stream to its north named Flying Rainbow Ravine (*Feihong jian*), and a pond to its east called 'Heaven Eye (*Tianmu*),' from which the mountain derived its name.[7] Lion Cliff, protruding from the waist of the mountain, and Lion Grove Temple, built in the cliff's vicinity, were two symbols of the sect of Mount Heaven Eye. Two Jade Ravines (*Yu jian*) flowed down the valley next to these scenic spots (Fig. 2.6a).

Gaofeng Yuanmiao (1238–1296) initiated the construction of scenes in Mount Heaven Eye.[8] As an heir of Linji's line, he arrived at Lion Cliff during the Song-Yuan transition and obtained enlightenment Under Lion Cliff in 1279. The gigantic Lion Cliff protrudes from the mountain range abruptly, resembling a roaring lion raising its head from the woods and overlooking the fathomless abyss underneath (Fig. 2.1b). This oblique cliff formed a cave underneath inside, which was called 'Lion Mouth (*Shizi kou*)' (Fig. 2.3a).[9] Gaofeng made a 'Death Pass (*Si guan*)' chamber in the cave (Fig. 2.3b,c). The chamber resembled a little boat, and its name 'Death Pass' was inscribed on a plaque at the cave's entrance. This name demonstrated Gaofeng's determination to get enlightened by meditating in the cave—otherwise, he would better fall into the fathomless abyss.

Years later, as the number of visitors increased, Gaofeng built a pavilion called 'Three Gates (*San guan*)' below the cave, serving as the front gate leading to the 'Death Pass' chamber (Fig. 2.3d). The pavilion possibly featured three door panels, representing the three questions Gaofeng asked the arriving Chan seekers to test their attainment.[10] Supposedly, without a satisfied answer, the Chan seekers would not be allowed to pass the gates and enter the chamber as Gaofeng's disciples. Very few monks finally passed Three Gates, such as monk Zhongfeng Mingben (1263–1323) and Duanya Liaoyi (d.1293). To seek answers to Gaofeng's three questions, Duanya meditated at the edge of the Lion Cliff and once fell from it, almost losing his life. Passing Three Gates, literally and metaphorically, Duanya received instructions from Gaofeng and later became a leading master.

The peaks, cliffs, and gigantic rocks became favoured by monastic meditation. The lofty views and expansive vistas afforded by the rocky terrains fostered monks' philosophical contemplation (Fig. 2.4a). Connected by a path to the southeast of Lion Cliff was Lotus Platform (*Lianhua tai*), Gaofeng's favourite meditation place (Fig. 2.4a, also see Fig. 2.2(7)).[11] To

Figure 2.2 Detail from 'Complete Map of the Excellent Realm of Western Mount Heaven Eye (*Xi Tianmu shan shengjing quantu*),' woodblock print, from Chen, *Excellent Scenery of Mount Heaven Eye*, 1933. Labelled scenes in the map: 1. Lotus Peak; 2. Heaven Eye Pond; 3. Lofty Clouds Peak; 4. Lion Grove Temple; 5. Entombment Cloister; 6. Lion Mouth; 7. Lotus Platform; 8. Heap Jade Peak; 9. Erected Jade Pavilion.

the west of Lion Peak was Lotus Peak (*Lianhua feng*), and to the east was Heap Jade Peak (*Cuanyu feng*) (see Fig. 2.2(1, 8)). At its summit, Erected Jade Pavilion (*Liyu ting*) commanded the view of Lotus Peak and other peaks in distance (Fig. 2.4). The pavilion, where Zhongfeng often engaged in meditation and philosophical discussions with his students, became a metonym of him. Buddhist meditations inscribed meaning to these places and incentivised further constructions to commemorate. For example, Zhongfeng spent months meditating on Fallen Sitting Rock (*Diezuo shi*). Later a Korean prince built a cloister there named 'Entombment Cloister' (*Huomai an*), symbolising Zhongfeng's enduring stillness in the meditation as if he was buried alive (Fig. 2.2(5)).[12]

Monks' Chan discussions transformed sceneries in Mount Heaven Eye into scenes inscribed with philosophical ideas. Initially, Erected Jade Pavilion and its surrounding features were named and perceived in relation to jade-like rock forms. The peak of the pavilion was named 'Heap Jade Peak' before monks' arrival because the angular joints of rocks on its surface resemble jade overlapping together. In correspondence, this pavilion perched at its top was named 'Erected (at the) Jade (*Li yu*),' and the ravine under it was named 'Jave Ravine (*Yu jian*).' The jade-associated imagery

Figure 2.3 Artefacts around Lion Mouth under Lion Cliff. (a) Lion Mouth cave. Source: Author's photo, 2018. (b) (c) Gaofeng's tomb and statue in Lion Mouth cave, the location of previous Death Pass chamber. Source: Author's photo, 2018. (d) Chen Pan-Show, *Lion Cliff, Double Clouds Pagoda (Gaofeng's tomb), Three Gates pavilion*, 1933, in *Excellent Scenery of Mount Heaven Eye*.

was illustrated by the gazetteer map *Marvelous Sights Within Seas* vividly (Fig. 2.4b)

Zhongfeng's discussion on clouds transformed the semiotic meaning of Erected Jade Pavilion, tying it to both the cloudy views and the philosophy of none-distinction. When the sky is clear, looking outwards from the pavilion, one would see Lotus Peak resembling a lotus in blossom emerging amongst clouds, as recorded in Tianru's writing, *Sayings of Clouds*:

> I once accompanied master Zhongfeng in Erected Jade Pavilion. We've seen numerous clouds appear as the sea, and peaks nearby or in distance arose among clouds. The peaks were like caps left, shoes floating, and vacant boats drifting among waves. The clouds changed their forms into spectacular peaks, jade trees, swimming dragons and dancing phoenixes.[13]

The writing continues to describe the dialogue between Zhongfeng and his guest evoked by this scene of clouds:

> A guest then said: 'when observing the illusory forms presented by the sea of clouds, I started to realize that myriad things between the great heaven and earth are a sea of clouds.'
>
> The master replied: 'You [used to] understand that illusions are illusive but fail to recognize that non-illusions are also illusive; you [now] understand that non-illusions are illusive but have not noticed what compose the non-illusive. Clouds gathered together to form the sea and represent numerous forms, but what constitutes the clouds?'[14]

The dialogue between Zhongfeng and his guest, concise yet allusive, exemplifies a typical gong'an, an 'encountered dialogue,' a teaching method developed in Chan Buddhism since the eighth century.[15] The ever-changing cloud forms inspired the guest to reflect on the subjective nature of human perception. However, Zhongfeng deepened this realisation by challenging the boundary between subjectivity and objectivity. Since then, Erected Pavilion had come to be synonymous with both the expansive view of clouds and the philosophy of transcending duality, marking a departure from its original associations with jade forms. It also became a metonym of Zhongfeng. These new imageries eclipsed the old as being circulated in the literature of Mount Heaven Eye.

The way monks in Mount Heaven Eye constructed philosophical meanings by interpreting and shaping the landscape in the 14th century marked a development from the traditional gong'an method, which used to be prevalent from the 11th to 13th centuries. Gong'ans, literally meaning 'public cases,' were intellectual dialogues and anecdotes of Chan masters, which reveal their philosophical insights or attitudes in Chan practice.

Figure 2.4 Meditation spaces in Mount Heaven Eye. (a) Chen Pan-Show, 'Lotus Platform,' from *Excellent Scenery of Mount Heaven Eye*, 1933. (b) Erected Jade Pavilion on Heap Jade Peak. Yang Er'zeng, section of 'Painting of Mount Heaven Eye (*Tianmu shan tu*)' from *Marvelous Sights Within Seas*, 1609, woodblock print. Berlin: Berlin State Library.

Most cases were circulated in the Tang dynasty (618–906), such as Huike (487–593) stood in snow and Zhaozhou (778–897) pointed to the cypress.

Along with the expansion of Chan Buddhism from the 11th century onward, gong'ans were anthologised and annotated in printed books, forging a distinctive Chan literature genre. For example, the aforementioned two anecdotes were compiled in books such as *The Blue Cliff Record* (1128) and *The Gateless Barrier* (1228). Their authors further expounded upon the meaning of the two gong'ans as 'one needs to possess ardent determination to achieve enlightenment' and 'words cannot fully express things,' respectively.

Gong'ans were also interwoven into Chan Buddhism practices. Masters discoursed gong'an cases in their lectures, and monks read, discussed, and meditated on gong'ans. Through practising with gong'ans, monks were expected to internalise their masters' philosophy, rejecting the original mental blueprints and reaching enlightenment. D. T. Suzuki explained the cognitive transformation intended in gong'an practice: 'when contemplating gong'ans, the practitioner reproduced the same psychic condition out of which the Zen ('Chan' in Chinese) master uttered these gong'ans.'

Traditional gong'an practices in Tang and Song periods relied on the agency of words to instruct enlightenment. Words from patriarchs, recorded in gong'an literature, and taught by teachers were the medium encoded with insights. Contemplating these words was a cognitive exercise for practitioners. By taking this mental load to work out the 'true meaning' hidden by enigmatic words, a practitioner achieved enlightenment. The enclosed monastic buildings were places for these literary gong'an practices.

In this context, Gaofeng's constructions beneath Lion Cliff and Zhongfeng's dialogue in Erected Jade Pavilion were instances of gong'an, as both embodied Chan philosophy with the goal of transforming students' mindsets. However, they differed from traditional gong'ans in the spatial medium used. In the gong'an cases developed by monks in Mount Heaven Eye, albeit words remained, scenes played a constructive role in illustrating masters' insights and facilitating practitioners' comprehension of teachings. The places where Gaofeng and Zhongfeng practised this scenes-assisted pedagogy were amongst the landscape, where sceneries could be easily picked up to encode philosophical ideas. This new mode of gong'an bolstered by scenes could be called 'scenic gong'an,' of which the landscape served both as the environment and catalyst for metaphorical constructs. Conversely, monks' religious practices inscribed scenes in Mount Heaven Eye with Zenith meanings, thus the scenes could be called 'gong'an scenes.'

Why did Gaofeng and Zhongfeng need to renovate the traditional gong'an method? The reason lies in its literary format, once capable of leading practitioners to enlightenment in the 11th–13th centuries, and was no longer effective in the 14th century for Gaofeng's generation. As pointed out by Alan Cole, gong'an anecdotes featured extremely reduced form–it omitted the background and context of the original story.[16] This deliberate omittance aims to challenge the practitioner. To work out the true meaning of the masters' answers, a practitioner needs to recruit all faculties and experiences to investigate gong'an cases persistently. It is by taking the internal conflicts that a practitioner's psycho-condition is transformed.

The social changes happened in the dynastic transition of the 13th century, and basic contexts of gong'ans gradually diminished from the societal knowledge. The ancient cases gradually lost resonance with the contemporary culture, becoming enigmatic and less appealing. Many students of the 14th century lacked the drive to engage with these complex riddles, and even those who were willing to often struggled to penetrate the surface of gong'ans' paradoxical language. Consequently, teachers like Zhongfeng had to provide extensive explanations.[17] This inadvertently fostered a teacher-dependent learning, which contradicted the self-reliance principle of Chan Buddhism since its inception in the 6th century. Furthermore, the widespread gong'an anthologies familiarised practitioners with answers to questions in gong'ans, reducing master-student dialogues to mere rote exchanges.

In response to the limitations of traditional gong'an practices, Gaofeng and Zhongfeng explored alternative methods. Scenes in Mount Heaven Eye were found as a spatial medium more effective for engaging Dharma seekers.[18] Zhongfeng wrote to promote scenic gong'ans. He criticised former literary gong'ans as 'artificial constructs' and claimed nature as a 'manifest gong'an,' advocating that 'aside from this manifest gong'an, there is no Buddha-dharma to continue the transmission of the lamp.'[19]

Why comparing with traditional literary gong'an, scenic gong'an was more effective for the aim of 'mind-to-mind transmission'? I would argue that it is because of the scenery's active role in engaging practitioners by deictic shift. For instance, when a practitioner visited the Erected Jade Pavilion in Mount Heaven Eye, the sea of clouds visually inspired him to comprehend the non-dualistic view in Zhongfeng's dialogue with his guest. Moreover, he could sit there for a long meditation until this insight was fully internalised. Similarly, the daunting view of the abyss beneath the Lion Mouth cave brought to life the determination practised by Gaofeng and others. The landscape provided a context that integrated the visual, sensory, and emotional engagement of the practitioners, rendering the philosophical practice as tangible, impactful, and relatable.

Spatial Translation

After his teacher Zhongfeng left Mount Heaven Eye and led a drifting life in the lower Yangtze river basin, Tianru migrated to several places and finally settled in Suzhou. When establishing his new temples in the city, Tianru faced two challenges. The first was how to practice the scenic gong'an teaching approach inherited from Mount Heaven Eye. If the relatively flat urban landscape of Suzhou did not feature inspiring mountain peaks as Mount Heaven Eye did, how could the teaching of Chan philosophy derived from interpreting scenes be feasible? The second challenge was to demonstrate his lineage from Mount Heaven Eye in a competitive environment of numerous temples. As discussed in the Introduction, the city of Suzhou had accumulated 372 temples upon Tianru's arrival during the mid-14th century, with 119 constructed in the intramural area where Shizi Lin was located (Fig. 0.1, 0.3). While some temples possessed substantial properties and could sustain on their own, many others relied on social and court patronages to flourish.[20] As a temple fitting into the latter category, it was imperative for Tianru to demonstrate his prestige lineage and pedagogy to make his new temple competent in amassing patronage. In response to these issues, Tianru recreated landscape features and gong'an scenes of Mount Heaven Eye in Shizi Lin.

The land plot selected by Tianru resembled a typical swampy condition of Suzhou during the Song-Yuan periods.[21] Albeit the site was described by Tianru as adorable and resembling that in the mountains, the topographical differences between the two places were dramatic: the land inside Suzhou city was almost flat in contrast to the gigantic mountain ranges of Mount Heaven Eye, and it did not feature soaring peaks and cliffs (Fig. 1.1a).

A hydro-topographical model was implemented to refurbish the land. The depressed terrain was dredged to form a river encircling the central mound. The inner river was linked to the main urban canal and enabled visitors to boat to the temple. The surplus silt, mud, and soil from dredging was repurposed to pile up the existing central mound into a hill. Bamboos were planted along the riverbank. They reinforced the bank, conserved and fertilised the soil for other plants to grow, meanwhile were also a source of food supply. Further complementing the site as both productive and habitable, a well was dug for drinkable water, a pond was created with fished raised, and paddy fields were plotted for cultivating vegetables.[22]

Although hydraulic and agricultural modifications were commonly practiced, Tianru's deployment of Lake Tai rocks was unusual and controversial during that period. After the capital Lin'an (current Hangzhou) of the Southern Song dynasty was occupied by the Mongols in 1279, the Lake Tai rock, once an imperial luxury, was now condemned as inauspicious and blamed for incurring the misfortune of the overturned dynasty. Kong Qi (d. 1367) noted that Lake Tai rocks were largely abandoned in cities of the lower Yangtze river.[23] Nevertheless, the discarded Lake Tai rocks around the original site were reused by Tianru. He dispersed them among the central hill and along the creek (Fig. 2.5).

These rocks, although only about one and a half meters in height, were referred to by Tianru as peaks; the hill where they grounded, around five meters high, was called a mountain. The newly dredged creek surrounding the mound was named 'Flying Rainbow (*Feihong*),' as indicated by its bridge called 'Flying Rainbow Bridge.' Tianru scattered the Lake Tai rocks on the hill and along the creek, as if they were peaks spreading among the mountainous terrain in Mount Heaven Eye. He named his temple 'Shizi Lin'—the same pronunciation as 'Lion Grove Temple' (獅子林寺) in Mount Heaven Eye. He also strategically removed the left radical of the term 'Shizi 獅子' which indicates lion, changing it into 'Shizi 師子,' which means 'the master's disciple,' adding a new layer of meaning to identify himself as from the sect of Mount Heaven Eye.[24] By inscribing the new temple with a metaphorical name and miniaturising features such as peaks, mountains, and ravines, Tianru made Shizi Lin resemble Mount Heaven Eye of Lion Grove Temple.[25]

Apart from configurating the site with mountainous features, Tianru reproduced several scenes from Mount Heaven Eye in particular. Inside the temple were six scenic objects named directly after those scenes in Mount Heaven Eye, including three rocks, a chamber, a bridge, and a pond (Fig. 2.6), whose names were either engraved on architecture plaques or stony surfaces. On the one hand, Salient geographical features were chosen to be translated: Lofty Clouds Peak (C), the peak at the summit of Mount Heaven Eye, was represented by a rock with the same name (c); Flying Rainbow Ravine (D) was translated into a creek called Flying Rainbow, and in correspondence its crossing bridge was called Flying Rainbow Bridge (d); Eastern and Western Jade Ravines (*yujian* 玉澗) (E1 and E2) were translated to the newly dredged square pond named Jade Mirror Pond (*yujian* 玉鑑) (e).[26]

On the other hand, famous gong'an scenes were also translated: the scene of Lion Cliff (A) at Mount Heaven Eye was represented by a rock resembling a lion, and was called 'Lion Peak (a)'; the scene of Erected Jade Pavilion (B) on the top of Heap Jade Peak was translated to a rock called 'Erected Jade Peak (b1),' and the view of clouds it commanded was translated into the view framed by a chamber called 'Lying Clouds Chamber (b2).' The scattered configuration of these scenes at Mount Heaven Eye was imitated: the gong'an scenes in Shizi Lin were dispersed among the hill and creek—the imagined mountain and ravine.

I borrow the cultural translation theory in poetry studies to investigate Tianru's reproduction of the landscape, because his craft of scenes parallels a translator's reworking of a poem—both of which aim at transplanting imageries and meanings across regions. Susan Bassnett makes a vivid analogy of translation: imageries and meanings are like seeds; Through a

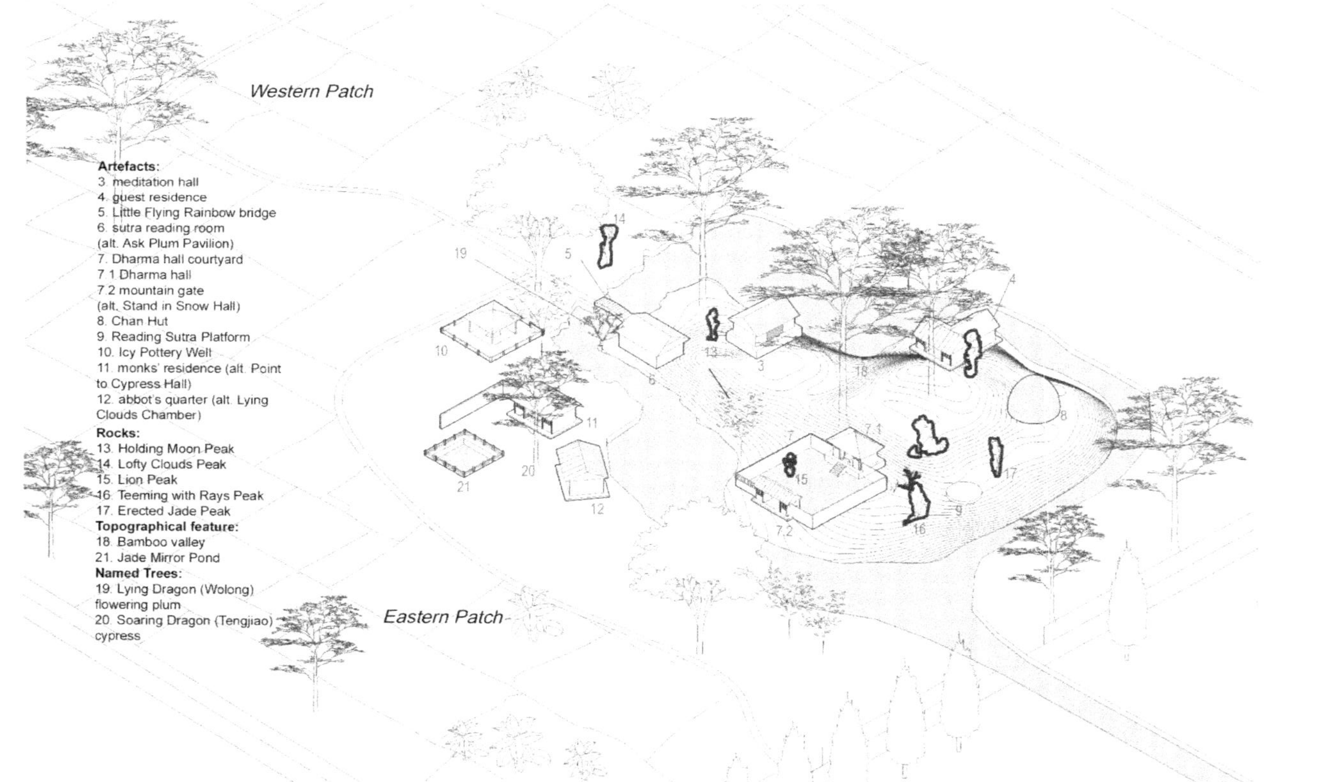

Figure 2.5 Shizi Lin Temple in the Late Yuan Dynasty. Source: Author's drawing and modelling.

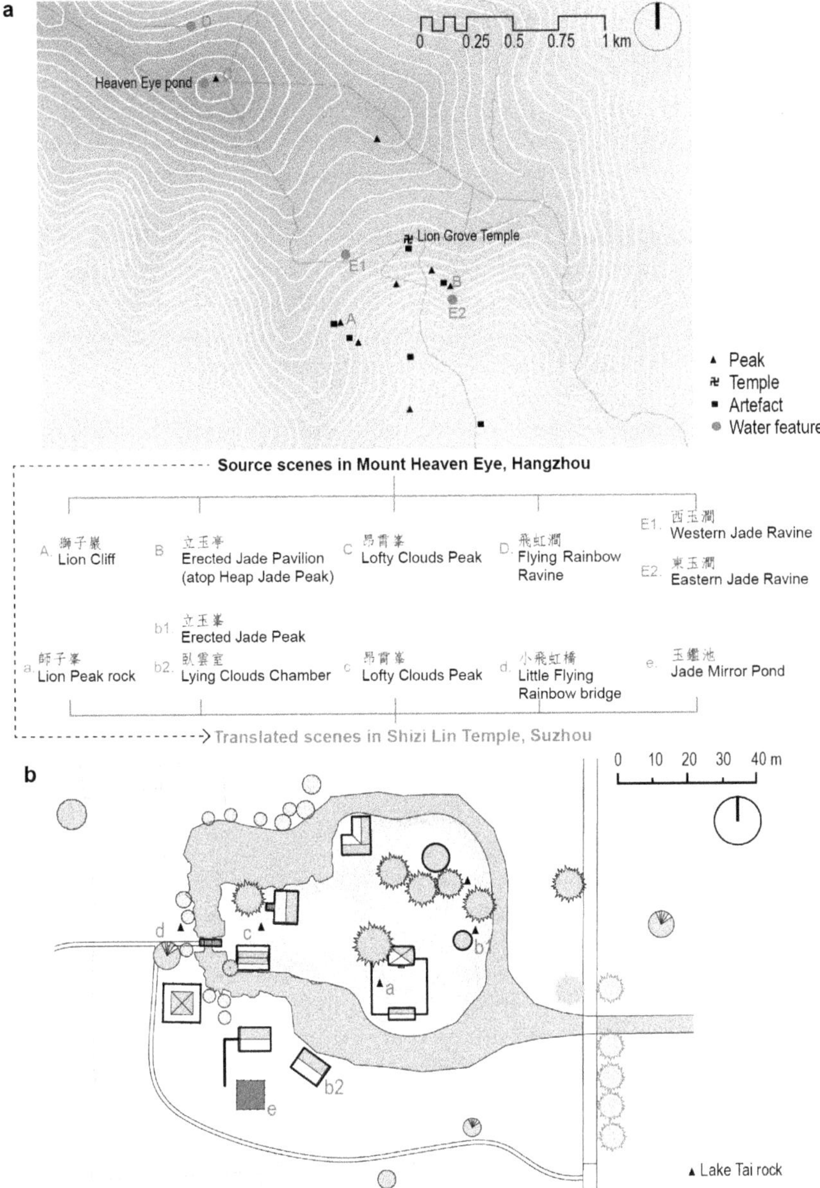

Figure 2.6 Scenes translated from Mount Heaven Eye to Shizi Lin Temple. (a) Source scenes at Mount Heaven Eye. (b) Translated scenes at Shizi Lin Temple. Source: Author.

translator's re-creation, they are transplanted to another region and cultivated into new forms.[27]

Three aspects are framed in the cultural translation. First, the translation reflects the need to fill the gap between the source culture and the target culture. Second, skills are required to choose a counterpart form in the target culture. The translator might select a mimic form, or analogical form with partial resemblance. The third dimension pertains to readers' comprehensibility. To fully appreciate the allusions in the translated subject, the reader needs to be equipped with some contexts of the original culture.[28] In the following paragraphs, these aspects will be applied in examining Tianru's translation of Lion Cliff to Lion Peak rock mimically, and the translations of Erected Jade Pavilion to Erected Jade rock and Lying Clouds Chamber analogically.

Among the many Lake Tai rocks collected by Tianru, the highest one mimicking a lion was chosen to translate Lion Cliff at Mount Heaven Eye (Fig. 2.6), as indicated by the name 'Lion Peak' engraved (Fig. 0.4a). Tianru located it at the southern slope of the hill overlooking the creek, making it a prominent feature for coming visitors. Wang Yi described that when he started to climb the central hill, Lion Peak was the first spectacle coming to his sight.[29] Centring on the rock, a Dharma hall was constructed at the rear with a mountain gate (*shan men*) at the front facing the river. They assumed a similar function and meaning as Gaofeng's two artefacts under Lion Cliff: The Dharma hall housed Tianru's sermons to students, while Gaofeng instructed students in Death Pass chamber. The mountain gate, inscribed with the name 'Stand in Snow Hall (*Lixue tang*),' symbolises determination, echoing the theme of Three Gates pavilion at Mount Heaven Eye. The name referred to the Chan gong'an of Huike, who stood in the snow for days demonstrating resolution.

Furthermore, the spatial sequence of the two artefacts surrounding Lion Peak rock imitated that at the mountain. In Mount Heaven Eye, monks needed first to demonstrate their resolution at Three Gates pavilion, and then pass through to reach Death Pass chamber in the cave, where instructions from Gaofeng's could be received (Fig. 2.3d). Analogously, in Shizi Lin, visitors first walked through Stand in Snow Hall, being reminded of the required determination, then entered the courtyard to attend Tianru's sermon (Fig. 2.7a). Furthermore, both Lion Cliff and Lion Peak rock were genius loci of the Buddhist sites.

Lion Peak rock conjured up images of the Lion Cliff in visitors. For instance, Zheng Yuanyou (1292–1364), a local official and a frequent guest of Shizi Lin, envisioned that the Lion Cliff was a lion dwelling in Mount Heaven Eye, leaping to Suzhou and transforming into the Lion Peak rock:

In Mount Heaven Eye stands the imposing Lion Cliff,
Its golden mane shimmering in the sunlight, amidst the mountain's
mist and haze.
It turns and roars towards the Central Wu Region,
As if challenging and joining the clouds that soar high in the sky.[30]

While Zheng's poem imagined a movement and interplay between the two
regions, the poem from Tu Zhen (fl. late 14th c.) closed the distance and
difference between the cliff and the rock. He depicted Lion Peak rock as
majestic and towering as Lion Cliff, merging the imageries of the two.[31]

Other visitors lauded Tianru's transmission of Dharma across regions.
Their poems typically begin by recounting Tianru's teacher Zhongfeng's
practices under Lion Cliff, highlighting Tianru's original lineage. The narrative then transitioned to Suzhou, centring on Lion Peak rock as a symbol
of Tianru's teaching.[32]

Visitors' poems reveal the dual purposes served by a translated scene:
it carried the imageries and meanings developed in Mount Heaven Eye
meanwhile demonstrated the lineage of Mount Heaven Eye. The associations evoked deictic shifts in visitors: while being in Shizi Lin, their minds
were transported to Mount Heaven Eye. The two spiritual realms merged
in perception.

Erected Jade Pavilion, perching atop Heap Jade Peak at Mount Heaven
Eye, underpins two imageries: the pavilion as a metonym of Zhongfeng,
and the cloudy view the pavilion commanded. To translate the first
imagery, Tianru used a Lake Tai rock to symbolise the peak, inscribing
'Erected Jade' on it to reference Erected Jade Pavilion. He positioned the
rock at the eastern slope of the mound near the creek (Fig. 2.6b(b1), also
see Fig. 2.5(17)), mirroring the way Erected Jade Pavilion overlooking the
ravine beneath (Fig. 2.6a(B)).[33]

Erected Jade Peak rock evoked visitors' imagination of Erected Jade
Pavilion, which standing aloft at the top of Heap Jade Peak, as illustrated
in He Zhen's (fl. late 14th c.) poem:

How the stone peaks rise and fall,
touching clouds next to bamboo so slender.
As if nodding in agreement,
standing lofty like jade above the clouds.[34]

The poem depicts a towering presence, which he perceived as lowering its
head in greeting. This lofty portrayal aligns more closely with the grandeur
of Heap Jade Peak rather than the modest height of Erected Jade rock,
which stands only one to two metres tall. The image of the stone soaring
into the clouds also defies the low altitude and flat terrain of Suzhou, where
such a cloud-engulfed scene is less likely. Thus, it can be deduced that He

Figure 2.7 Stand in Snow Hall and Lying Clouds Chamber. (a) Xu Ben, leaf no. 7, 'Stand in Snow Hall.' (b) Xu Ben, leaf no. 8, 'Lying Clouds Chamber.' This chamber is Tianru's abbot quarter. Source: Taipei: Collection of the National Palace Museum.

Zhen's poem superscribed the majestic scene of Erected Jade Pavilion onto the rock in Shizi Lin.

The translated imagery, indicated by its name, would lose its resonance with visitors unfamiliar with Mount Heaven Eye. This issue became apparent in the Early Ming period when the visitors of Shizi Lin changed to a group of literati. Among them was Wang Yi, who was unfamiliar with the site's origin, found the rock's name and form puzzling. He felt the rock resemble less to the form of jade and more to sword-shaped leaves. He thus proposed renaming it to 'Earth Lung (*Di fei*).'[35] While the rock's imagery evoked He Zhen's imagination during the Yuan dynasty, it lost its cue to Wang Yi in the Ming. This shift illustrates that a full appreciation of a translated scene requires the knowledge of its origin.

To translate the scene 'sea of clouds' commanded by Erected Jade Pavilion, Tianru employed a chamber as an analogical form (Fig. 2.5(12)). He strategically located the chamber across the river, facing the central hill, to replicate the viewing experience of Erected Jade Pavilion. From this chamber, one could see the 'mountain' represented by the hill and other 'peaks' symbolised by Lake Tai rocks. The presence of clouds was invoked through the chamber's name—'Lying Clouds Chamber.'

Once in the chamber, the scene of the 'mountain' and 'peaks' evoked Tianru's memory of sitting with Zhongfeng in Erected Jade Pavilion. Lamenting the decease of his beloved teacher and reflecting on challenging years after leaving Mount Heaven Eye, Tianru wrote:

> The sea is the clouds, which reside in my chamber.
> Clouds are the sea, covering mountains and rivers.
> As a person of the Way, he does not care it rains or shines.
> Just sit down and observe the clouds transforming, rising and moving.
> Buddha Dharma would not perish after three thousand years.[36]

Tianru's imaginary seeing was visualised in Xu Ben's album leaf of 'Lying Clouds Chamber' (Fig. 2.7b). The chamber, depicted as a small icon at the bottom left, was a metonym of Tianru—Tianru was addressed as 'lying clouds teacher' by visitors of Shizi Lin. The chamber oriented towards a towering mountain at the centre of the painting: it was emerging from the lofty clouds, with peaks hidden and revealed. The painting presented the scene Tianru used to appreciate with Zhongfeng at Mount Heaven Eye, which Tianru aimed to recreate in Shizi Lin.

This imagery of a Chan master within cloudy mountains and peaks circulated among poets of the late Yuan dynasty. Tu Zhen, Zhou Ji (fl. late 14th c.), Zheng Yuanyou, and Yang Zhu (fl. late 14th c.) all imagined the chamber as being surrounded by clouds. Chen Qian even romantically imagined Tianru as 'walking atop the white clouds in morning and rest within clouds in evening.'[37]

Tianru's translation of scenes from the mountain transformed Shizi Lin into a landscape imbued with the Buddhist culture of Mount Heaven Eye. The spatial translation requires the landscaping skills extending beyond finding counterparts as in the cultural translation. The first step of landscape translation involved addressing the practical requirements of a temple, such as water management, agricultural use, habitability, and the monastic programme.

Second, spatial translation requires the configuring, orientating, and choreographing scenes. After selecting a symbolic object to represent the source scene, Tianru needed to integrate it with other objects. The artefacts commanding the views were strategically placed to realise Tianru's seeings at Mount Heaven Eye. The strategies of 'facing the view' and 'framing the view,' frequently discussed in the modern scholarship of Chinese gardens, were observed in Shizi Lin as Tianru oriented Lying Clouds Chamber to command the view of the central mound. Moreover, Tianru reproduced the spatial sequence from Mount Heaven Eye, allowing visitors to derive meanings not only visually but also in movement. These three techniques distinguish spatial translation from miniature art, which is appreciated through visual inspection.[38]

The way people experienced the translated landscape is through 'deictic shift': scenes in Shizi Lin prompted visitors to mentally conjure up the landscape of Mount Heaven Eye, allowing them to feel present both in Shizi Lin and in the distant mountain simultaneously. Since translated scenes did not fully resemble the source scenes, inscribing the name of the latter forged a clear reference between. The agency of scene names in evoking a deictic shift, as identified by Lu Andong in Chinese gardens, was similarly observed in the translated landscape.[39] What distinguishes a deictic shift induced by a translated scene from that in a non-translated setting is that in the former, the viewer's mind is transported to a different region, whereas in the latter, it travels to the historical past of the same location.

Translating Literary Gong'ans into Scenes

Apart from the spatial translation, Tianru also translated gong'ans from classical Chan records into scenes, aligning his temple with Chan Buddhism broadly. The three translated gong'ans were 'Point to Cypress (*zhibai*),' which was translated into a cypress tree in front of the monks' residence (Fig. 2.8a); 'Ask Plum (*wenmei*),' which was translated to a plum along the creek facing the sutra reading room (Fig. 2.8b); and 'Stand in Snow (*lixue*),' translated as the mountain gate in front of the Dharma hall (Fig. 2.7a).

For 'Point to Cypress,' an ancient cypress tree at the original site was preserved. Beneath its verdant canopy, a square pool was made, described by visitors as reflecting the tree's beautiful branches (Fig. 2.8a).[40] This pond was named 'Jade Mirror,' translated analogically from 'Jade Ravine' in Mount Heaven Eye. A Sangha hall for monks' meditation and residence faced towards the tree and Jade Mirror Pond. The plaque of the hall was inscribed with the name of the gong'an—'Point to Cypress Hall (*Zhibai xuan*).' Walls extending from the hall formed a semi-courtyard enclosing and focusing attention on the cypress tree (Fig. 2.5(20)).

The gong'an 'Point to Cypress' originated from the dialogue between a practitioner and the Chan master Zhaozhou.[41] The practitioner posed a traditional question to Zhaozhou: 'why did the first patriarch come from the West (India)?' The underlying enquiry was 'What is Buddha's dharma?'[42] It is a loaded question, with the asker expecting the master to demonstrate the profoundness of the Buddha's teaching. To break the

Figure 2.8 Point to Cypress Hall and Ask Plum Pavilion. (a) Xu Ben, 'Point to Cypress Hall,' leaf no. 9, in the album *Painting of Shizi Lin*. (b) Xu Ben, 'Ask Plum Pavilion,' leaf no. 10, in the same album. Taipei: Collection of the National Palace Museum.

premise, Chan masters tailored their responses to the situation.[43] Probably in the story of Zhaozhou, there was a cypress tree in the courtyard, thus Zhaozhou replied: 'Cypress tree in the garden':

> Then the enquirer doubted: 'Master, please don't teach using an object.'
> Zhaozhou said: 'I am not showing an object to you.'
> The enquirer said: 'What is the meaning of the Ancestor's coming from the West?'
> Zhaozhou said: 'The cypress tree in the courtyard.'[44]

This case exemplifies the illogical linguistic feature of a gong'an. Dōgen Kigen (1200–1253) interpreted this gong'an as showcasing the Chan principle that 'words cannot express things and speech does not convey the spirit.'[45] Modern scholar Cheng Liao's psychological explanation is more specific: Zhaozhou directed the practitioner's attention to the cypress tree (the external) and then immediately negated his pointing to the tree. Thus, he was indeed encouraging the practitioner to realise that, instead of seeing the tree, the practitioner should see himself (the internal), who is now looking at the tree. Thus Zhaozhou instructed the practitioner to understand that the essence of Buddha's teaching did not exist outward but within oneself at the current moment.[46]

Tianru translated the cypress tree from Zhaozhou's answer into a tangible scene at his temple. Xu Ben's album leaf no. 9 depicted this scene: a robust cypress tree leans slightly, its sprawling branches forming a canopy over the courtyard. Beneath the cypress, the square pond 'Jade Mirror,' bordered by stone railings, was placid and reflective, enhancing the contemplative atmosphere. The pond and the tree create a focal space that draws viewers in, inviting them to reflect on the gong'an 'Point to Cypress' (Fig. 2.8a).[47]

The inscribed poems on the opposite leaf explicate the painter's introspective reaction to the scene:

> The lush cypress tree in the garden clearly shows the intention of why Buddha came from the West.
> Dark green cypress before the courtyard stands,
> Its vivid intent hails from the West.
> The Chan master points, instructing the seeker,
> Yet the deeper meaning lies within the second layer.[48]

The poem resonates with Dōgen and Wumen's observation about the indexical nature of the Chan language. Furthermore, other visitors engaged with this gong'an by cross-referencing other gong'ans, with some even pondering whether the cypress tree itself possesses Buddha nature.[49] The poems reveal that the translated scene created an immersive and learning-inducive environment. The multi-sensory spatial engagement of the landscape facilitated visitors' personal insights, rather than their passively waiting for teachers' instructions.

The second translated gong'an is 'Ask Plum,' an anecdote between Mazu Daoyi (709–788) and his disciple Damei Fachang (752–839):

Fachang initially approached Mazu and asked: 'What is the Buddha?' Mazu answered: 'One's mind is the Buddha.' Fachang immediately understood and then retreated to Great Plum Mountain (*Damei shan*), where he got his toponym as Damei (Great Plum)

Mazu sent a monk to test Fachang: 'Recently, the master's teaching on the Dharma has changed; he now says neither mind nor Buddha.'

Fachang replied: 'Let him say neither mind nor Buddha. I maintain that the mind is the Buddha.'

When the monk returned and relayed this to Mazu, Mazu declared, 'the plum is ripe!'[50]

This story illustrates the function of gong'an dialogue to assess a student's level of enlightenment.[51] To translate this gong'an, Tianru preserved an ancient plum tree by the riverbank and oriented the sutra reading room towards it (Fig. 2.8b, Fig. 2.5(6, 19)). The translation reflects Tianru's intention to use gong'an dialogue for both teaching and assessing disciples, emulating Mazu's approach. Tianru expressed this purpose in his verses: 'In Ask Plum Pavilion dwells Mazu, examining and discerning his disciples. In Point to Cypress Hall resides Zhaozhou, with the mats laid out open for discourse.'[52] This sentiment was echoed by Zheng Yuanyou, who wrote, 'When plum blossoms bear fruit, sprint is met once again,' implying that Tianru cultivated many disciples.[53]

The scene 'Ask Plum Pavilion' also prompted practitioners' reflections on the other two Chan Buddhism concepts. The first one is the 'undistinguished mind,' as reflected in Tu Zhen's poem. He juxtaposed the experience of discovering white plum flowers in Shizi Lin with reading sutras in the hall:

Snow fills the temple, where should I seek the plum blossoms?
I recognize the place where the fragrance drifts; in the western
garden, a tree leans.
The mountain hermit opens a small pavilion, facing it, he reads the
Lankavatara Sutra.[54]

The poem depicts a setting where the temple is blanketed in snow. The pervasive whiteness obscures plum blossoms, making them indistinguishable to the eye; yet their lingering fragrance subtly indicates the presence.[55] The poem suggests that by ceasing to search for visible distinctions—as one tries to discern the blossoms in the snow—and instead embracing the state of self-awareness, symbolised here by sitting in the sutra reading room, one can reveal the essence of Dharma. The poem also hints at a true understanding in Buddhism not merely from the intellectual but also integrating embodied experience and perception. He Zhen's experience was visualised in Xu Ben's painting leaf no. 10 (Fig. 2.8b), in which a monk was sitting towards the plum tree. The poem on the facing leaf similarly interplayed between the plum and the snow.[56]

Visitors' second enlightenment, prompted by the plum tree, revolved around the playful action of 'asking the plum about spring.' Xu Ben's group's poem wrote:

> Spring, whence do you depart and whence do you return?
> Holding the intent of departure and return, I ask of the plum before the pavilion.[57]

Spring is an abstract concept, akin to the idea of emptiness in Buddhism, while the plum blossom serves as a tangible manifestation of spring. The act of 'seeking the plum blossom to find the spring' further suggests that personal experiences could lead to sudden enlightenment. This idea is echoed in Dōgen's verse: 'you cannot paint the spring, but you can paint the plum to know spring.'[58] The poems of the scene 'Ask Plum Pavilion' demonstrate how gong'an scenes could inspire visitors to investigate Chan Buddhism from multiple angles, rather than constraining to the teacher's instruction.

The gong'an scenes at Shizi Lin established on-site deictic centres to immerse visitors in historical Chan moments. This is exemplified by the third translated gong'an 'Stand in Snow Hall.' 'Stand in Snow' recounts a tale in Song Mountain between Bodhidharma (d. 536), the first patriarch, and his disciple Huike:

> Bodhidharma sat facing the wall. The second patriarch [Huike] stood in the snow. He cut off his arm saying, 'Your disciple's mind still has no peace! I beg you, master, please pacify my mind!'
>
> Bodhidharma said, 'Bring your mind here and I'll pacify it for you.' Huike replied, 'I've searched for my mind and there is nothing to be had.' Bodhidharma answered, 'Then I have pacified your mind.'[59]

This influential story, originally depicted by Yan Ciping (fl. 1163–1189), portrays Huike standing in a lower position, partially buried in snow, earnestly seeking guidance from Bodhidharma.[60] Bodhidharma, depicted sitting aloof on a stone platform, faced the cave with his back to Huike, showing a stance of disdain (Fig. 2.9a).

Tianru translated this gong'an to the mountain gate at Shizi Lin, inscribing the gate's hanging plaque 'Stand in Snow Hall.' In Xu Ben's painting leaf no. 7, the setting is adapted to the winter landscape of Shizi Lin, marked by barren and withered trees, revealing the harshness of the season. The main characters are changed to the master and his disciples in Shizi Lin (Fig. 2.9b). The posture between them became more interactive, with the master seated inside the hall facing the disciple standing outside.[61]

In his poems, Tianru imagined Bodhidharma sitting in 'Stand in Snow Hall,' praising the patriarch's straightforward style, high standards for students, and insistence on 'mind to mind transmission.'[62] In other verses, Tianru and Li Xiaoguang (1285–1350) compared Zhuofeng (fl. early 14th c.), Tianru's principal disciple, to Huike, portraying Zhuofeng as standing resolutely in the snow seeking enlightenment.[63] Inspired by these poetic reinterpretations, visitors to Shizi Lin reflected on their own commitment to Chan practices.[64]

The analysis above demonstrates how scenic gong'ans immersed practitioners with historical Chan Buddhism. Participants not only relived the legendary tales of Bodhidharma and Huike but also connected history to personal spiritual journeys. The research also finds that gong'ans selected for translation reflected Tianru's principles: the 'mind to mind transmission' initiated by Bodhidharma, the gong'an dialogue model initiated by Mazu, and Zhaozhou's approach of integrating Chan practice with daily life.[65] Scenic gong'ans functioned as versatile vehicles, adapting to mentalities of individuals and leading towards enlightenment. The scenic translation effectively addressed the limitations of literary gong'ans.

Buddhist Architecture Incorporating Scenes as Pedagogy Evolved

Tianru's spatial translations transformed Shizi Lin into a landscape full of inspiring scenery, suitable for Tianru to unfold his inherited pedagogy. Shizi Lin soon attracted numerous followers of various kinds and grew into an influential Chan temple. Hundreds of people came to attend lectures, and the temple was always full of visitors.[66] Its influence extended nationwide—the Yuan court bestowed national titles to the temple and to the teacher Tianru, putting it on par with the esteemed Lion Grove Temple at Mount Heaven Eye.[67] This section examines the evolution of monastic activities, comparing those at Southern Song dynasty monasteries during the 12th–13th centuries, Mount Heaven Eye in the early 14th century,

Figure 2.9 The scene 'Stand in Snow' in paintings. (a) Attributed to Yan Ciping (fl. 1163–89), *Bodhidharma Meditating Facing a Cliff* (Damo mianbi). Section of a hanging scroll, ink on silk, painting only: 116.5 x 46.4 cm (45 7/8 x 18 1/4 in.). The Cleveland Museum of Art, John L. Severance Fund 1972.41. https://www.clevelandart.org/art/1972.41. (b) Xu Ben, section of leaf No.7, 'Standing in Snow Hall.'

and Shizi Lin in the mid-14th century. It explores how the scenic gong'an pedagogy reshaped ritual programmes and spaces.

The layout of Tiantong monastery showcases how monastic activities in the 12th–13th centuries were organised. Three ritual activities were arranged along the north-south central axis sequentially: Buddha worships (A) were conducted in the Buddha hall, the temple's most prominent building; sermons (B) were delivered in the Dharma hall behind it; and individual instructions (C) took place in the abbot chamber at the rear. Perpendicular to this axis at the Buddha hall, the Meditation hall (D) to the west and the Reading room (E) to the east facilitated communal meditation and scriptural studies respectively (Fig. 2.10a).[68]

Each activity also occupied a regular time slot in the temple's daily, monthly, and annual schedules.[69] The regulation of monastic ritual activities in both space and time disciplined trainees' bodies towards achieving two transcendences: transcending to the world of Buddha and receiving enlightenment from former patriarchs. Although many of these monasteries were situated among beautiful mountains, the five activities were primarily confined within the compact temple complex, with few interactions with its surrounding landscape, as if the wall of the temple complex strictly demarcated the internal monastic life from the environing landscape.[70]

This rigid boundary began to loosen, as witnessed at Mount Heaven Eye in the early 14th century. Gaofeng constructed his chamber under Lion Cliff facing the fathomless abyss, and monks' meditation and discussions spread out onto the mountain landscape. The intersection between monastic rituals and the landscape progressed further at Shizi Lin in the 14th century, where the architectural boundary diminished. Traditional Buddha worship was replaced with the practice of gong'an scenes. Architectures were adapted to open up to these views, fusing the interior and exterior into a continuum.

Figure 2.10b maps out the monastic programmes in Shizi Lin. It shows that activities were not organised in a cruciform layout centred around the Buddha hall but were distributed amongst the hill and along the river. The eastern slope of the hill near the temple entrance was designated for sermons and individual instructions by the abbot, meanwhile the serene western half, nested in the quiet bamboo valley, accommodated meditation, reading, and guests' residence.

During the Song dynasty, sermons in temples were held in the Dharma hall, which was architecturally designed to emphasise the centrality and sanctity of abbot's teachings. As shown by the interior settings of Kennin-ji Temple (Fig. 2.11a, Fig. 2.11b), the abbot's seat was the focal point, enclosed by the decorated dais below, a prominent ceiling above, and a backdrop screen. An inner ring of columns delineated the central space reserved for the abbot, with an outer ring of columns defining the area for students, who had no permanent seating. In the sermons, the abbot would ascend to the central seat to address the assembly below. The wall of the Dharma hall screened off outside distractions and focused attention to the abbot's speech.[71]

In contrast, Shizi Lin featured a courtyard for sermons (Fig. 2.11c, Fig. 2.11d). The courtyard was encircled by low bamboo walls, with Dharma

a

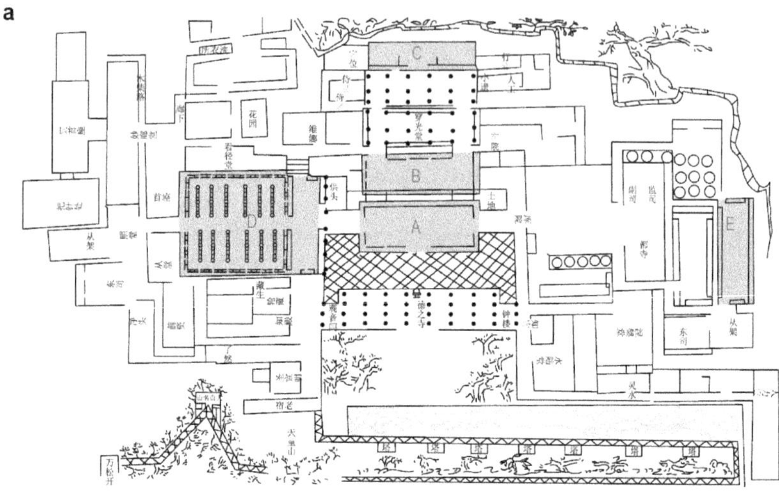

b

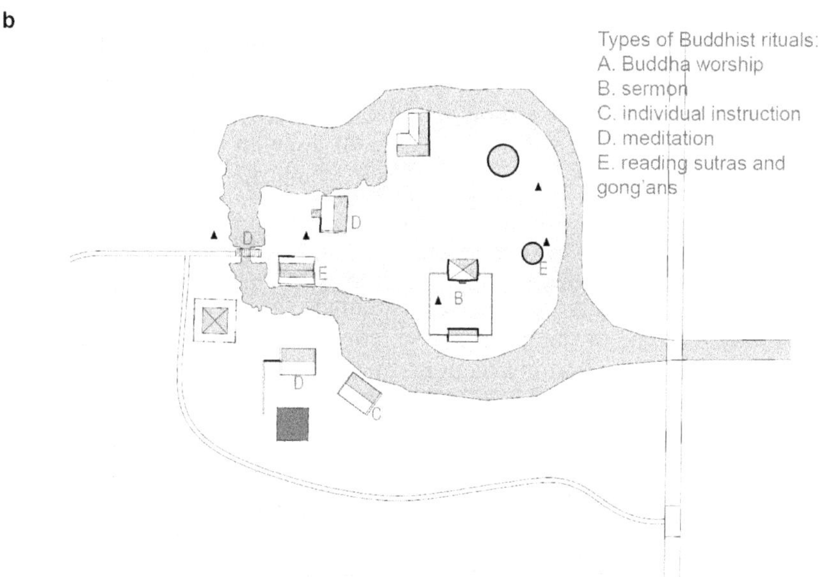

Figure 2.10 Spaces for Buddhist rituals in Tiantong Monastery (12th–13th c.) and in Shizi Lin Temple (14th c.). (a) Monastic rituals in Tiantong monastery. Author's drawing following the map from Sekiguchi, 1991. (b) Rituals in Shizi Lin Temple. Author's drawing.

hall at the back and the mountain gate at the front. Lion Peak served as the monumental centre—a scene translated from Mount Heaven Eye. The door panels of the Dharma hall were removed to offer an unobstructed view of Lion Peak rock. It could be deduced that during sermons, hundreds of attendees would gather around Lion Peak rock, listening to Tianru's

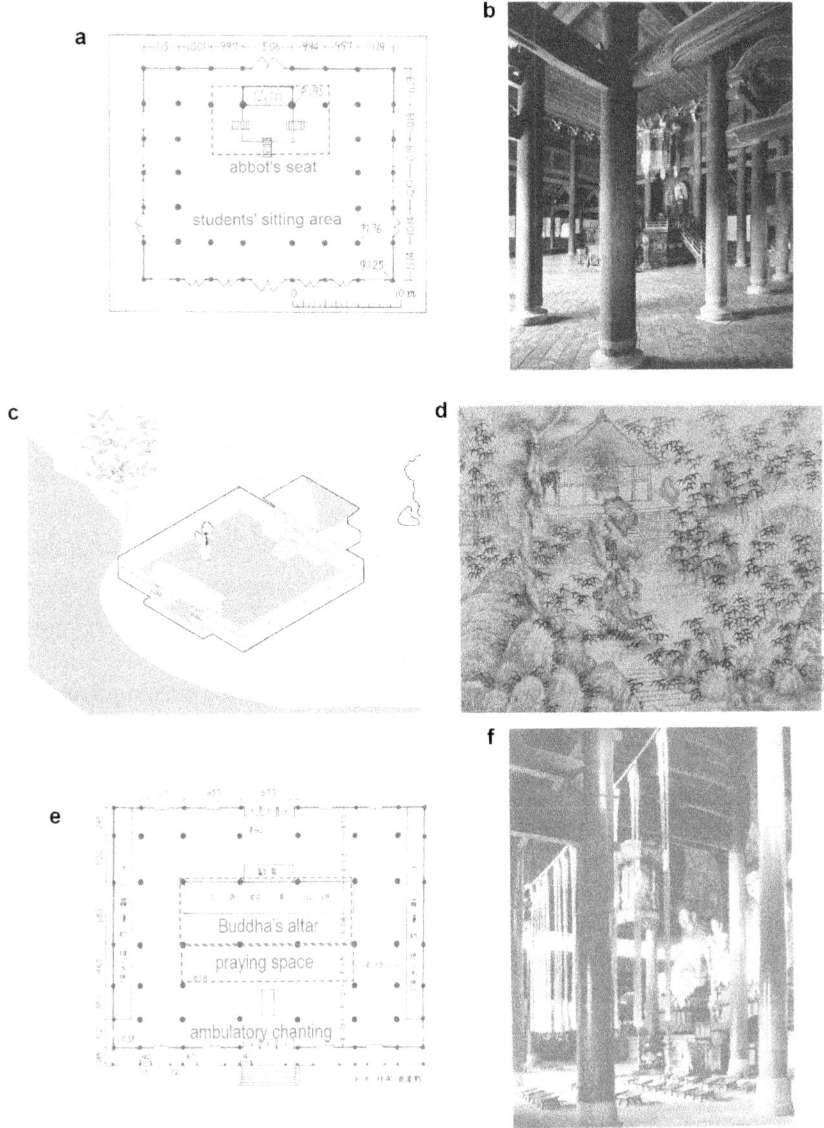

Figure 2.11 Comparison of the Dharma hall and Buddha hall in the 13th century with the Dharma courtyard in Shizi Lin in the 14th century. (a) (b) The plan and photo of Dharma hall in Kennin-ji Temple. Source: Sekiguchi, *Monasteries of the Five Mountains*, 30, 91. (c) Dharma courtyard in Shizi Lin Temple. Author's drawing. (d) Dharma Hall as depicted by Xu Ben, 'Lion Peak.' Taipei: Collection of the National Palace Museum. (e) (f) the Buddha hall of Tiantong Temple. Source: Sekiguchi, *The Origins of Chan Temples in Jiangnan and the Development of Goryeo*, 14-15.

delivery of teachings from the Dharma hall at the rear.[72] This spatial arrangement replaced the centrality of the abbot with the rock, creating a scenic sermon environment.

Lion Peak rock, centring at the courtyard, imparted spirituality to the locale and served as a frequent metaphor in the sermon. In one lecture, in order to emphasize personal determination over teacher's instruction, Tianru made a vivid metaphor that 'a lion could not teach its son, it just pushes it to the edge of a cliff.'[73] Attendants often likened the roar of a lion to the insights of Tianru.[74] Other scenes in Shizi Lin also served as metaphorical constructs enriching Tianru's teaching and attendees' communication.

Tianru adeptly transformed everyday experiences at Shizi Lin into poems with philosophical meaning. For instance, when Flying Rainbow Bridge (Fig. 2.12a) was reconstructed by a patron, a commemorative sermon was held. Tianru's poem inscribed the new bridge with Chan philosophy:
Little Flying Rainbow resides in Shizi Lin.

Below the Bridge, the bamboo breeze is as cool as water.
The old rainbow flew, the new one arrives.
Crossing the rainbow, I begin my journey.
Everyone freely travels between the two banks,
with no distinctions between them.
Stepping firmly on the ground like an elephant king turning over,
each step is like that of a lion.[75]

The poem begins with the sensory experience of crossing the bridge, and feeling the coldness of the water. The reference to old and new rainbows embodies the concept of impermanence and the cyclical nature of endings and beginnings. In the fifth and six lines, Tianru underscored an undistinguished mindset which transcends the temporal (past, present, and future) and spatial boundaries (here and there). The last two lines suggest that even ordinary actions, such as walking, can be transformative if taken with mindfulness.[76]

In Song dynasty monasteries, individual instructions were conducted in the abbot's quarter. The abbot's quarter in Eihei-ji Temple was designed with door panels that enclosed an intimate space for dialogues between the abbot and the student (Fig. 2.12b). Such a setting blocked outside distractions and oriented the student's attention towards the teacher's oral instruction.[77]

However, at Mount Heaven Eye, masters like Gaofeng and Zhongfeng chose natural settings for individual guidance. In Shizi Lin, Tianru's abbot's quarter commanded outdoor scenes. This abbot's quarter, named 'Lying Clouds Chamber,' encompassed a panoramic view of the mound and rocks, as simulated in Figure 2.12c. This view served multiple roles in

Figure 2.12 The Meditation space and abbot's quarter in Shizi Lin (14th c.). (a) Xu Ben, 'Little Flying Rainbow,' leaf no. 4, in the album *Painting of Shizi Lin*. Taipei: Collection of the National Palace Museum. (b) Individual instructions in the abbot's quarter of Eihei-ji temple. Source: Shinohara, 67. (c) Simulated view of Lion Peak rock as seen by Zhu Derun and Tianru from the abbot's quarter. Author's drawing.

a practitioner's psycho transitions, as exemplified by the dialogue between Tianru and the court official Zhu Derun (1294–1365).[78]

Lion Peak rock, standing out among other rocks and visible from the abbot's quarter, prompted Zhu Derun to inquiry about the power of a lion's formidable appearance: 'Master, you named this temple 'lion,' is that because of the formidable look of a lion could conquer all demons?' This question reflects Zhu's perspective, shaped by his political experiences, where apparatus such as official titles and social status were deemed powerful.

Challenging Zhu's notion, Tianru replied: 'No. It is by chance that this rock resembles a lion, not because I want to select one with a formidable look thereby suppressing the evil.'[79] Tianru gave an analogy regarding Manjushri riding a demonic lion, suggesting that Zhu confused the lion's appearance with the Buddha's inherent power to pacify delusions.

Zhu, however, entrenched in the original schema, was inflicted to ask whether it is the lion's intimidating 'roar' that could eradicate delusions. To break Zhu's impasse of form and sound, this time Tianru directly negated: 'If you are looking for the sound and appearance, there are none of them. Could you understand the hidden meaning unspoken?'

Confronted by Tianru's challenging response, Zhu entered the third stage of breaking the impasse. He likely experienced what Pelowski has described as a 'sudden aloneness and loss of expectations,' a state where confusion and frustration peak. This cognitive dissonance led to Zhu's introspection, moving from external validation towards the assessment of internal beliefs. Ultimately, he grasped the essence of Tianru's teaching, realising the power of an undistinguished mind to remain undisturbed amidst external chaos. He used the rock as a metaphor to demonstrate this new mindset which finally won the teacher's agreement:

> Although the teacher and the rock are of different substances, with a single thought on the teacher, all rocks can embody the teacher; with a thought on the rock, the teacher too can manifest as the rock. Wouldn't it be better to forget the distinction between the teacher and the rock?

The scene of Lion Peak rock played multiple roles in the three stages of Zhu Derun's psycho-transition. In the questioning stage, it served as both a catalyst for Zhu Derun and the teacher's metaphorical constructs and a mirror reflecting different inner schemas. Zhu's outward, form-based perspective contrasted with Tianru's inward, mind-focused philosophy. As Zhu navigated through the second breaking impasse stage, he may have gazed at the mound and rocks. The serene landscape settled his mind and brought a state of focus leading to insight. Finally, as Zhu transcended his previous mentality, the scene again facilitated new metaphors, revealing his spiritual progress for the teacher's re-assessment.

Zhu Derun's enlightenment at Shizi Lin had a lasting impact on him. Years later, he could vividly recall details of this cognitive transition, feeling compelled to capture it in a painting.[80] Unlike the confined abbot's quarters of Song monasteries, which focused a practitioner's attention solely on the abbot's words (Fig. 2.12b), the abbot quarter of Shizi Lin framed scenes to facilitate practitioners' realisation, aligning their minds and feelings with the landscape (Fig. 2.12c).

Meditation in Song dynasty monasteries was scheduled in the enclosed meditation hall. Part of the meditation is to contemplate on gong'ans. In Eihei-ji, the windows of the meditation hall were raised beyond monks' eye level to block outside views, ensuring an inward attention (Fig. 2.13a).[81] Meditators sat facing the wall, a tradition known as 'wall meditation,' inherited from Bodhidharma around the 5th century (Fig. 2.9a). At Mount Heaven Eye, meditations were often practiced in the landscape amongst gong'an scenes (Fig. 2.4a).

In Shizi Lin, indoor and outdoor meditation spaces were arranged around Flying Rainbow bridge (Fig. 2.10b). The Meditation hall, with its central bay door panel removed, commanded the view of Holding Moon Peak in front (Fig. 2.13b). In Xu Ben's painting leaf no. 3, a monk was depicted sitting inside the hall, gazing out at the rock. The poem on the opposite leaf reveals that the rock, with moonlight casting through its aperture onto the ground, inspired gong'an-like reflections during meditation. The meditation in Shizi Lin, where sceneries seamlessly integrated into the practice, contrasts sharply with the traditional 'facing wall' meditation.

Stone beds were arranged around Flying Rainbow Bridge for outdoor meditation.[82] The soothing sound of cascading water, the gentle breeze weaving through bamboo groves, and the evocative metaphors associated with the bridge idealised this area for meditation (Fig. 2.12a). Tianru wrote a verse describing practitioners' meditations there:

> People coming for seeking the doctrine of Shaolin.
> Officials and attendants were crowded in the temple, blocking the path.
> They spread into the deep bamboo valley near Reside Phoenix Pavilion,
> and sat in stone beds around Flying Rainbow.[83]

It could be deduced that upon arrival, visitors deposited their personal items in Reside Phoenix Pavilion, the guest house, and then went to the nearby bamboo valley to meditate. Tianru and monk Zhuofeng meticulously managed the landscape of Shizi Lin to optimise visitors' experiences.

It is recorded that they added stony stools under the cypress tree near the square pond to facilitate meditation.[84]

The sutra reading room in Eihei-ji Temple resembled the introspective architectural style of other monastic structures (Fig. 2.13c). Its raised

Figure 2.13 Comparing the sutra reading rooms and meditation halls in Eihei-ji Temple (13th c.) and Shizi Lin Temple (14th c.). (a) The meditation hall in Eihei-ji Temple. (b) The meditation hall in Shizi Lin Temple. Xu Ben, detail from Leaf no. 3, 'Holding Moon Peak.' (c) The sutra reading room in Eihei-ji Temple. (d) the sutra reading room in Shizi Lin Temple. Xu Ben, detail from Leaf no. 9, 'Ask Plum Pavilion.' Source: (a), (c) Shinohara and Satō, *Illustriation on Zen Life*, 32, 101; (b), (d), Taipei: Collection of the National Palace Museum.

window only allowed adequate light for reading, preventing outside views. Similar to the 'wall-meditation' practice, monks also read in a wall-facing manner.

However, in Shizi Lin, there was a large opening on the gable wall of the reading room, which was unprecedented in earlier monastic buildings (Fig. 2.13d). It framed the view of a slanted plum tree in blossom for the monk reading inside. As previously discussed, textual evidence suggests that meditators were contemplating the plum blossom while reading.

Scenes in Shizi Lin served as a platform for cultivating Buddhist realisation. The scenic gong'an thinking pattern, taught by Tianru, was extensively practiced by visitors, and they transformed their see-ings into philosophical constructs. Eight scenes (Fig. 3.8) were canonised through practitioners' poems. The learning experience at Shizi Lin was interactive; visitors engaged not only with Master Tianru but also with peers. They read previous visitors' interpretations of scenes and inscribed new insights and reflections. Their collective poems and writings were compiled by monks of Shizi Lin in *Anthology of the Recorded Scenery of Shizi Lin* and *Recorded Sayings of Master Tianru Weize*, circulating to wider audiences.

The above analysis identifies an evolution of Chan ritual practices in the 14th century, epitomised by Shizi Lin temple, where scenes were integrated into Buddhist ritual programmes. Scenes permeated into sermons, individual instructions, meditation, reading, becoming the primary subject of seeing and discussion. Scenes exerted agency in practitioners' enlightenment. The Chan Buddhism pedagogy transitioned from disciplining the body in enclosed space to engaging the mind in the landscape. Interacted with this pedagogical change, the walls dissolved in Shizi Lin—architecture were no longer enclosed to avoid outside views but facing and framing scenes. This incorporation of landscape by architecture further echoed the Chan principle of non-distinction.

Ideological Transition and the Temple as a Garden

The most striking feature in the architecture of Shizi Lin is the absence of Buddha hall, an absence which indicates the evolution of Buddhist visuality in the early 14th century. The traditional Buddha-centred viewing in Buddha halls originated from the Pure Land belief. It is believed that through visualising Buddha, the Pure Land, and performing worshipping rituals, such as reciting Buddha's name and chanting sutras, the practitioner would gain merit and reborn at the Pure Land afterlife.[85] To achieve this transcendence, monastic spaces centred on Buddhist representations and reined worshippers' attention to them. This Buddha-centred viewing space persisted in Chinese Buddhist sites from the antiquity to the medieval period, and into the Late Song dynasty.[86]

By contrast, in the scenic viewing mode developed in the Yuan dynasty, as witnessed in Shizi Lin and Mount Heaven Eye, the practitioner was placed to be surrounded by scenes. Zhongfeng and Tianru believed that through crafting and interpreting scenes, philosophical ideas could be derived and developed. This landscaping process assisted practitioners in realising inherent Buddhahood and achieving awakening.

The objects in focus served as visual stimuli in both viewing modes, albeit leading to contradictory visions: in the Buddha-centred viewing, the representation of Buddha's paradise guided to the other world and facilitated its imagination; while in the scenic viewing mode, scenes directed to the current world and assisted practitioners in introspecting their inner schema.

These contrasting ideologies could be elucidated through two types of mirror halls: one invented by Fazang (643–712) in the 7th century for Empress Wu, and the others built at Mount Heaven Eye and Shizi Lin in the early 14th century (Fig. 2.14). Fazang's mirror hall featured a central Buddha statue circled by eight vertical mirrors, with two horizontal mirrors placed above and below, creating an enclosed space. Illuminated by a lamp, the Buddha's reflections multiplied into infinity. This seeing visualised Fazhang's idea that Buddha's realm is boundless and pervasive.[87] The multiplying reflections further demonstrate the interdependence in Avatamsaka Buddhism: aspects of Buddha's teachings incorporate and reflect each other.[88] The design of Fazang's mirror hall centred around the Buddha, magnifying his presence and power. Spectators could only see from outside, observing but not partaking—mirrors demarcated the sacred and the mundane.

The mirror hall at Mount Heaven Eye, later adapted by Tianru in Shizi Lin, also utilised ten mirrors to encircle a space (Fig. 2.14b). The practitioner centred the hall instead of the Buddha statue. The practitioner was circled by ten Buddha statues, each in front of a mirror. The practitioner saw both his and the Buddha's reflections multiplied into infinity. The reflections of the practitioner and the Buddhas also intermingled, merging into a continuum.[89] As the spectator represents the secular and the Buddha as the sacred, this mirror hall argued for a unity of the secular and sacred.[90] The secular world was not only valued in view but also regarded as intersecting with the sacred.[91]

Zhongfeng's and Tianru's mirror halls were archetypal to their gong'an scenes at Mount Heaven Eye and Shizi Lin. Both mirror reflections and natural scenes oriented the practitioner as the centre. According to Zhongfeng, mirror reflections equalled natural scenes—both were 'see-ings' by a practitioner, reflecting inner schemas. It also merits consideration of why

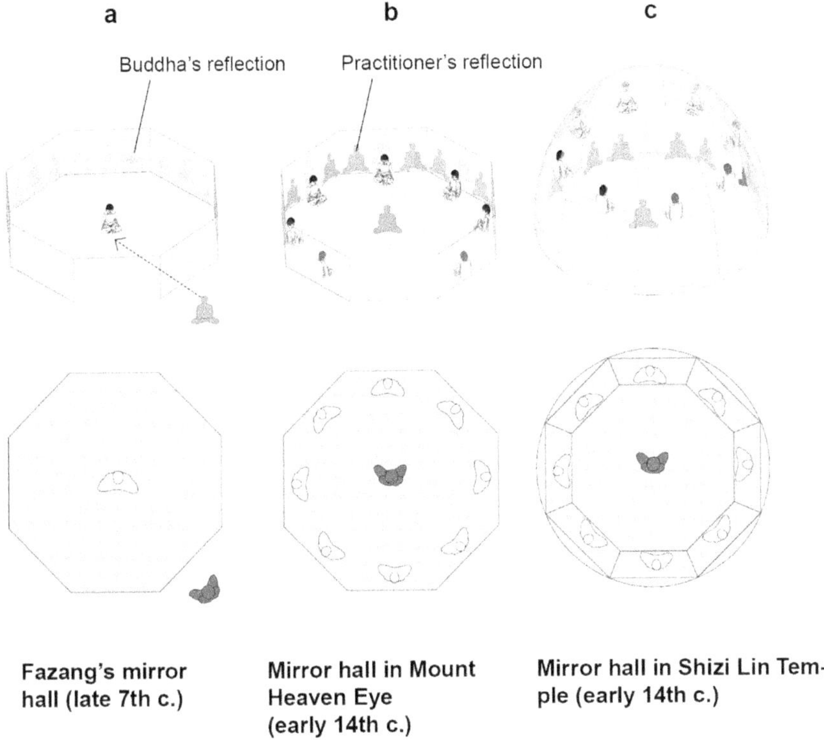

Fazang's mirror hall (late 7th c.)

Mirror hall in Mount Heaven Eye (early 14th c.)

Mirror hall in Shizi Lin Temple (early 14th c.)

Figure 2.14 Evolution of mirror halls from the 7th to 14th centuries. Source: Author.

natural scenes replaced Buddha portrayals as the subject of viewing. Iconic representations are monumental, eternal, and awe-inspiring, while scenes are illusive, ephemeral, and common to find. While iconic representations directed devotees' viewing of Buddha's sacred world, scenes encouraged practitioners to introspect their inner world and find meanings in the current world.

While mirror hall spaces of the sect of Mount Heaven Eye were fully manifested in their gong'an scenes, Fazang's mirror hall was archetypal to the five monastic architectures in medieval monasteries of China and Japan, especially the Buddha hall. The thick walls of the Buddha hall eliminated visual inputs from the external, asserting a similar dualistic view from Fazang. These walls allowed limited light, casting the interior in shadow to create a sacred realm (Fig. 2.11f).[92] As worshippers passed from the bright courtyard into the dimly lit hall, the lighting contrast conveyed the idea of the sacred realm inside as distinct from the mundane world outside. Shafts of light from clerestory windows illuminated the sublime expression of the Buddha statues, which gaze downward, penetrating the worshipers' desires and sins and beckoning them towards the Pure Land.

The centrality of Buddha advocated by Fazang was realised in the spatial configuration of the Buddha hall (Fig. 2.11e). The inner ring columns formed the central volume for three gigantic Buddha statues on the elevated plinth; the outer ring columns defined the worship space for devotees' prayer or ambulatory chanting as the periphery.[93] The numerous constructions of Buddha hall during the Song dynasty, as diagrammed in Figure 2.14, attested to the obsession of Buddha-centred viewing and the aspire to transcend to the Buddha's realm.

However, the traditional Buddha-centred viewing was not performed in Shizi Lin. Textual and pictorial evidence indicates no presence of a Buddha hall or related rituals in the temple. This absence should not be attributed to the temple's hill-like topography since it is Tianru's intentional landscaping. Furthermore, the west and south areas of the temple were comparably flat and suitable for constructing a grand Buddha hall complex. Indeed, when the temple was reconstructed around the 17th century, the southern part was chosen to realise an axial layout with a central Buddha hall. Financial constraints were also unlikely, given the temple's generous patronage.[94] Rather, I would argue that the absence of Buddha hall was rooted in an ideological transition from Buddha-centred to self-valued. The goal of transcending to the sacred realm gave way to realising one's innate Buddha nature.

Similarly, the enclosed Dharma hall was transformed into a courtyard in Shizi Lin (Fig. 2.11c, Fig. 2.11d). The walls separating the interior and the exterior were opened up, as exemplified in the abbot's quarter (Fig. 2.12c), meditation hall (Fig. 2.13b), and sutra reading room (Fig. 2.13d) in Shizi Lin. The abbot quarter of Lying Clouds Chamber, as previously discussed, commanded an exquisite view of the hill and rocks (Fig. 2.12c).

The axial and condensed layout characteristic in former Song dynasty monasteries was not adopted in Shizi Lin—the conceptualised 'Shakyamuni's body' disappeared.[95] This Late Yuan architectural transformation epitomised by Shizi Lin is salient when comparing its layout with Song monasteries and even Early Yuan temples such as Tiantong Temple, Wannian Temple, Eihei-ji, Kencho-ji, and Grand Clouds Cloister in Figure 2.15. In contrast to the traditional axial and cruciform monastic layout, buildings in Shizi Lin were oriented towards, represented by, and named after scenes. The temple's dispersed artefacts featured the 'dominant beautiful disorder' as described by Jean Denis Attiret: 'The Chinese garden characterized by an elaborate composition of hills, valleys, lakes, sinuous streams, and architectural elements, scattered with masses of trees and traversed by winding paths.'[96] The natural curved paths connecting scattered

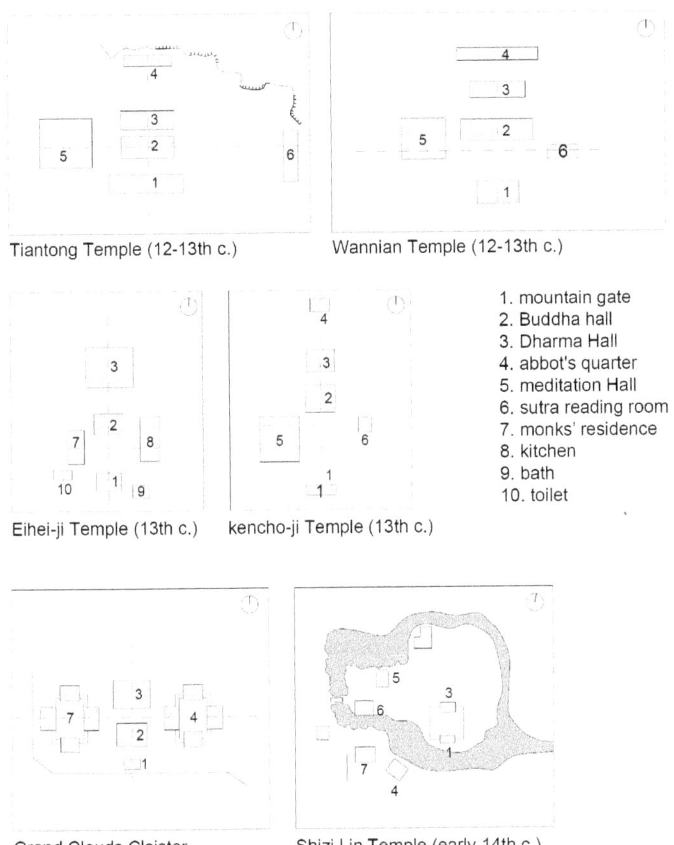

Figure 2.15 The scattered distribution of monastic buildings in Shizi Lin, in comparison to other axial monastic layouts. Source: Author's drawing.

buildings in Shizi Lin also aligns with the meandering experience described by Chen Congzhou.[97]

Monk Tianru's landscaping would surely meet Chen's standards for great gardening, which judges a garden's artistic achievements by exemplifying a break from the convention to optimise the 'trademark' scenery. What are deemed to be key techniques of Chinese garden making, 'facing a scene (*dui jing*),' 'borrowing a scene (*jie jing*),' and 'framing a scene (*kuang jing*),' were observed in Tianru's craft. This configuration associates Shizi Lin more with a garden rather than a temple. The blurred boundary between the indoor and outdoor at Shizi Lin gave it one of the much-praised garden features as 'intermingling the interior and exterior.'[98] The temple was a garden.

Summary

Tianru's spatial translation of scenes from Mount Heaven Eye and literary Chan gong'ans to Shizi Lin Temple illustrates Buddhist's garden practices in China. The crafted scenes were gong'ans manifesting Buddhist philosophy. Meditative engagement within scenes facilitated deictic shift in practitioners, aiding their enlightenment. Shizi Lin Temple represents an architectural evolution of the late medieval Buddhism: the absence of a Buddha hall, walls opened up to frame scenes, and architectural buildings dispersed along rivers and the mound, actively integrating the surrounding landscape. The shift of visuality from Buddha worship to scenic gong'ans underpinned a pedagogical and ideological change towards a self-centred and undistinguished mentality.

Monks

The Yuan dynasty saw the construction of 109 new temples, with the existing temples well-preserved, as mapped out in Figure 0.1 and Figure 0.3. The substantial increase in monastic construction paralleled the revival and expansion of diverse Buddhist sects in Suzhou, including Chan, Tiantai, and Huayan Buddhism. Temples oriented towards those sectorial practices are mapped out in Figure 2.16 and Figure 2.17. Who constructed these temples and how did the gardens inside integrate with religious beliefs? The current and subsequent sections outline the contributions from monks and patrons respectively, based on an in-depth analysis of the construction records of temples.[99]

Chan Sect: Network and Urbanisation

The development of Chan Buddhism intertwined with the urban development of Shantang district, situated northwest of the Chang gate. The establishment of Tiger Hill Chan Temple (*Huqiu chan si*) in Tiger Hill marked the earliest Chan practice in this region. During the early Southern Song dynasty, the renowned monk Huqiu Shaolong (1077–1136) migrated to this temple and taught for three years. He actively disseminated the teachings of Linji lineage in Suzhou.[100] His toponym as 'Tiger Hill' shows his profound influence within the locale.[101]

In the Song dynasty, Shantang district was relatively undeveloped with only sparse settlements. Yet, during the 13th–14th centuries, this area witnessed the establishment of numerous cloisters, temples, and estates, as illustrated in Figure 2.16. Amongst these new establishments, several monasteries belonged to the lineage of Mount Heaven Eye, which was the most prominent Chan sect during that period. The expansion of monastic

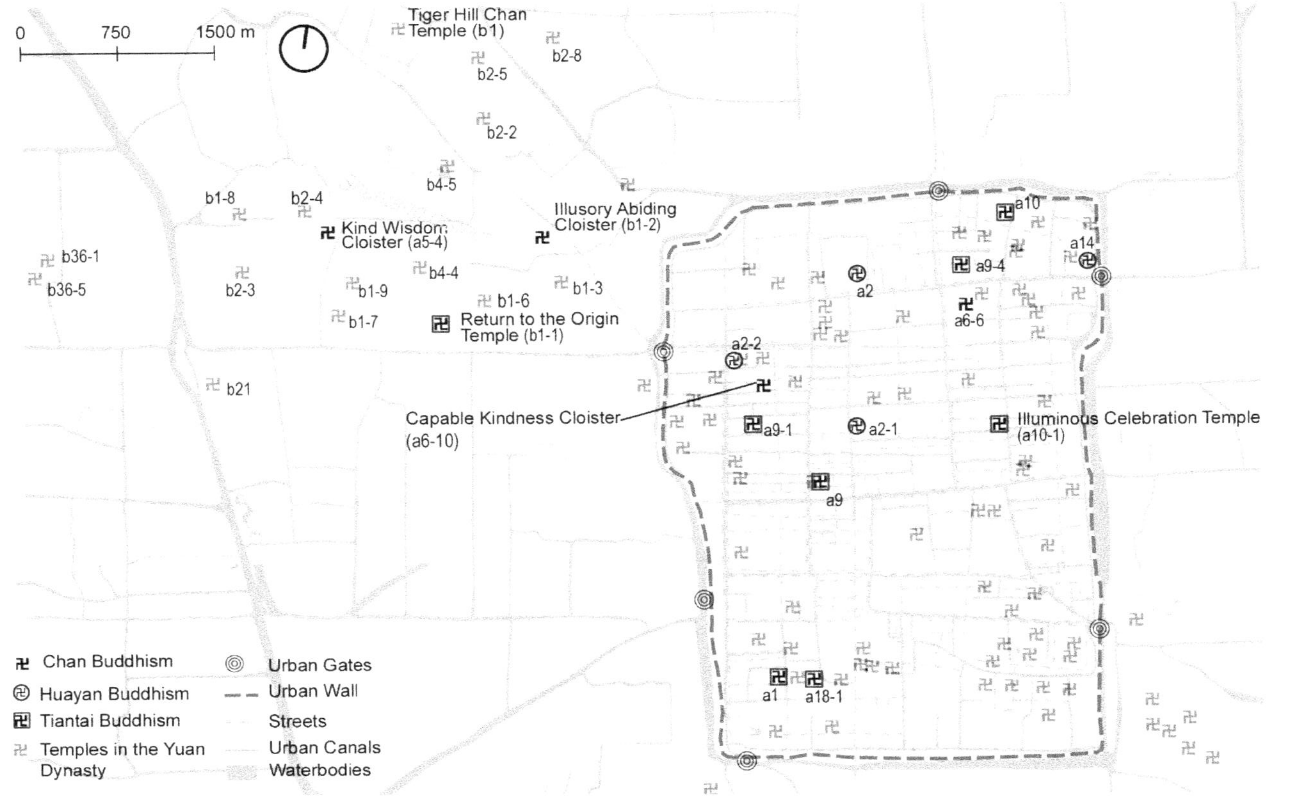

Figure 2.16 GIS map of temples oriented in Chan, Tiantai, and Huayan Buddhism in Suzhou during the Yuan dynasty. For names of the temples, see Table A.1 and A.6 in Appendix. Source: Author.

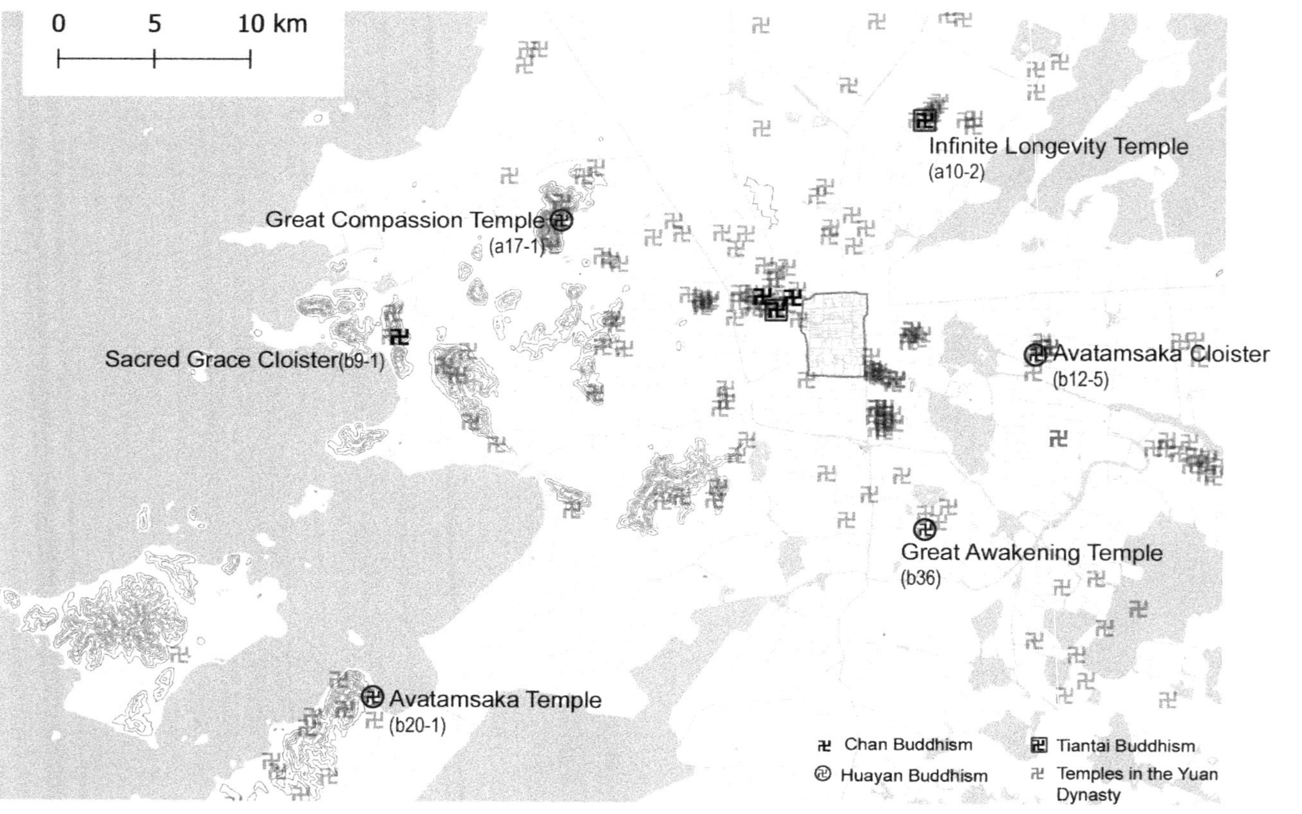

Figure 2.17 GIS map of temples oriented in Chan, Tiantai, and Huayan Buddhism in Suzhou's extra-mural area during the Yuan dynasty. For names of the temples, also see Table A.6 in Appendix. Source: Author.

communities influenced the hydraulic and religious landscape through the construction of temple estates.

The introduction of Mount Heaven Eye's lineage to Shantang district was marked by Zhongfeng Mingben's founding of Illusory Abiding Cloister (*Huanzhu an*) around 1300.[102] During his stay in the then-underdeveloped and swampy Shantang area, Zhongfeng received lands of several *mu* from a generous lay follower. Situated on higher ground and dotted with wild pine groves, this land became the site for Zhongfeng's modest thatched hut, which he named 'Residing Clouds Chamber (*Qiyun shi*).' The name was a reference to Residing Clouds Peak at Mount Heaven Eye, a site where Zhongfeng and his disciples often meditated on the transient nature of perception, inspired by the view of drifting clouds. The renowned Yuan literati Zhao Mengfu (1254–1322), in close connection with Zhongfeng from their time at Mount Heaven Eye, inscribed the cloister's name on its plaque in exquisite calligraphy.[103]

Initially, Zhongfeng only constructed three tiny thatched houses as a shelter for ten monks. Their impoverished life was depicted in Zhongfeng's poem:

> In a thatched hut of three rooms, as cold as ice,
> A dozen monks with ash-smeared faces and earthy countenances.
> They sweep away idle branches and leaves of their own,
> No bothering the tangled vines of other regions. [104]

In the face of these stark living conditions, Zhongfeng and his disciples embraced a life of asceticism and introversion, viewing it as a conduit to spiritual awakening.[105] Their frugal life and meagre situation formed a vivid contrast to the often extravagant lifestyles of monks in the Yuan dynasty.[106]

Zhongfeng soon attracted hundreds of followers, necessitating the construction of a cloister. With collaborative efforts from both monks and laypeople, a monastic compound was completed with halls, rooms, gates, and corridors within three months. Zhongfeng named it 'Illusory Abiding Cloister,' marking the second such cloister he built in his drifting life.[107] Part of this land was reclaimed into paddy fields, although Zhongfeng noted that the agricultural yields were sometimes insufficient to meet monks' needs. Referencing the hydraulic landscaping model commonly used in Yuan temples, it could be deduced that the original marshy land was transformed into a monastic estate with a well-established hydraulic infrastructure. The temple also attracted literati, whose poetic compositions were inspired by the natural scenery and inscribed Chan metaphors.

While serving as abbot at Illusory Abiding Cloister, Zhongfeng played a pivotal role in expanding Mount Heaven Eye's lineage in Suzhou. He was instructive in founding new temples in this lineage, notably 'opened the mountain' for Kind Wisdom Cloister (*Shanhui an*) and Follow the Mind Cloister (*Shunxin an*).[108] This indicates that these two cloisters became new nodes within the monastic network of Mount Heaven Eye's lineage.

Zhongfeng also supported other sectorial temples in Suzhou. He authored stele inscriptions for temples such as Western Avatamsaka Temple (*Xi huayan si*) within the city and Plant Benevolence Cloister (*Zhongfu an*) on its outskirts. Zhongfeng's record validated and endorsed the religious practices carried out within these temples and enhanced their recognition.

Illusory Abiding Cloister, under Zhongfeng's guidance, became a nurturing ground preparing disciples to establish their own temples. One such disciple, Falei Zuzhen (fl. early 14th c.), founded two cloisters, Follow the Mind Cloister and Capable Kindness Cloister.[109] In 1311, the Chen family donated their family graveyard to Zhongfeng, inviting him to open the mountain of this cloister. Zhongfeng named the cloister 'Follow the Mind Cloister' and appointed Zuzhen as its abbot. The cloister was nested in a serene environment at the foot of Wu Mountain, surrounded by pines and cypress trees, and frequented by cranes.[110]

Zuzhen adeptly managed Follow the Mind Cloister, fulfilling Chen family's wish to alleviate the suffering of their deceased relatives and secure blessings for the family's longevity. Under his leadership, the cloister saw a notable increase in both visitors and patrons, leading to a significant growth in its monastic community. Additionally, the cloister also expanded its role to include accommodations for visiting monks. As a disciple of Zhongfeng, Zuzhen's record of Capable Kindness Cloister also received the calligraphy from Zhao Mengfu, further enhancing its prestige.

Zuzhen's effective management of the cloister likely accumulated him wealth and social recognition. A decade later, in 1322, Zuzhen purchased land within the city to establish Capable Kindness Cloister. This time, he attracted patronage independently.[111] Among the supporters was a lay Buddhist named Wu, who donated for crafting Avalokitesvara and Arhat statues, as well as items to adorn the Buddha and Bodhisattvas and make offerings. Other supporters also fulfilled the cloister's various needs.

Meanwhile, Zuzhen received the donation of a mountain villa from Chan master Daoweng, located on the city's outskirts. Zuzhen transformed it into Celebrate Benevolence Cloister (*Qingfu an*), affiliated to Capable Kindness Cloister. He appointed his disciple Zhengxing (fl. early 14th c.) to oversee the temple.

Following Zuzhen's establishment of cloisters, Tianru Weize founded Shizi Lin Temple inside the city in 1326. Before founding his own temple, Tianru resided in Zhongfeng's Illusory Abiding Cloister for approximately one year.[112] This temporary residence likely facilitated Tianru's acquaintance with Suzhou, helped him establish connections with potential patrons, and ultimately motivated his decision to settle there. It can be inferred that other temples in Mount Heaven Eye's lineage also played a similar role as stimulus and supporters for new monks' migration and establishments in Suzhou.

The development of five temples of Mount Heaven Eye's line within one century showcases monks' mutual support in establishing temples. During the Yuan dynasty, the area of Shantang district saw the founding of approximately 17 new temples (Fig. 2.16), a significant increase compared to the four temples in previous dynasties. These newly founded temples supported each other and catalysed further establishments. With each new temple, the existing network of temples grew stronger and more expansive, creating a virtuous cycle that fuelled the establishment of even more temples.

These new temple estates played a driving role in the urbanisation of Shantang district. By the end of the Yuan dynasty, the main road of this area teemed with passengers, bustling shops, and thriving settlements. Monks, as the largest landholder in the district, actively contributed to public service, exerting cultural influence in the local community.[113]

Tiantai Sect: Integrating Seeing and Scenery

During the 1080s to 1090s in the Northern Song dynasty, monk Jingfan (d. 1128) migrated to Suzhou and became the abbot of Northern Chan Temple (*Beichan si*).[114] He delivered lectures on three major Buddhist canons, and his teachings gained wide acclaim. Legendary accounts recorded that Jingfan's sound was as loud as a striking bell. His lectures were often accompanied by extraordinary occurrences, including miraculous appearances of Samantabhadra and other Bodhisattvas, flying snow, and scattered flowers. The local gazetteer depicted these spectacles as 'celestial beings moved in procession and demon kings paid homage, many miraculous phenomena appeared.'[115] Monks and lay followers flocked to his guidance, and the people he ordained filled the city.[116] The local governor recognised him as the prime religious authority in the region.[117]

Jingfan also introduced the essential Tiantai practice of contemplation and repentance to Suzhou. In the 1120s, he constructed the Infinite Longevity Temple (*Wuliangshou si*), situated west of the Northern Chan Temple. This new temple featured a contemplation hall. The hall was surrounded by a pond known as the 'Forbidding Frog Pond (*Jinwa chi*),' named after a legend where Jingfan silenced a croaking frog during meditation by dotting it with a vermillion pigment, causing it to cease making noise.[118] Another anecdote described that during Jingfan's meditation, Weituo (alt. Skanda) kneeled in front of his seat.[119]

Unfortunately, the Jianyan war in 1127 devastated the contemplation hall and other monastic structures at the temple, marking the beginning of the decline of Tiantai Buddhism in Suzhou.[120] Although Huishen (fl. 12th c.), a disciple of Jingfan, managed to rebuild the contemplation hall, many other buildings remained in ruins.[121] The monk community struggled to support themselves as food donation from patrons became unreliable. Huishen constantly feared that 'one day the food supply would be suspended.' The financial struggles of Northern Chan Temple and Infinite Longevity Temple continued for two centuries, until Tianquan Yuze (fl. 14th c.) led their revival in the early Yuan dynasty.[122]

Yuze transmitted Buddhism from Ningbo to Suzhou. Born in a sub-county of Suzhou, he took a vow to Buddhism at the age of 17. He then travelled to Ningbo and trained in prestige temples, studying under Wo'an Benwu (1286–1343) in Extended Celebration Temple (*Yanqing si*) and then Shishi Zuying (1291–1342) in King Ashoka Temple (Ayu Wang si).[123] After completing his training under these eminent Tiantai masters and growing matured, Yuze returned to Suzhou and taught for 50 years.[124] Numerous followers came to partake in his lectures on Tiantai canons and practiced contemplation and repentance under his guidance.

Yuze prioritised the reconstruction of the contemplation hall in his revival of Tiantai Buddhism. The contemplation hall reconstructed by Yuze was recorded as:

> At the beginning of Huangqing's reign (1312–1314), abbot Yuze thought that if the site were derelict now, it would vanish forever. He was afraid that the tradition of Tiantai Buddhism would diminish, and the foundation of teaching would be shattered. He planned the restoration and designated his disciple Depu to help to erect the inclined columns. They spared no effort. The Lu family was glad to devote strings of coins. [In the Lu family's support] they repaired the decayed beams and rafters, amended damaged tiles and bricks, and refurbished those decorative polychrome paintings. Goods and furniture were arranged according to the ritual. Sixteen monks were appointed to practice contemplation of Tiantai Buddhism in this hall. Each one of them occupied one chamber. The most excellent practitioner would be elected as the leader.[125]

Deducing from the textual record, the design of Yuze's contemplation hall followed a pattern established in the 12th century. This archetype was exemplified by the Sixteen Contemplation Halls (*Shiliu guantang*) at Extended Celebration Temple in Ningbo and Shang Tianzhu Temple in Hangzhou, where Yuze received his training.[126] Xie Hongguang's research shows that this hall originated from the sixteen contemplations in the Lotus Sutra. The hall featured a central volume that was typically three or four stories high, housing a giant Buddha at the celestial centre. Surrounding the central volume, 16 chambers were arranged, each accommodating a monk to gaze at the Buddha while meditating. The corridor linking the central volume with these chambers facilitated walking meditation (Fig. 2.18). This hall was placed above a pond filled with lotus. Looking in the distance, the hall seemed to rise from the sea of lotus, registering the sense of purity and spirituality. Upon entering, the dark and enclosed interior focused viewers' attention on the central Buddha and his magnificent realm.

Apart from restoring the contemplation hall, Yuze reconstructed ruined artefacts comprehensively and crafted several scenes in Northern Chan Temple and Infinite Longevity Temple. Later Qing writers depicted that the temple was well landscaped, with fields in front and rivers surrounded it.[127] The new splendid scenes include Great Transmission Pavilion (*Datong ge*), Rain Flower Hall (*Yuhua tang*), and Forbidding Frog Pond.[128] Great

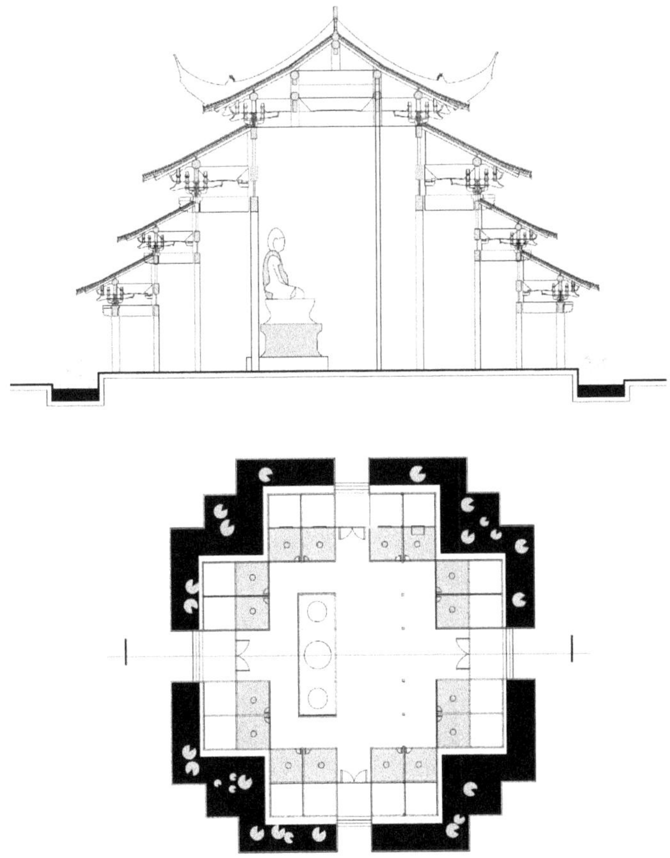

Figure 2.18 Plan and section of Sixteen Contemplation Hall. Source: Author's drawing following Xie Chongguang.

Transmission Pavilion was likely the Sixteen Contemplation Hall, which was historically described as a multistorey building with an integrated central volume for a giant Buddha statue.[129] Rain Flower Hall was the Dharma hall where Yuze lectured. The name encoded the vision that when Tiantai masters lectured, flowers fell like raindrops. Yuze named one of his anthologies *Collection of Rain Flowers (Yuhua ji)*.[130] Additionally, Yuze constructed a chamber and a well in Great Propagation Temple, the well still exists today in the eastern part of Unsuccessful Politician's Garden.[131] The naming of these scenes indicates that Yuze, a monk versatile in poetry, infused the landscape of his temple with metaphors from Tiantai Buddhism, similar to how Tianru crafted scenes in Shizi Lin Temple with Chan Buddhist meanings.

Another prominent monk who specialised in Tiantai Buddhism was Miaosheng Jiugao (1309–1384). Similar to Yuze, Miaosheng was born in Suzhou and was versed in Tiantai teachings and composing poems. Instead of pilgrimaging to Hangzhou as Yuze did, Miaosheng received most of his training in Suzhou.[132] Then he took positions in Northern Chan Temple and Bright Virtue Temple (*Jingde si*) in his middle ages, and finally settled in Forever Pacification Temple for the rest of his life.

Miaosheng revived Forever Pacification Temple as another centre of Tiantai Buddhism. In the Tang dynasty, the temple was noted not for its Buddhist practice but Wei Yingwu's (737–792) poems. Wei was demoted from the court in Chang'an to Suzhou as a magistrate. Frustrated, Wei sought solace in Forever Pacification Temple and composed several poems. Two lines in his poem *Touring the Northern Pond and Monks' Chamber in Forever Pacification Temple* were well received: 'The green pond was suffused with the fragrance of grass. Idling Chamber (*Xian zhai*) was shaded by the tree in the spring.'[133]

During the Southern Song dynasty, Forever Pacification Temple did not flourish and was even subjected to conversion.[134] It was when Miaosheng took the abbacy, that the temple started to be a popular Buddhist Centre. Disciples and followers increased, and the temple became eminent among other Buddhist sites, as recorded by Yao Guangxiao (1335–1418):

Until the reign of Zhizheng (1341–1378), when master Jiugao Sheng took the abbotship of this temple, disciples of this temple flourished and the place for discoursing Buddhism prospered. Miaosheng's appearance was tidy and graceful. He was versed in the 'inside study' (Buddhism) and 'outside knowledge' (non-Buddhist cultures). He was extremely talented in literature and gained respect during that time. He was able to promote Buddhism. When he ascended the dharma seat and lectured on the Lotus Sutra, the monastics and laymen, the elderly and the children all came to listen. The number of attendants reached thousands. Supplies in the kitchen were sufficient, meanwhile, incense and lamps were well-equipped. Forever Pacification Temple became a supreme temple that outranked others.[135]

Valuing the temple's history, Miaosheng picked up the phrase 'Idling Chamber' from Wei Yingwu's poem to name his abbot chamber and inscribed it on the plaque. This chamber, with views of the outside trees and sceneries, provided Miaosheng inspiration for composing poems. Miaosheng wrote two poems in this chamber which echoed Wei Yingwu's poem. In the first poem, he reminisced about a deceased friend and transposed the season of spring in Wei's poem into autumn. He wrote: 'In spring, Idling Chamber was silent. In the wind of autumn, slim shadows of trees are cast into the chamber. [Reflecting on] my life, my tears fall in front of the lamp.'[136] In the second poem dedicated to an official, he wrote: 'In the evening spring, trees cast moving shadows in Idling Chamber. The rain falling down again, when I only think of you.'[137] Miaosheng also entertained guests in this chamber. It was recorded that during vocations, officials gathered in this chamber, discussed Buddhism with Miaosheng, and made poems the whole day.[138]

Miaosheng also constructed the lecture Hall. This hall was titled 'Ocean Seal Hall (*Haiyin tang*).' The metaphor 'Ocean Seal' from the Flower Ornament Sutra (alt. the Avatamsaka Sutra, the Flower Garland Sutra) refers to the image of the sea where, when no wind gives rise to waves, the placid surface reflects and illuminates everything in the world.[139] Thus, naming the lecture hall 'Ocean Seal' alluded that Miaosheng's lectures and instructions would help practitioners to clarify their understanding of the world and reach the stage of Samadhi.

Officials came to take Miaosheng's lecture frequently. One of these officials, Taibuhua (fl. 14th c.), inscribed the name of this lecture hall on its plaque for Miaosheng.[140] During the Yuan dynasty, visiting temples and masters was a part of officials' life, especially for temples of versatile masters such as Miaosheng and Tianru. Although existing archives do not record who financed Miaosheng's constructions, given the generous Buddhist patronages during the Yuan dynasty, it is likely that a renowned monk such as Miaosheng had no difficulty in obtaining donations.

From Jingfan's introduction of Tiantai Buddhism to Suzhou during the Northern Song dynasty to the devotion and advancement of the sect by local monks like Yuze and Miaosheng in the Yuan, Tiantai Buddhism was acculturated and firmly established in Suzhou. This revival was realised by the reconstruction of previous Buddhist artefacts, and the establishment of new Temples oriented towards Tiantai practice. Yuze and Miaosheng, both versed in poetry and well-acquainted with Suzhou's local culture, integrated the viewing practice of Tiantai Buddhism with Suzhou's local tradition of crafting and appreciating scenes. Suzhou became a competent centre of Tiantai sect. Yuze and Miaosheng produced several leading disciples such as Yongzhen Fuliang (1317–1371) and Qi Zongyou (1334–1407). Unlike during the Song dynasty, when Suzhou mainly received Tiantai teachings from other regions, the Yuan era saw the city nurturing its own Tiantai masters who extended influence beyond the local. By the Early Ming, when the emperor recruited Tiantai masters, lots of them were from Suzhou.[141]

Huayan Sect: Monastic Economy and Career

Repay Kindness Temple used to be a flourished centre of Huayan Buddhism led by master Fori Qisong (1007–1073) in the Northern Song dynasty.[142] The temple was one of the oldest temples in Suzhou, prominently located at the north end of the city's central avenue. Its nine-story pagoda, dating back to the Liang period (502–557) and rebuilt around the 1130s, is a significant urban landmark.[143] During the Tang dynasty, the temple comprised five sub-temples, each with a different name on *Pingjiang Map*, indicating their distinct sectorial practices. In the Northern Song period, master Qisong realigned these sub-temples under the Huayan lineage. He publicly lectured on the Flower Ornament Sutra, greatly enhancing the temple's social standing and leading to its renaming as Avatamsaka Teaching Temple (*Xianshou jiao si*).[144] However, the temple's prosperous era was short-lived; most of its artefacts suffered damage during the Jianyan period war (1127–30). Subsequently, the succeeding abbot only managed to restore several key artefacts.[145]

In the late 13th century, the economic prospects of the temple improved significantly, as evidenced by its frequent reconstructions. Within a century during the Yuan dynasty, the temple underwent two comprehensive reconstructions and one extensive refurbishment. The rate of rebuilding—approximately every three decades—speaks to the temple's remarkable economic resilience. The following discussion examines the stele inscriptions recording these reconstructions, investigating how the temple accumulated wealth and capital through the building projects. Analysing the rhetoric used by the two Hanlin academician authors also provides insights into the national agenda of promoting grand temples.

The first reconstruction was directed by abbot Nanxuan Xun (fl. 13th c.), a successor in Qisong's line. Nanxuan restored the whole temple comprehensively and expanded it, as recorded by the Hanlin academician Yan Fu (1236–1312):

At the beginning of the unified imperial dynasty, [Nanxuan] took abbacy of the teaching position…he supervised the selection of materials and training of workers. Over the years and months, he established three outer gates, constructed the eastern and western cloisters, built the central Dharma hall, and adorned the pagoda courtyard. Whether it be the guest quarters, the repentance rooms, the Earth Deity shrine, the Huayan ancestral hall, the monks' living quarters, the well pavilion, the bathhouse, the lion's seat, or various ritual implements, what was worn was replaced, what was lacking was completed, and what had fallen was raised. The columns and beams are numbered in the hundreds, the reclaimed land amounts to thousands of *mu*, and the labour costs tens of thousands of strings of cash.[146]

There were several notable aspects of Nanxuan's reconstruction. First, the project was resource-intensive, indicated by the substantial number of structures built, the expansive land reclaimed, and the significant labour and material costs involved. Nanxuan's diligent supervision of material selection and labour training indicates a commitment to high-quality construction. The sheer scale of reconstruction and substantial financial investment were comparable to the imperial family's construction of Illuminous Celebration Temple, which will be discussed later. Though the source of funding was not specified, there is a sense that financial constraints were not a concern. Notably, Nanxuan added a hall dedicated to Huayan patriarchs, reinforcing the temple's orientation on Huayan Buddhism.[147]

In his closing remarks, Yan Fu commended Nanxuan's reconstruction as an adeptness to transmit the Dharma. Yan underscored the temple's role in fostering the moral development of citizens. He passionately called upon citizens and officials to reward their parents and country, to thus realising the true meaning of 'repaying the kindness.' Buddhist temples were not seen as retreats from social obligations, but rather as places that actively promoted and supported the social order.[148]

The second reconstruction guided by Abbot Biechuan (fl. 14th c.) three decades later was prompted by a devastating typhoon that swept through Suzhou. Located at the end of the central avenue, the temple bore the brunt of the typhoon, suffering extensive damage to most of its buildings. Monks were forced to suspend all teachings and focus on restoration. Biechuan donated his personal wealth to reconstruct the bell tower and Nirvana pavilion, and to repair monks' residences and abbots' quarters. The outcome was extolled by Huang Jin (1277–1357) as 'everything that once existed was fully restored, and even things that were not there before were fully prepared. The broad eaves and spacious rooms, the thick beams and large pillars, all displayed a grandeur and solemnity that exceeded the old.'[149]

Apart from a full restoration of the numerous artefacts initially constructed by Nanxuan, Biechuan also established sixteen hectares of permanent fields to bolster the temple's financial stability.[150] He personally financed the purchase of three hectares, while the remaining thirteen were acquired through patronage. Biechuan set up rules for managing these fields, appointing virtuous monks to oversee the generated income. This revenue was primarily used to cover practitioners' living expenses, with a reserve set aside for future monastic needs.[151] Huang Jin commended Biechuan's significant investment of wealth and effort towards enhancing the temple. Echoing his colleague Yu Ji (1272–1348), Huang Jin advocated that the practice of Huayan Buddhism could benefit the whole country: 'learners of this temple should concentrate on studying Buddhism, instead of being satisfied with a comfort life. The longevity of our country could thus be ensured, and our people would be rewarded. The merit is infinite.'[152]

Two decades later, the third refurbishment was carried out under the supervision of abbot Guangxuan Wuyan (fl. 14th c.), again recorded by Huang Jin. Huang's portrayal of Guangxuan highlights his appeal to disciples, noting, 'the abbots, following the ancestral teachings, have widely received all beings. With a wave of the fly whisk, students gathered like clouds.'[153]

Observing the deterioration of many artefacts, tilting statues, and fading golden hues, Guangxuan advocated a renovation by devoting his personal wealth as a start. People from all social strata, responded to Guangxuan's appeal and donated generously, including noble clans, distinguished families, and common households. The renovation transformed the temple into an architectural marvel, resembling a paradise on earth:

> The entrance stands tall and wide in front, impressive and open, allowing unobstructed passage for humans and heavenly beings alike. The ease of access to the towers and pavilions is such that one need not even lift a finger.
>
> Two corridors alongside were profound serene and displaying intricate patterns. Five hundred arhats manifest, emerging vividly among them. The bodies of these Śrāvakas were wearing robes untainted by dirt. The [Buddha] hall is adorned with serene and dignified images, while the [Dharma] hall expounds the true essence of teachings, providing a sanctuary for purity and gathering. Various embellishments enhance its sanctity, radiating splendour and beauty. The seat of Bodhi is no different from the heaven of Trāyastriṃśa. The granaries are full, and the kitchens are abundant, reminiscent of the fragrant lands where the remnants of the World-Honoured One's meals are used for Buddhist rituals.[154]

Repay Kindness Temple's three reconstructions showcased the substantial economic strength of a grand monastery during the Yuan dynasty. First, the management of monastic buildings was efficient. Whenever they were damaged in natural disasters or showed signs of wear, immediate reconstruction efforts were initiated, reflecting strong financial backing. Monks were vigilant in monitoring the condition of the buildings, ensuring the refurbishments were timely and effective.

Second, the destruction or wear of artefacts did not spell crisis for the temple but turned out to be opportunities for enhancement. Each reconstruction phase not only repaired the damage but also improved the architectural splendour and expanded the landholdings. The first restoration led by Nanxuan amplified the number of monastic artefacts, the second reconstruction directed by Biechuan equipped the temple with permanent fields, and the third renovation led by Guangxuan promoted the temple to the peak of its architectural grandeur. This contrasts with the temple's declining financial condition in the succeeding Ming and Qing dynasties: when the temple required reconstruction and refurbishment, monks only managed to renew a few primary artefacts, leaving other artefacts to fall into ruin and be lost. Abbots were poor, and patrons were inactive. The monk community shrank, and lands slipped away from the temple's possession.

The thrive of Huayan sect in the Yuan dynasty could also be glimpsed from the diverse career paths available for monks aspiring to establish themselves, as exemplified by Rushan (fl. 14th c.) and Guting Shanxue (1307–1370). Rushan, though not a direct descendant in the Huayan lineage, gained local acclaim through his efforts in constructing a new temple. In comparison, Shanxue earned his reverence as a Huayan Buddhism leader due to his profound interpretations of sutras and intellectual contribution.

Rushan, who lacked a detailed biographical record, was chiefly remembered for his construction of Heavenly Benevolence Avatamsaka Teaching Temple (*Tianhui xianshou jiao si*).[155] The literati Chen Ji (1314–1370)'s record offered a succinct introduction to Rushan. Hailing from a well-established family in rural Suzhou, Rushan was raised in a household with Buddhist beliefs. In his youth, he engaged with worldly classics externally while focusing on the true vehicle of Buddhism internally. He sought out renowned teachers and became obsessed with the Flower Ornament Sutra.[156]

Inspired by the 'Great Repaying Kindness' chapter of Huayan scriptures, Rushan was motivated to build a temple to repay his parents.[157] He led a life of simple food and humble clothing, accumulated wealth through diligence and frugality, and bought a small plot of land inside the city. Initially, the land was just enough to build a modest cloister for his dwelling. Over time, he expanded this humble cloister into a full-fledged temple, as portrayed by Chen:

> The temple gate is grand and splendid. Palatial halls are deep and majestic, and the Dharma hall is tall and spacious, all arranged orderly from east to west. The sutra storage pavilion, meditation hall, the abbot's quarter, and areas such as the dining hall and kitchen, all perfectly equipped for religious activities.[158]

Rushan guided monks in studying scriptures in the temple, teaching both Sanskrit and Chinese texts, gaining admiration from scholars and commoners. A former minister of the Commission for Buddhist and Tibetan Affairs (*Xuanzheng yuan*) petitioned the emperor to grant the temple the prestigious title of 'Temple of Ten Directions' and an official name. A disciple of Qingliang Chengguan (738–839) was invited to inaugurate the temple and offer teachings. From then on, the temple became a legitimised Huayan Buddhist site in Suzhou city.

Three aspects are noted in Rushan's construction. First, Rushan was neither a holder of Dharma lineage nor an enlightened monk. The temple where he took his tonsure might not be a recognised one, whose name was not even specified by Chen Ji. Despite founding the temple through persistence, Rushan was not eligible to 'open the mountain' for it, necessitating the invitation of a monk from the Huayan lineage for the inauguration. Nevertheless, Rushan successfully built a cloister from the ground up. He managed the cloister efficiently, eventually accumulating enough wealth to expand, which was later legitimised by the court as a temple. This reveals the feasibility of commencing a monastic career by constructing a temple. Third, both society and the court showed inclusive attitude towards new temples. Once a temple was equipped of considerable scale, it attracted many to join the sangha community, and local officials were keen to support and promote.

Differing from Rushan's trajectory, monk Shanxue gained esteem through contributions to Huayan philosophy rather than temple construction. He received stele records from four personages after his death, all attested to a high recognition.[159] Inquiry for a deep understanding of the Dharma, Shanxue embarked on a journey to renowned temples and teachers. He was ordained in Great Awakening Temple (*Dajue si*), and then moved to Light Benevolence Temple (*Guangfu si*) to follow master Linwu Qinggong (1272–1352). Feeling the aspiration to improve further, he sought the guidance of monk Baojue Jian (fl. 14th c.) at Caoxi. Under Baojue's tutelage, Shanxue delved into scriptures such as Avatamsaka Sutra, Sutra of Perfect Enlightenment, Surangama Sutra, and The Awakening of Faith, and finally reached enlightenment.[160]

At the age of 37, Shanxue rose to prominence and held positions in various temples. In 1344, he was nominated by Abbot Biechuan Jiao to the chief seat at the Huayan Buddhist centre, Repay Kindness Temple. Then he was appointed by the Commission for Buddhist and Tibetan Affairs to establish Jianfu Temple in Kunshan and subsequently became the abbot of Great Compassion Temple (*Daci si*) at the Sun Mountain (Yangshan). Acknowledged as a 'leading scholar of Avatamsaka teaching (*Xianshou Xuekui*)' by people in Suzhou, Shanxue was invited to assume the abbotship of Light Benevolence Temple.[161]

Shanxue contributed to the scholarship of Buddhism by integrating the philosophy of Huayan sect and Tiantai sect. His influential works, such as *Odes on Ten Mysterious Gates* (*Shixuan men fu*), elucidated the core principles of Avatamsaka Buddhism. These works were widely circulated and advanced Buddhist thinking. As Yao Guangxiao commented: 'I used to criticise how scholars of the Xianshou (Avatamsaka) and Tiantai schools, each staunch in their doctrines, seemed to be in a conflict akin to the Chu-Han contention. Recently, among those who managed to bridge the gap and unify the teachings, Shanxue stood out prominently.'[162]

Rushan and Shanxue demonstrated diverse paths to becoming an esteemed monk. Rushan gained recognition by building a temple and engaging with the local community, whereas Shanxue competed with Buddhist masters through scholarly research and publication. Their stories shed light on the Buddhist landscape in Yuan dynasty Suzhou, characterised by accessible sutras copies, prominent teachers, and temples. This accessibility was conditioned by the flourishing publication industry and the government's promotion of Buddhism. Many citizens showed a keen interest in Buddhism, aspiring to monastic careers. Buddhist imagery, texts, and temples were ubiquitous in the city, and Buddhist practices permeated everyday life. It was common for people to read Buddhist canons, discuss Buddhism, and pursue a Buddhist path.

Patrons

The Imperium and National Agenda

One of the temples in Suzhou patronised by the Yuan Imperium was Illuminous Celebration Temple, located in the underdeveloped northeast quadrant of the city.[163] The temple's construction was initiated by the court official Azala (fl. 14th c.), with authorisation from the princess Sengge Sagi. The temple received sutras and financial donations from the imperium and glorifying records from Hanlin academicians. As a high-ranking Buddhist official, Azala, who was ordained by the national teacher Rinchen Gyaltsen (1256–1305) and trained under Gyaltsen's disciple Shajian (fl. early 14th c.), had previously held significant Buddhist positions in northern China. When emperor Renzong (1285–1320) allocated extensive fields in Suzhou to Princess Sagi, Azala was dispatched to the region to oversee the fields. Upon his arrival, Azala purchased the land and established the temple with the princess' approval.

Azala's construction of the temple adhered to the conventional seven-hall layout. Arranged along the central south-to-north axis were the mountain gate, two stone towers, a Buddha hall, a Dharma hall, and the abbot's quarter. To the east were a storage room and a kitchen, and to the west was the shrine of the Tiantai patriarchs, underscoring the temple's dedication to Tiantai Buddhism. Azala also allocated fields to ensure a steady income for the monastic community. Yu Ji described the temple as being fully equipped with all necessities, noting its rapid construction within four years (1325–1328). Upon its completion, Azala reported to the emperor, earning commendation and an official certificate legitimising the temple. Monk Huaishou (fl. early 14th c.), likely an influential figure in the Yuan court, was appointed as the temple's abbot. Two years later in 1330, the emperor granted the temple 5000 copies of Tripitaka, and the empress acquired the temple additional fields. She instructed the monks to recite the scriptures daily to pray for prosperity and longevity.

In 1349, Azala commissioned the construction of a grandiose two-story pavilion positioned behind the Buddha hall at the rear of the temple's central axis. Huang Jin described this pavilion as spanning three bays and reaching a height of approximately 18.7 metres. Considering the width of one bay as five to eight metres, it can be inferred that the pavilion's columns and beams were notably robust. Huang Jin also detailed the pavilion's elaborate decorations:

> Its layered brackets and eaves, square lattices, and curved railings are meticulously measured on all four sides. Painted in vermillion, gold, and azure, it shines brilliantly both inside and out, adorned in various ways, exquisitely and wonderfully.
>
> The second floor of the pavilion housed a Buddha statue, exquisitely carved from precious sandalwood, flanked by Manjushri and Samamtabhadra Bodhisattvas. Thousands of Buddhas and the landscape of Mount Wutai were painted on the wall. The sutras previously endowed by the imperium were aligned along the wall.[164]

On the first floor, Azala configured the space to enshrine himself, his teacher, and his wife. In the central bay, a statue of Avalokitesvara Bodhisattva from Mount Potalaka was placed, alongside a statue of Azala's teacher. The left bay served as a living shrine for Azala, while the right bay was dedicated to his wife. Huang Jin noted that the pavilion, completed swiftly within a year, was entirely financed by Azala himself, without drawing on local government or patronage resources. The grand pavilion evoked awe and admiration from all who saw it.

The construction of Illuminous Celebration Temple showcased the imperial families' financial strength to establish grand temples in regional cities. The construction was officially praised by the court as accruing merit for the imperial family and the entire empire. This architectural feat represented the imperial advocacy of Buddhism in local regions. Interestingly, the temple's construction process did not assimilate into the local culture. Instead, it symbolised the imperial court's effort to transplant its northern

Buddhist culture to the south. The temple featured paintings of Mount Wutai, a revered site in northern Buddhist tradition, underscoring this cultural transmission. The enshrinement of Azala and his wife and teacher in the temple clearly asserted the authority of a court official in a regional city.

Pious Uyghur Officials

Uyghur officials, migrating to Suzhou on administrative posts, formed the second powerful patron group. Originating from Kucha region (*Gaochang*), the Uyghurs had a long-standing tradition of temple patronage, notably exemplified by their contribution to Mogao caves of Dunhuang during the tenth to 11th centuries.[165] The Song-Jin wars and Mongol conquests prompted their migration into central and southern China, and their proficiency in both Mongolian and Chinese positioned them as key intermediaries between the Mongol rulers and Han Chinese.[166] In Suzhou, Uyghur officials formed a substantial part of the local administration.[167] They patronised Shizi Lin by purchasing land and setting up artefacts, meanwhile leveraging their social networks to promote the temple to officials.

Tuglug (fl. early 14th. c.), a Uyghur official overseeing the lower Yangtze river basin (*jiangzhe pingzhang*), was a prime donor in the establishment of Shizi Lin Temple. This is revealed from the three letters Tianru addressed to him. In one letter, Tianru credited Tuglug's frequent visits and generosity for transforming the temple: Since [I] laid out the ground of Shizi Lin and made thatched houses, your repeated kind visits have made the grass, trees, bamboo, and rocks shine with a hundredfold brilliance. My disciple and I were not talented enough to make them so. It was all thanks to your endowment.'[168] In another letter, Tianru conveyed his deep gratitude: 'A single bowl beneath the trees, a life outside of worldly concerns, quietly dwelling in peace without external disturbances—all these are illuminated by your residual influence, and I dare not forget them for even a day.'[169]

As a devout Buddhist, Tuglug maintained thoughtful exchanges with Tianru. Their correspondence illuminates a deep bond and How Tianru instructed Tuglug towards enlightenment. In one letter, Tianru lauded Tuglug's deeds for the citizens while pursuing the Buddha's path, congratulating him on his recent promotion and urging him to harmonise worldly laws with Buddha's teachings.[170] Tianru's other letter addressed Tuglug's aspiration for 'the ultimate state of great rest and cessation,' advising him to seek enlightenment within worldly entanglements rather than in seclusion.[171] In a third letter, Tianru congratulated Tuglug's successful southern campaign against bandits. He draws a parallel between Tuglug's suppression and the Buddha's vanquishing of demons, and further inspired Tuglug to transcend the dichotomy between sage and demon, maintaining a nondualistic mentality.[172]

Tuglug's colleague Dotung (fl. 14th c.) provided literature support to Shizi Lin Temple and promoted it to other officials. Following the temple's completion, Dotung inscribed 'Bodhi Villa (*Puti lanruo*)' on the temple's entrance plaque. Serving as an administrator of Pingjiang Lu, Dotung was stationed in Suzhou for eleven years. While not as deeply invested in Chan Buddhism as his colleague Tuglug, he was well-acquainted with Buddhist masters and actively supported their promotion, such as appointing Xuechuang Guang (fl. early 14th c.) as the abbot of Heavenly Bestowed Benevolence Temple. Dotung also facilitated connections between the Buddhist community and the political sphere. He introduced Tianru, the abbot of Shizi Lin, to the court official Zhu Derun. When Zhu expressed his distress over complex political matters and a desire to retreat, Dotung recommended consulting with Tianru for guidance.

Maiju (fl. 14th c.), the third patron of Shizi Lin, was also a Uyghur official and one of Suzhou's largest landholders.[173] His father, Atai Toyia, was a devout follower of Tianru's teacher Zhongfeng. Maiju described himself as 'a descendant of Kucha, devoted to Buddhism as a life command. When seeing an esteemed monk, I show serious reverence as if a silent cicada in winter.'[174] Residing near Shizi Lin Temple, Maiju had ample opportunities to visit Tianru. He inscribed the plaque name of the Dharma hall as 'Stand in Snow Hall' and Tianru's abbot's quarter 'Lying Clouds Chamber.' The naming conveyed Maiju's steadfast dedication to Chan practice and his familiarity with Chan gong'ans and the culture of Mount Heaven Eye.

As analysed above, Uyghur officials in Suzhou formed a group of devout and reliable patrons. As intermediaries between the imperial court and the locals, they were capable of introducing temples and monk masters to higher echelons of the society. Financially well-off, many Uyghur officials owned extensive fertile lands in the Yangtze river basin, generating income for Buddhist patronage. They frequented Buddhist teachers for spiritual guidance. As immigrants adapting to a new cultural environment in Suzhou, monastic spaces provided them with a sense of community and belonging, where they could gather, communicate, and engage in their Buddhist tradition.[175]

Locals' Investment

Local citizens of Suzhou, similar to loyal families and officials, were fervently dedicated to the construction and patronage of temples, generously endowing lands and fostering Buddhist communities. A prime example of this devotion is Return to the Origin Chan Temple (*Guiyuan chan si*, alt. *Guiyuan xingguo chan si*), established by Cao Ruli. The Cao family were long-time residents of Tiger Hill County (now the Shantang area in Suzhou). They led a modest life, accumulating wealth through agriculture. In 1281, Cao Ruli experienced a spiritual awakening, claiming to have received a divine response from the Bodhisattva Mañjuśrī at Mount Wutai. This led him to Beijing, where he obtained a monk's certificate from the emperor Kublai Khan. Upon returning to Suzhou, he took monastic vows at the Clouds Cliff Chan Temple on Tiger Hill near his home. Following him, his wife and two elder sons also adopted monastic lives.[176]

Residing in Clouds Cliff Chan Temple, the Cao family invested their personal wealth into the temple's upkeep. Cao Ruli served in several positions and endowed money for the refurbishment of the temple's buildings. His generosity, however, incurred jealousy among some, compelling Cao Ruli to leave and construct his own temple.[177] Chen Lü (1288–1343) detailed this venture:

> He returned to the ancestral home in Plum Grove (*Meilin*). The place, with its green fields and surrounding waters, was ideal for a Buddhist retreat. He built a grand hall with long corridors and imposing gates, a spacious and elegant Dharma hall and a Buddha pavilion with intricate gold and jade decorations. He established monastic residences, labourers' quarters, and all other necessary building, and acquired over three thousand *mu* of fertile lands for the temple's permanent income.[178]

After Cao Ruli's death, his sons took over to manage the temple, continuing its expansion and development. Honouring the father's wish, they constructed an abbot's quarter and invited Chan master Zhongfeng Hai (fl. early 14th. c.) to take the abbacy.[179] In 1316, they received the imperial decree which elevated the temple to a primary one in Suzhou. After the demise of the two elder sons, the youngest son Cao Ju (fl. 14th c.) took over the temple. He constructed more structures, including the Bell tower, storage tower, Bodhisattva pavilion, rear hall, eastern and western abbot halls, monks' quarters, meeting hall, and guests' hall. Furthermore, he established an administrative office inside the city to enhance the temple's reach to intramural residents. He also increased the temple's land holdings with another 300 *mu* (around 192 km²) of fertile lands.[180]

Cao family's endeavour encapsulates three aspects of local citizens' monastic construction. Unlike royal families who were able to rapidly establish large temples in three years, Cao family, with limited financial means, took sixty years of two generations to complete the project.[181] The family's fortune grew as they converted private property into a monastic estate. The youngest son Cao Ju bought 192 km² of fields to the temple, a hundredfold increase compared to the land his father acquired thirty years ago. He was even capable of buying expensive urban properties. Similar patterns are observed in the cases of monks Zuzhen and Rushan, who gained substantial wealth by constructing temples. The three examples prompt an intriguing question: did wealth facilitate monastic patronage and construction, or did investment in monastic estates lead to the accumulation of wealth?

Second, the temple's establishment secured Cao family's status as monks, a household status they previously struggled to maintain at Clouds Cliff Chan Temple. This suggests that during the Yuan dynasty, monkhood was highly desired, and it was attainable through constructing temples. Upon completion, the temple quickly gained legitimacy from the Yuan court, which, unlike its Ming successors, did not impose suspicious eyes and quotas on temples and ordinations. The Yuan court seemed to adopt a Laissez-Faire policy towards the ever-increasing number of temples and monks. Monks could fully invest in estates and other commercial activities, becoming great landlords and elites.

Wang Jinping has argued that taking a Buddhist career was a favoured choice during the Yuan period, and many social elites were monks at that time.[182] None of Cao family's members was a prominent Buddhist leader with substantial influence on the religion. Cao family's establishment of Return to the Origin Chan Temple leads to speculation about whether other cloisters in the Yuan dynasty emerged from strategies to convert private assets into Buddhist properties while forging a Buddhist identity.

Conclusion

This chapter investigated the temple-scape in the Yuan dynasty Suzhou both from the physical establishment of temples and the cultural construct of Buddhist ways of seeing. Monks expressed religious ideologies by crafting temples and gardens. The construction of monastic estates, Buddhist pursuits, and the economic benefits were mutually reinforcing. The hundreds of temples served as spatial nodes for exchanging Buddhist ideas and formed a social network between monks, patrons, and officials. The analysis of construction records shows that temple estates were profitable land investments. Patrons from multiple ethnicities donated to monastic construction projects, converted their lands into temples, and aspired to become monks.

Notes

1 Discussions on Lion Grove Garden as a typical Chinese garden could be seen in works like Keswick, *The Chinese Garden*, 172–79, 224; Henderson, *The Gardens of Suzhou*, 34–59; Rinaldi, *The Chinese Garden*, 79.
2 Initiated by Qianlong emperor's replication of Shizi Lin in his Mountain Estate to Escape the Heat in Chengde (Chengde bishu shanzhuang), this 'water surrounding central hill' configuration was popularised as an iconic

Chinese garden layout. This layout was later re-identified by modern scholars like Tong Jun as the typology of Chinese garden. See Tong, *Gazetteer of Jiangnan Gardens*, 16, 34.

3 The site's chronological changes has been outlined in researches such as Tian and Fang, 'Research on the Historic Appearance of the Lion Grove from the Yuan Dynasty to the Republic of China'; Suzhou shi Yuanlin Lühua guanliju, *Gazetteer of Shizi Lin*; Wei, *History of Classical Gardens of Suzhou*, 167–181.

4 Recently the Buddhist history of Shizi Lin has called scholars' attention. See Sensabaugh, 'The Lion Grove in Space and Time'; Zhang, 'The Lion Grove Garden: The Only Yuan Dynasty Garden of Rinzai Zen in Regions South of the Yangtze River.' I express my gratitude to David Sensabaugh for his discussion with me regarding the identification of a Shizi Lin painting attributed to Zhu Derun in the Artron.net 2011's auction as a forgery of the 17th century.

5 禪, 'Chan' in Chinese and 'Zen' in Japanese, was a sect of Buddhism taken shape in China around the 6th century. It was then transferred to and flourished in Japan around the 12th century.

6 For the geographical features like peaks, cliffs, and ravines, see *TMSZ*, 44, 59–80. Those soaring peaks were formed due to glacial movement. I would like to thank monk Miaozhi from Chanyuan Temple for his warm hospitality during our fieldtrip to Mount Heaven Eye, and for generously providing us a copy of the mountain's gazetteer.

7 For the description of Lion Cliff in Mount Heaven Eye, see *TMSZ*, v1: 63–64, v2: 102–105; For Lofty Clouds Peak and poem on it, see *TMSZ*, v1: 62, v5: 357. For Eastern Jade Ravine, see *TMSZ*, v1: 74, v4: 273–274; For Flying Rainbow Ravine, see *TMSZ*, v1: 74; v5: 369.

8 Zhongfeng's inscription for the portrait of Gaofeng is 'Opened the Heaven Eye, sat on Death Pass. The peak is lofty, ten thousand fathoms—Dangerous and hard to climb. 揭開天目, 坐斷死關. 峰高萬仞, 險絕難攀.' This verse nicely combines geographical features in Mount Heaven Eye with Gaofeng's toponym, eulogising Zhongfeng's achievements in Chan history.

9 '東獅子口, 高峰居之. 複閣四層, 長列數楹. 其所自名舫室, 死關.' In *TMSZ*, 278.

10 For the description of Three Gates pavilion, see *TMSZ*, v2: 92–93, 103–104; for the description of Death Pass, see *TMSZ*, v2:94, 104–105; for the circulation of 'Three Gates' as a public case, see *TMSZ*, 395–396, 580–581.

11 See *CXTMSZ*, 285.

12 '活埋庵在師子巖南. 岡巒竹樹, 窅然深秀. 前為香爐峰, 後為趺坐石. 幻住和尚嘗禪寂石上. 瀋王真際謂 "我師其活埋於此乎?" 遂以名庵. 今廢." See *TMSZ*, v2: 85.

13 'Sayings of Clouds 雲海說 (Yunhai shuo)', in *TMSZ*, v7: 492.

14 Ibid.

15 Poceski traces the origin of this dialogical teaching to Mazu, and Mcrae discusses the context of its emergence. See Poceski, *The Records of Mazu and the Making of Classical Chan Literature*, 51–54. John R. Mcrae, 'The Antecedents of Encounter Dialogue in Chinese Ch'an Buddhism,' in *The Kōan*, ed. Heine and Wright, 46–74.

16 Cole, *Patriarchs on Paper*, 252–274.

17 Musō, *Dialogues in a Dream*, PDF pgs. 197–199.

18 Gaofeng also experimented with other media for conveying Chan ideas, such as symbols. See Cai, *An Outline of the Lineage of Tiger Hill in Linji Chan Buddhism*.

19 Zhongfeng's many poems transformed natural scenes into philosophical ideas. In one of these poems, he claimed the natural scenery as 'manifest gong'an':

'除此現成公案外, 且無佛法繼傳燈.' See Heller, *Illusory Abiding*, 163. Also see Heller's discussion of Zhongfeng's idea that natural landscape served as a medium for Chan philosophical interpretation in pgs. 158–172.

20 For the recorded temples inside and outside Suzhou city, see *GSZ*, v29: 1–89; also see Gao Qi, 'Preface to twelve Odes of Shzi Lin.'

21 This marshy condition in the intra-mural area of Suzhou during the 14th century could be seen in Shen Zhou's, 'Painting Album of Eastern Estate (*Dong Zhuang tuce*)' and his *Painting of Thatched Cloister* in the Early and Middle Ming.

22 This hydrologic process was known by reading the scattered records of Tianru's landscaping in *JSJ* and *YL*.

23 The emperor Huizong was notorious for his Flower Rock Net project (Huashi gang), in which Lake Tai rocks produced in Lake Tai area were ordered to be collected and transported to Hangzhou for imperial garden construction. This incurred many social criticisms and was blamed for leading to the decline of the Southern Song dynasty. See Yang and Kong, *New Sayings on Mountain Living, Straight Records of Zhizheng*, 75–76. Tian Ye suggests that those Great rocks were discarded near the site of Shizi Lin in the late Southern Song. See Tian and Fang, 'Research on the Historic Appearance of the Lion Grove from the Yuan Dynasty to the Republic of China.'

24 Hu Zheng 胡震 (fl. 14th c.) wrote '從人喚作師子林, 不忘吾師舊宗旨. People called it 'Shizi Lin,' never forgetting the old principles of the masters.' in 'Short Song Composed for Shizi Lin 短歌行為師子林賦 (Duange xing wei Shizi Lin fu)' in *JSJ*, v1.

25 Ouyang Xuan 歐陽玄 (1283–1358), 'Record of Shizi Lin Bodhi True lineage Chan Temple 師子林菩提正宗寺記 (Shizi Lin Puti Zhengzongsi ji)': '因地之隆阜者, 命之曰山. 因山有石而崛起者, 命之曰峰. 曰含暉, 曰吐月, 曰立玉, 曰昂霄者, 皆峰也. Where the land rises in prominence, it is named a mountain. Upon these mountains, where rocks protrude and ascend, they are named "peaks". Names like "Teeming with Rays", "Holding Moon", "Erected Jade", and "Lofty Clouds", all refer to peaks.' in *JSJ*, v1: 1. Tianru, 'Fourteen Immediate Verses on the Sceneries in Shizi Lin 師子林即景十四首 (Shizi Lin jijing shisi shou)' in *JSJ*, v1. Duan Tianyou 段天佑 (fl. 1324), 'A Poem for Cherished Memories and Sceneries 紀胜寓懷一首 (Jisheng yuhuai yishou),' in *JSJ*, v1.

26 '玉鑑 yujian' is a pun of '玉澗 yujian,' pronouncing similarly with difference in the second characters.

27 Bassnett, *Constructing Cultures*, 11.

28 Pellatt and Liu point out that for translating cultural-specific terms in Chinese poems, an author needs to give more explanation in order to equip the reader with contexts. Pellatt and Liu, *Thinking Chinese Translation*, 162–164.

29 Wang, 'Record of Touring Shizi Lin,' in *JSJ*, v2: 10.

30 Zheng Yuanyou, 'Eight Scenes in Shizi Lin 師子林八景 (Shizi Lin ba jing)' in, *JSJ*, v1.

31 This motif could also be found in Tu Zhen's poem, 'Eight Verses on the Eight Scenes of Shizi Lin, Lion Peak 分詠林中八景凡八章, 師子峰 (Fenyong linzhong bajing fan bazhang)': '我聞師子峰, 來自師子巖. 騰驤勢欲吼, 萬籟振松杉. 為問同虓虎, 聽經到幾函. I have heard of Lion Peak, coming from Lion Cliff. With a surge as if to roar, it stirs the myriad whispers among the pines and cedars. I want to ask the tiger, who is also reverential here, which section of the sutra he has heard?' in *JSJ*, v1; also see poems from Zhou Ji and Chen Qian, in *JSJ*, v1.

32 For example, Hu Zheng wrote: '天目之山師子巖，中峰今代人師子…衣鉢之傳付屬誰？楚士天如得其髓. The Lion Cliff of Mount Heaven Eye, the present generation in lineage of the master Zhongfeng... To whom are the robe and bowl transmitted? Tianru from the Chu region has received the essence.' in 'A Short Ode in praise of Shizi Lin 短歌行為師子林賦 (Duange xing wei Shizilin fu).' Also see poems from Zeng Jian and Duan Tianyou in *JSJ*, v1.

33 The second imagery about the prospect commanded by Erected Jade Pavilion could be found in the paintings of Xu Beihong and Lu Shaoyan. They depicted the view of Lotus Peak commanded by Erected Jade Pavilion.

34 '石峰何低昂, 連雲傍修竹. 豈無點頭意, 凌雲如立玉. Why does the stone peak bow low or rise high, with clusters of clouds lingering beside the slender bamboo? Could it be without a nodding intention, soaring through the clouds like erected jade?' In *JSJ*, v1.

35 See Wang, 'Record of Touring Shizi Lin.'

36 Tianru, '雲海說 Sayings on the Sea of Clouds,' in *TMSJ*.

37 In *JSJ*, v1: 10–20; v2: 7–8, 12.

38 Stein, *The World in Miniature*, 23–48.

39 Lu, 'Lost in Translation.'

40 Wang, 'Record of Touring Shizi Lin.'

41 Zhaozhou was a Tang dynasty Chan master as the second generation of Mazu's disciple. Zhaozhou's anecdotes were widely cited as gong'ans by other Chan practitioners. See Buswell Jr. and Lopez Jr., *The Princeton Dictionary of Buddhism*, 1053.

42 The first patriarch refers to Bodhidharma, who came from India to China.

43 During the Tang Dynasty, recorded answers to this question were hundreds, see Zhengguo, *Principles of Chan Buddhism*, 129–134.

44 Dōgen, *The True Dharma Eye*, 162–163.

45 Wumen Huikai (1183–1260)'s interpretation of this gong'an is similar to Dōgen's, see Cole Alan, 'Kōans and Being There,' in *Patriarchs on Paper*, ed. Alan Cole, 262.

46 Cheng, *There is One Thing in Your Heart*, 33–38.

47 This viewing mode in which a central subject confronting the viewer is identified by Wu Hung as 'non-self-contained' in the pictorial composition of religious artworks. In this non-self-contained mode, the significance of a represented deity relies on the viewer's gaze to be realised. See Wu, *The Wu Liang Shrine*, 133–134.

48 '蒼蒼庭前柏, 明明西來意. 禪翁指示人, 又在第二義.' in Xu Ben's album leaf no. 9.

49 One of the visitors' poem was '此柏佛性全, 天寒神自王. 青青柏樹枝, 累累柏樹子. 此意已自知, 如待分明指. The cypress fully embodies the Buddha-nature. In the cold days, its spirit reigns supreme. The branches of the cypress are ever green, its seeds numerous and clustered. The meaning I already know within myself, as if waiting to be clearly pointed,' in *JSJ*, v1. This poem cross-referenced another gong'an whether a dog has Buddha nature.

50 See *Guzunsu Yulu* 古尊宿語錄, in CBETA 2024, R1, X68, no. 1315, 290a6-20.

51 Poceski, *The Records of Mazu and the Making of Classical Chan Literature*, 51.

52 Tianru wrote in his verses: '問梅閣裏著馬祖, 勘辨參徒. 指栢軒中容趙州, 放開鋪席. In Ask Plum Pavilion there resides Mazu, he tests the discernment of his disciples. In Point to Cypress Hall there resides Zhaozhou, he laid out the cushion [for Chan discussion].' in '雲南尊講主請 (像在林屋之間)' in *YL*.

53 Zheng Yuanyou, '謾道龍駒踏殺人, 梅花成實又逢春. It is absurd to say drag-on's steed tramples men to death, when plum blossoms bear fruit, sprint is met once again.' in *JSJ*, v1.

54 Tu Zhen, '分詠林中八景八章, 問梅閣: 滿林都是雪, 何許問梅花? 暗識飄香處, 西園一樹斜. 山人開小閣, 相對讀楞伽.' in *JSJ*, v1.

55 This dialectical relation between the visual and olfactory perceptions was expressed in the famous Chan verse from Tani Sōyō (1526–1563): 'The plum blossom is covered with snow, but its fragrance covers all.' in Weiss, *Zen Landscapes,* 64.

56 The inscription reads: '雪中疎蕊開, 不知暗香發. 幽人試問時, 正值黃昏月.'

57 '春從何處去, 復從何處來. 持此去來意, 一間閣前梅' in *JSJ*, v2, 8.

58 Loori, *The Eight Gates of Zen*, Ch. 22, pg. 1.

59 Alan, 'Kōans and Being There,' 252–274.

60 After Yan Ciping, this gong'an was painted by Liang Kai (d. 1210) in the Southern Song dynasty, and by Sesshū Tōyō (1420–1506) during the Middle Ming.

61 Many Chan teachers were rather harsh to students and even beat them. Contrary to this severe image of a Chan teacher, Tianru was praised as taking a genial attitude towards students. See *JSJ*, v1.

62 In Tianru's one inscribed poem on a portrait of him commissioned by a monk from Yunnan, he wrote '老胡多意氣, 大坐立雪堂. 呆漢少機鋒, 安居臥雲室. The old foreigner (Bodhidharma) was full of spirit, grandly sitting in Stand in Snow Hall. The foolish man lacked sharp wit, contentedly residing in Lying Clouds Chamber,' imagining Bodhidharma sitting in Stand in Snow Hall. Tianru, 'Yunnan respect invites the lecture (the portrait is between the grove and the chamber) 雲南尊講主請 (像在林屋之間),' in *YL*.

  In Tianru's another poem dedicated to Bodhidharma, he wrote: '師子林中, 立雪堂裏. 佛也不著, 因甚著你. 不重你面壁八九年, 不重你西來十萬里. 只重你心直口直, 不順帝王情. 眼高氣高, 不失宗師體. 坐斷天下人舌頭, 一花五葉香風起. In Shizi Lin, in Stand in Snow Hall. Neither Buddha nor anything else attaches—so why do you? It is not about facing the wall for eight or nine years, nor about your journey from the west of a hundred thousand *li*. What matters is your straightforward heart and speech, defying the emperor's will with no regard. With lofty eyes and spirit unyielding, maintaining the dignity of the Chan master. Seated, you silence the tongues of all under heaven. As a single flower with five pedals blooms, a fragrant wind arises,' in *YL*.

63 Li Xiaoguang's poem dedicated to Zhuofeng wrote: '道人久住師子林, 師旁立雪覓其心. A person of the way, long resides in Shizi Lin. Beside the master standing in snow, seeking heart within.' In Tianru's one poem eulogising Zhuofeng, 'Song of Keting (Zhuofeng),' he wrote: '可師立處一庭雪, 金剛脚跟凍欲裂. 覓心不得便心安, 敢保老兄猶未徹. Where master Ke stands, a courtyard covered in snow. The sole of his vajra feet freezing to the verge of cracking. Unable to find the mind, yet thereby the mind finds peace. Dare I assure, elder brother, that you have yet not fully comprehend.' In ibid.

64 Another poem portrays people of various ages who came to Shizi Lin for Tianru's teaching and could endure the cold and the snow. They remarked that unlike Bodhidharma, who had only one disciple, Tianru had many. Tu Zhen's poem on Stand in Snow Hall wrote, '歲寒守孤硬, 小大立盈門. 卻笑嵩山下, 齊腰一箇僧. In the cold of years, guarding solitude steadfastly. Both young and old fill the gate, standing firm. Yet laughing beneath Mount Song, a lone monk, snow up to his waist.' In Zhou Ji's poem on Stand in Snow Hall, he wrote: '問法身忘倦, 堂前積雪深. 飛花知落處, 弟子已安心. Asking about the

Dharma body, forgetting fatigue. In front of the hall, the snow piles deep. Flying pedals know where to fall. The disciple heart has found its peace,' in *JSJ*, v1 .

65 Tianru frequently discussed Zhaozhou's teaching methods in his sermons in Shizi Lin. Zhaozhou's idea was referenced at least 27 times in *Recorded Sayings of Master Tianru Weize*.

66 Tianru, '相君來扣少林宗, 官從盈門隘不通. When the prime minister came to enquire the Shaolin school, officials crowded the gate so much that it became impassable,' in *JSJ*, v1.

67 in *JSJ*, v1.

68 Tiantong Monastery was ranked as the third in the Five Mountain Ten Monastery system of the Song dynasty and represents a typical monastic layout. For Chan rituals and activities, see Zongze, *The Origins of Buddhist Monastic Codes in China*; Heine and Wright, eds., *Zen Ritual*.

69 For the schedule of rituals carried out in Song monasteries, see Heine, *From Chinese Chan to Japanese Zen*, 195–196.

70 I would like to thank Steven Heine for his insights during our discussion on the separation between Chan monasteries and the landscape. He emphasised to recognise that many early Chan meditations took place within natural landscapes.

71 The idea of the abbot as an enlightened person equivalent to a living Buddha has been discussed in Heine's study of Chan masters during the Song and Yuan periods in *From Chinese Chan to Japanese Zen*, 157–158.

72 Ouyang Xuan's stele inscription for Shizi Lin wrote: '師每説法, 糸問多至數百. 隨其悟解開導誘掖, 有所質疑, 剖析至當. 莫不虛往寔歸.' in *JSJ*, v1: 1. The numerous attendees were also portrayed in *YL*, v1: 761b20, v2: 758a15: '結夏前十日, 遠近禪侶共集. 立書記詎藏主請為普說,' and '結夏示眾. 今晨四月十五. 諸方浩浩譚禪, 不是結却布袋, 便是同泛鈷船…打鼓上堂, 入室據座. 舉揚般若, 提唱宗乘.'

73 Tianru: '師子無法教兒, 直向懸崖一捺. 要從捺處翻身, 死活不容眼眨.' in *YL*.

74 '師云上天無路入地無門. 進云今日多幸得聞師子吼也.' in *YL*, v1: 761b20.

75 '江陵周善宗為重建小飛虹橋示眾.' in *YL*, v1: 0760c16.

76 Tianru's adeptness in instructing students was praised by Ouyang Xuan, *JSJ*, v1: 1: '師每説法, 糸問多至數百. 隨其悟解開導誘掖, 有所質疑. 剖析至當, 莫不虛往寔歸.'

77 Eihei-ji Temple was constructed by Dōgen who transmitted the sitting meditation (Zazen) ritual to Japan in the 13th century. The temple was modelled after Tiantong temple and other Song dynasty temples thus provide a glimpse of the Song dynasty monastery layout. Heine and Wright, eds., 'Is Dōgen's Eiheiji Temple "Mt. T'ien-t'ung East"? Geo-Ritual Perspective on the Transition from Chinese Ch'an to Japanese Zen,' in *Zen Ritual*, 139–166.

78 Pelowski, 'Satori, Gong'an and Aesthetic Experience.'

79 Zhu Derun, 'Preface to the painting of Shizi Lin 師子林圖序 (Shizi Lin tuxu),' in *JSJ*, v2: 1–2.

80 For a 17th-century forgery of Zhu Derun's painting of Shizi Lin, see Tian and Fang, 'Research on the Historic Appearance of the Lion Grove from the Yuan Dynasty to the Republic of China.'

81 Shinohara and Satō, *Illustration on Zen Life, the Living Zen Buddhism*, 32.

82 Tianru wrote his meditation on the bridge, where the flow of water evoked diverse feelings and thoughts in *YL*.

83 Ibid.

84 Tianru's letter to his primary disciple Zhuofeng reveals that Zhuofeng was in charge of managing the landscape of Shizi Lin. Considering visitors' needs, they also enhanced the other outdoor facilities for meditation. See Tianru, '又答可庭藏主. 外檜行之側各作矮塼墻, 開二小門入東西圃. 小池四岸疊以蠻石, 而青石蓋之. 繞池石闌之外作小街道, 池南種竹數箇. 池北撤去舊籬. 古柏臨池如蓋. 樹下洒掃列瓦鼓為數客坐處. 禪餘飯罷儘有負暄散步之所矣.' in *YL*,v8: 826b10, 823a21.

85 The Buddha-centred viewing and its associated beliefs and rituals are discussed in Wu Hung's study of the Amitayurdhyana Sutra painting in Cave 172 of Dunhuang. See Wu, 'Reborn in Paradise.'

86 Zhang argues that Buddha Hall possesses the highest significance in the Song dynasty temples. See Zhang, *Buddhist Temples of Jiangnan in China*, 72.

87 See a modern reconstruction of the medieval mirror installation by Boston artist Victoria I, in Wang, *Shaping the Lotus Sutra*, 256–259.

88 For a detailed study of Fazang's mirror hall and the two principles he attempted to illustrate in Flower Ornament Scripture, see Chen, *Philosopher, Practitioner, Politician*, 183–187.

89 This mirror hall in Mount Heaven Eye, named 'Great Round Mirror Hall,' was constructed inside Great Awakening Temple. See Zhongfeng, 'Record of Infinite Light in Great Awakening Temple 大覺寺無盡燈記 (Dajue si wujin deng ji),' in *TMSZ*, 255–257.

90 The mirror hall in Mount Heaven Eye was described by Zhongfeng as '位置十面,面各一鏡.鏡各一佛.中然一燈,交光相攝.外以彰法界之無盡,內以標事理之不窮.' in ibid. The mirror hall in Shizi Lin was described by Wei Su (1303–1372) as '下設禪座, 上安七佛像. 間列八鏡. 鏡像互攝. 以顯凡聖交參. 使觀者有所警也.' The holistic thinking between the divine and the mundane advocated by the sect of Mount Heaven Eye was also extended to their concepts of time, such as Tianru's interpretation of Flying Rainbow Bridge, which emphasises a continuum of past, current, and future, and Zhongfeng's insights into the intersection between subjectivity and objectivity.

91 The dualistic view held by the Pure Land belief was discussed by Isozaki Arata in his analysis of Jodo-ji in *Japan-Ness in Architecture.*

92 Since in classical Chinese timber-framed architecture, the thick brick wall is non-structural, its application indicates religious purposes.

93 The rituals discussed by Isozaki in his study of Pure Land Hall apply to the Buddha hall in Song dynasty monasteries: 'chanting facing the Buddha' and 'circumambulating around Buddha icons.' See Isozaki, *Japan-Ness in Architecture.*

94 Tianru's many sermons, instructions, and letters addressed wealthy and devoted patrons, as revealed in *Recorded Sayings of Master Tianru Weize.*

95 Heine, *From Chinese Chan to Japanese Zen*, 184–186.

96 Rinaldi, ed., *Ideas of Chinese Gardens*, 91.

97 'The classical Chinese garden is centred on mountain and water…sceneries may be enjoyed when the onlooker stays put or moves along. Hence the terms, "in-situ viewing" and "in-motion viewing".' in Chen, *Chinese Gardens: Theory and Practice*, 82–83.

98 Research on gardening techniques could be found in Zhou, 'The Development and Composition of Borrowed Scenery;' Zhang, *Treatise on Crafting Chinese Garden*; Yang, *On Gardens in Jiangnan.*

99 This subchapter has been presented in Young Scholars' Forum in Chinese Studies 2021 titled as 'The Rise of Buddhism in Suzhou during the Yuan Dynasty: Transference of Cultures and Establishment of New Social Network 元代蘇州佛教的興盛: 外域文化的傳入與新社會網絡的形成.'

100 Shaolong was a master in Linji 臨濟's line. He was the outstanding disciple of Yuanwu Keqin 圓悟克勤 (1063–1135). See Cai, *An Outline of the Lineage of Tiger Hill in Linji Chan Buddhism*, 9–15.

101 Huang Jin 黃溍 (1277–1357), 'Reconstruction Record of Clouds Cliff Chan Temple in Tiger Hill in Pingjiang Prefecture [平江路]虎丘雲巖禪寺興造記 ([Pingjiang Lu] Huqiu Lingyan Chansi Xingzao Ji).'

102 Zhongfeng Mingben, 'Record of Illusory Abiding Cloister in Pingjiang 平江幻住庵記 (Pingjiang Huanzhu An ji),' in *Complete Anthology of Zhongfeng Mingben*, 315.

103 Song Lian, 'Record on the Reconstruction of Illusory Abiding Chan Cloister in Suzhou 吳門重建幻住禪庵記 (Wumen Chongjian Huanzhu An Ji),' in *Complete Works of Song Lian*, 1801–1802. '棲雲峰, 在藏雲塔東. 層巒窈窕, 如雲生態.' in *TMSZ*, 61.

104 Zhongfeng Mingben, 'Yangdang Chuye 雁荡除夜' in *Complete Anthology of Zhongfeng Mingben*, 372–373.

105 Heller, *Illusory Abiding*, 147–154.

106 Chen, 'Buddhism and Society in the Yuan Dynasty'; Ye, 'Famous Yuan Dynasty Monk Zhongfeng Mingben and His Deep Connection with Buddhism in Suzhou.'

107 Heller, *Illusory Abiding*, 423.

108 *GSZ*, vol. 29, eight, 10. As noted by Steven Heine, the act of 'opening a mountain' signified the establishment of a teaching lineage, with the master who inaugurated the temple being recognised as its first patriarch. Heine, *Opening a Mountain*.

109 Falei Zhen 法雷震, 'The testament of the founding patriarch of Capable Kindness Cloister 能仁菴開山祖師遺囑 (Nengren An Kaishan Zushi Yizhu),' in *WDWCXJ*, vol. 30, 47–49.

110 *GSZ*, vol. 29, 8.

111 Monk Zuying described Zuzhen's capability in attracting patronages as 'Where his heart desired, donors would respond 心之所施者必應.' in Shishi Zuying, 'Record of Capable Kindness Cloister in Suzhou 姑蘇能仁庵記 (Gusu Nengren An Ji),' in *Complete Anthology of Yuan Literature*, 122–123.

112 Long, 'Study on Chan Master Tianru Weize and His Chan Buddhism,' 36.

113 Two cases showcased monks' active participation in public services. Monks of Brilliant Wisdom Cloister (Shanhui an) constructed a pavilion along the dusty road of Shantang district to provide medical tea freely for passengers. Tianru, 'Record of the Tea Offering Fields of Brilliant Wisdom Cloister 善惠菴施茶田記,' in *YL*, 0809b03. Another example is Gravel Sand Teaching Temple (Qisha jiaosi), which was located in an island at the centre of Chen Lake. The temple provided accommodation for people transporting across the lake. Yuanzhi 圓至 (1256–1298), 'Record of Firewood Estate for the Bathroom of Ten Thousand Longevity Temple 平江府萬壽寺浴院柴莊記 (Pingjiang Fu Wanshou Si Yu Yuan Chai Zhuang Ji),' in *WDFC*, vol. 10, shang, 51–52.

114 In 'Anecdotes of two masters Jingfan and Qiyu 淨梵齊玉二師傳,' CBETA, X77, no. 1524, 0378–9, c23–a15.

115 Huang Jin, 'Record of the Contemplation Hall in Northern Chan Temple 北禪寺觀堂記,' in *Comprehensive Anthology of Huang Jin*, 326.

116 In 'Anecdotes of two masters Jingfan and Qiyu.'

117 Huang, 'Record of the Contemplation Hall in Northern Chan Temple.'

118 '禁蛙池.淨梵止觀, 以硃點蛙, 使之不鳴.' Xu and Zhang, *Smoke and Water of a Hundred Cities*, 449–450. Dotting someone's forehead with vermilion pigment was regarded as a ritual to impart Buddha spirit.

119 '師禪觀之處, 眾嘗見金甲神跪於座前.' in 'Anecdotes of two masters Jingfan and Qiyu.' 'Golden Armor God 金甲神' is a name for Weituo in folklore.
120 '建炎初, 毀於兵.' in Huang, 'Record of the Contemplation Hall in Northern Chan Temple.'
121 '北禪梵法主法嗣.' in CBETA, T49n2035.
122 Huang, 'Record of the Contemplation Hall in Northern Chan Temple.'
123 '慈忍法師' in *WDFC*, juan six, xia, 51.
124 Ibid.
125 Huang, 'Record of the Contemplation Hall in Northern Chan Temple.'
126 Xie, *Research on Tiantai Buddhism Architecture*, 89–91.
127 Refer to Pan Yijun, 'Record of Reconstructing Northern Chan Temple 重修北禪寺碑記 (Chongxiu Beichan Si Bei Ji),' in the database of Lidai Jiaowai Shefo Wenxian Shuju Ku 歷代教外涉佛文獻數據庫 集部佛論 (二) in the website ancientbooks.cn.
128 Zhu Yunming, 'Fund Raising Appeal for the Reconstruction of Flower Rain Terrance in Northern Chan Temple 北禪雨花臺脩造疏 (Beichan Yuhua Tai Xiu Zao Shu),' in *Anthology of Zhu Yunming*, 509. In *Map of Gusu City*, there was a pond to the west of the temple. It is unclear whether it would be Forbidding Frog Pond. There were vast fields to the temple's east.
129 Ge 閣 (translated as 'pavilion') refers to multi-storied building in temples, as represented by the Bodhisattva pavilion in Dule Temple 獨樂寺. See Steinhardt, *Liao Architecture*, 40–45.
130 Yuze's anthologies are not extant.
131 In *GSZ*, vol. 29, 16.
132 Gao, 'Study and Discussion on the Poet Monk Miaosheng from the Late Yuan to Early Ming Dynasty.'
133 In Wei Yingwu, 'Touring the Northern Pond and Monks' Chamber in Forever Pacification Temple 遊永定寺北池僧舍 (You Yongdingsi),' *GSZ* 29: 15–16.
134 Yao Guangxiao, 'Record of the Reconstruction of Sea Seal Hall in Forever Pacification Universal Compassion Tiantai Teaching Temple 永定普慈天台讲寺重建海印堂记' (Yongding Puci Tiantai Jiangsi Chongjian Haiyin Tang Ji),' in *WDFC*, juan 10, shang, 73–74.
135 Ibid.
136 Miaosheng, 'Funeral Eulogy For Maozhong from Shuiding 水定茂仲寶挽詞 (Shuiding Mao Zhongshi Wanci),' in *Anthology of Donggao*, 35.
137 Miaosheng, 'Letter to Cui Zongwen 寄崔宗文 (Ji Cui Zongwen),' in *Anthology of Donggao*, 55.
138 Yao, 'Record of the Reconstruction of Sea Seal Hall in Forever Pacification Universal Compassion Tiantai Teaching Temple.'
139 The term 'ocean-seal' originated with monk Fazang. See Yu, *Reimagining Chan Buddhism*, 126–127.
140 Yao, 'Record of the Reconstruction of Sea Seal Hall in Forever Pacification Universal Compassion Tiantai Teaching Temple.'
141 See Yao Guangxiao's multiple odes and biographies of these monks in *Anthology of Yao Guangxiao*, vol. 25, 335–362.
142 For Qisong's biography, see Buswell Jr. and Lopez Jr., *The Princeton Dictionary of Buddhism*, 303. I extend my gratitude to monk Zhongwei 中伟 who graciously accommodated my visits to Repay Kindness Temple and showed me archives.
143 See Yan Fu, 'Stele Record of Repay Kindness Ten Thousand Avatamsaka Teaching Temple 報恩萬歲賢首教寺碑 (Bao'en Wansui Xianshou Jiao Si Bei),' in *WDFC*, vol. 10, shang, 52–54.

144 In Song Lian, '重塑釋迦文佛臥像塔銘 Chongsu Shijia Wen Fo Woxiang Taming,' in *Complete Works of Song Lian*, vol. five, 1616–1618. 'Avatamsaka 賢首 (xian shou)' is an alternative name for Huayan Buddhism. Qisong's profession and influence in Huayan sect is discussed by Morrison, *The Power of Patriarchs*.

145 '建炎之難, 鞠為煨燼. During the Jianyan calamity, it was almost completely destroyed.' in Yan, 'Stele Record of Repay Kindness Ten Thousand Avatamsaka Teaching Temple 報恩萬歲賢首教寺碑 (Bao'en Wansui Xianshou Jiao Si Bei).'

146 Ibid.

147 The five patriarchs of Huayan sect were Du Shun 杜順 (557–640), Zhiyan 智嚴 (602–68), Fazang, Chengguan 澄觀 (738–839), and Zongmi 宗密 (780–841). See Hamar, *Reflecting Mirrors*, XV.

148 Yan, 'Stele Record of Repay Kindness Ten Thousand Avatamsaka Teaching Temple 報恩萬歲賢首教寺碑 (Bao'en Wansui Xianshou Jiao Si Bei).'

149 Huang Jin, 'Record of Permanent Fields of Repaying Kindness Ten Thousand Avatamsaka Teaching Temple 報恩萬歲賢首教寺長生田記 (Bao'en Wansui Xianshou Jiao Si Changsheng Tian Ji),' in *Comprehensive Anthology of Huang Jin*, 397–398.

150 Ibid.

151 Ibid.

152 Ibid.

153 Huang Jin, 'Record of Reconstruction of Repaying Kindness Ten Thousand Avatamsaka Teaching Temple 平江路報恩萬歲教寺興造記 (Pingjiang Lu Bao'en Wansui Jiaosi Xingzao Ji),' in *Complete Anthology of Huang Jin*, 337.

154 Ibid.

155 Chen Ji, 'Record of Heavenly Benevolence Avatamsaka Teaching Temple 天惠賢首教寺記 (Tianhui xianshou jiao si ji),' in *Yibai Zhai Gao*, vol. 26, 7–9, 1352. This temple was not inventoried in *GSZ*, suggesting that the number of temples during the Yuan dynasty might exceed those indexed in *GSZ*.

156 Ibid.

157 Ibid.

158 Ibid.

159 These four persons are: Song Lian, Minghe, Yao Guangxiao, and Miaosheng.

160 Miaosheng, 'Gu Guting Fashi Xingye Ji 故古庭法師行業記 (Record of the Practices of the Late Master Guting),' in *Anthology of Donggao*, juan zhong, 62–64.

161 Song Lian, 'Inscription on the Pagoda of Huayan Master Guting Xue 華嚴法師古庭學公塔銘 (Huayan Fashi Guting Xue Gong Ta Ming),' in *Complete Works of Song Lian*, 1619–1622.

162 In Yao Guangxiao, 'Record of the Pagoda of the Deceased Huayan Master Guting Xue 故華嚴法師古庭學和尚塔表 (Gu Huayan Fashi Guting Xue Heshang Ta Biao),' in *Comprehensive Anthology of Yao Guangxiao*, vol. 2, 317–318.

163 Yu Ji, 'Inscription of the Yuan Imperial Decree to Construct Grand Illuminous Celebration Temple 元敕建大昭慶寺碑 (Yuan Chijian Da Zhaoqing Si Bei),' in *WDFC*, vol. 10, shang, 38–40.

164 Huang Jin, 'Record of Sandalwood Pavilion of the Grand Illuminous Celebration Temple 大昭慶寺旃檀閣記 (Da Zhaoqing Si Zhantan Ge Ji),' in *Comprehensive Anthology of Huang Jin*, 796–797.

165 See Russell-Smith, *Uygur Patronage in Dunhuang*, 1–4.

166 See Chen, 'The Yuan Dynasty's Internal Migration of the Uyghur People and Buddhism.'

167 *Gazetteer of Gusu* inventoried 140 officials appointed to Suzhou during the Yuan dynasty, a majority of them were Uyghurs as indicated by name conventions. In *GSZ*, vol. 3, 34–39.
168 '師子林發地結茆, 屢枉惠顧. 草樹竹石, 精采百倍. 傅徒匪材, 荷玉成之賜.' in *YL*, X70n1403_008.
169 '林下一鉢, 生涯之外, 宴默安居, 外惱不至. 皆餘光之所照暎, 其敢一日忘邪.' in *YL*, X70n1403_008.
170 '竊嘗謂當今貴公子身處顯位而信慕佛法者, 固有之. 求如居士不肯以少自足, 直欲親到大休大歇之地者, 幾何人哉?' in *YL*, X70n1403_008.
171 Ibid.
172 In *YL*, 0825a17.
173 'In 1322, ten thousand mu fertile fields were granted to Maiju.' Wang et al., *Outlined History of Suzhou*, 134.
174 '余青出高昌, 依佛为命. 覩兹僧寶, 敢同寒蝉.' in Zheng Yuanyou, 'Record of Stand in Snow Hall 立雪堂記 (Lixue Tang Ji),' in *JSJ*, 5–7.
175 Officials' generosity in donating money to establish temples is also recorded by Zhu, 'Record of Imperial Endowed Repay the Nation Chan Temple in Suzhou.'
176 Chen Lü, 'Stele Record of Return to the Origin and Reviving the Nation Chan Temple in Pingjiang Lu 平江路歸元興國禪寺碑 (Pingjiang Lu Guiyuan Xingguo Chan Si Bei),' in *Anthology of Anya Hall*, vol. nine, 22–24.
177 Ibid. I would express thanks to monk Shanze 善择 in Western Garden Temple for bringing this stele record to my attention.
178 Ibid.
179 Ibid.
180 Ibid.
181 Ibid.
182 Wang, *In the Wake of the Mongols*, 118–165.

3   Converting Temples into Literati Gardens

During the High Ming period, the local gentry acquired Buddhist estates through various means, repurposing them into private gardens. They erased monastic traces on site meanwhile capitalised on the existing hydraulic landforms originally laid out by Buddhists. Leveraging the topography of former temples, the gentry established artefacts, crafted scenes, and commissioned paintings, fostering a distinctive literati garden culture. Artworks collected in temples were leaking out from temples and circulated within gentry's sphere, providing inspiration for the design and aesthetics of literati gardens.

This chapter uses 3D modelling to visualise the key transformative stages of Shizi Lin Temple into Lion Grove Garden and Great Propagation Temple into Unsuccessful Politician's Garden. The transformation of Shizi Lin Temple demonstrates how gong'an scenes were erased of Buddhist connotations and re-interpreted by the literati. The transformation of Great Propagation Temple showcases the gentry's configuration scenes amongst the mounds and rivers previously laid out by Buddhists. They redefined the meanings of scenes oriented in literati culture. The gentry also redefined garden aesthetics—the way to appreciate, gather, and stroll in the garden. Through adapting temples into literati gardens, the gentry rose not merely as the largest landholder but also as elites exerting cultural hegemony.

From Shizi Lin Temple to Lion Grove Garden

In the mid-19th century, scholar Jiang Yun (b.1816) visited Lion Grove Garden and created a painting (Fig. 3.1a). He envisioned the garden as a spherical rockery complex floating above the water. This rockery ball featured intricate caves and rugged Lake Tai rocks stretching outwards. Two gnarled pines grew out, with a pavilion nestled beneath. Pathways, hiding and revealing between rocks, guiding gentlemen to climb, as if were in a mysterious mountain. Jiang's portrayal of the site as a grove of lion-like rocks captured a prominent perception of Shizi Lin from the Qing dynasty

DOI: 10.4324/9781003387169-4

to today: the site is a literati garden of rocks (Fig. 3.1b). This image contrasts with the Buddhist landscape of Shizi Lin in the 14th century (Fig. 0.4a). How did a Buddhist temple transform into a literati garden themed on lions?

This section traces the evolution of Shizi Lin Temple through four stages. After the temple's transference to a wealthy family in the High Ming, the site was divided during the Qing dynasty into a northern part, owned by the gentry family as a garden, and a southern, managed by Buddhists as the temple. The wall separating the garden and the temple indicates a split between literati culture and Buddhism. Garden practices, both on-site and in artworks, were transferred from the Buddhist realm to that of the literati. As the site evolved in aspects of spatial layout, hydrology, and social network, the representation of the site came to overshadow the physical site in historical authenticity (Table 3.1).

A Temple of Literati

In the early Ming, around the 1370s, Ruhai Deyin (fl. 14th c.) assumed the abbotship of Shizi Lin, attracting visits from a new group of literati.[1] The group included elites such as Gao Qi, Xu Ben, Ni Zan, and Wang Yi.[2] During the Yuan era, they were based in Kunshan, north of Suzhou, where they gathered in each other's garden estates for artistic exchanges.[3] The outbreak of war during the Yuan-Ming transition devastated their estates and expelled them from hometown. Jade Mountain Thatched Hall (*Yushan caotang*), a primary site for their gatherings, suffered from invasion extensively, with many artworks looted. Consequently, these literati migrated to Suzhou, seeking refuge and new venues for cultural activities.

With the establishment of the Ming dynasty, they were appointed as court officials and assigned responsibilities by the emperor.[4] They frequently visited Shizi Lin, lingering, resting, and discussing Buddhism with Ruhai. Having lost their families and homes, the seclusive Shizi Lin Temple offered them a sense of tranquillity and peace. The Chan sentiments inscribed in the scenes also resonated with them, as the Buddhist notion of impermanence and illusion could provide solace during the ravages of war.

Xu Ben's album, *Painting of Shizi Lin*, portrayed twelve scenes of the temple, and together with seven other literati they created the linked verses *Twelve Odes of Shizi Lin*. Figure 3.2 visualised the location of these twelve scenes in the album's sequence. Eight of the scenes were acclaimed as gong'ans in Tianru's time. Following this scenic gong'an tradition, the literati engaged in interpreting scenes as paths to enlightenment. For instance, in the 7th scene 'Stand in Snow Hall,' Xu Ben's painting leaf transposed the historical gong'an of Huike and Bodhidharma from Song Mountain to Shizi Lin Temple (Fig. 2.7a). The linked verses in this scene demonstrate literati's collective investigation into Chan philosophy through poetry. Gao

Figure 3.1 Lion Grove Garden in the Late Qing and the current. (a) Jiang Yun, leaf no. 10, 'Lion Grove,' 1881, in album *Painting of Journeys in Jiangzhe Region (Jiangzhe jiyou tu)* ,colour ink on paper, 24.8×32.3cm. Nanjing Museum. (b) Aerial photo of current Lion Grove Garden in Suzhou. Source: Courtesy of 699pic.com.

Qi and Zhang Shi's verses vividly capture the severe cold Huike endured, with snow up to his waist at Song Mountain. Wang Xing (1331–1395) then shifted this narrative to the setting of Shizi Lin, while Xie Hui (fl. 14th c.) depicted personal experiences of visiting the master amidst the snow-covered temple. Shen Tuheng, Zhang Jian, and Tao Shen further demonstrated perseverance in seeking enlightenment—one became so absorbed that he remained unaware of the cold, despite being covered with snow

Table 3.1 Site Evolution Biography of Shizi Lin

Decade	Southern Part	Northern Part
Temple stage, Late Yuan to Early Ming		
1340s	1342, Tianru Weize constructed Shizi Lin Temple.	
1370s	1371, Gao Qi's group frequently visited Shizi Lin and composed *Twelve Odes of Shizi Lin*; Wang Yi wrote *Record of Touring Shizi Lin*; Ni Zan and Xu Ben created paintings.	
Decaying stage		
1520s–1560s	1522–1566, the site was purchased by Lu family and became dilapidated.	
Transformative Stage 2, Late Ming		
1590s–1700s	Monk Qing'an Mingxing (fl. 16th c.) reconstructed Shizi Lin Temple during the 1590s. The temple was named as Imperial Benevolence Temple (*Sheng'en si*). Monk Guangyun Fazhi (fl. 17th c.) carried on the reconstruction of the temple in 1642.	
Transformative stage 3, Early Qing		
1640s		1648, The northern part was transferred to Zhang Shijun (fl. 18th c.) and renovated as Lion Grove garden.
1700s	1703, the temple was endowed with the title The Ancient Shilin Temple (*Gu Shilin si*).	
Transformative stage 4, High Qing		
1730–1740s		Huang Xingren bought the garden part and renovated it as 'Garden of Five Pines' around the 1740s.
1750s	1751, The temple part was renamed as Painting Chan Temple (*Huachan si*).	1757, Qianlong emperor visited to Lion Grove Garden in spring and summer.
1760s	1762, the temple received its name plague from Qianlong emperor.	1762, Qianlong emperor's 3nd visit to Lion Grove Garden; 1765, Qianlong emperor's 4rd visit to Lion Grove Garden.
1770s		1771, Gao Jin's painting of Shizi Lin

overnight. The linked verses culminate at Yao Guangxiao's insight that a truly settled mind should realise that Dharma does not exist. Ruhai further anthologised the linked verses into a handscroll after Zhu Derun's painting of Shizi Lin and invited Gao Qi to write a preface, which was quoted in the Introduction.[5]

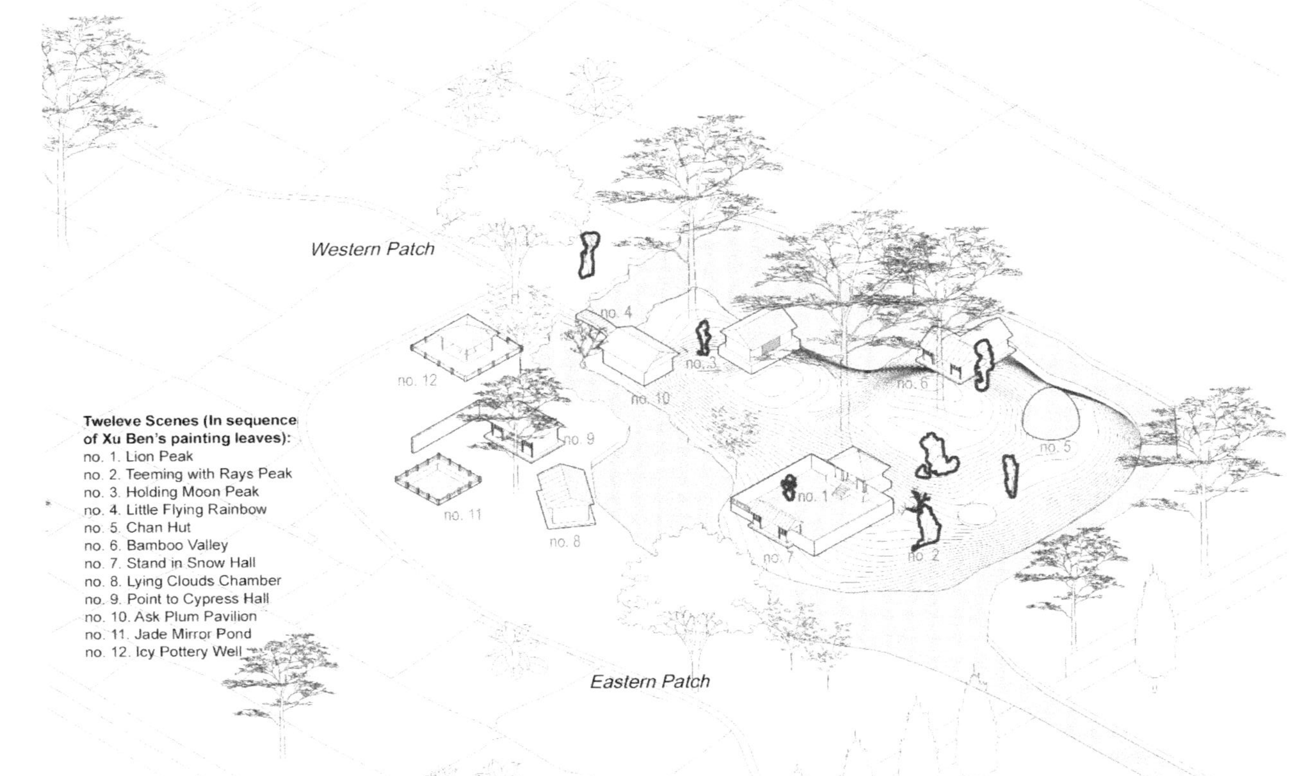

Figure 3.2 Twelve Scenes of Shizi Lin Temple in the Early Ming Period. Source: Author's modelling and drawing.

Moreover, the literati expanded the scenic repertoire by introducing four new scenes: Icy Pottery Well, Chan Hut, Holding Mood Peak, and Teeming with Rays Peak (Fig. 3.8). These additions and modifications reflect changes in the landscape during the early Ming period. The introduction of Chan Hut as a scene suggests its new accessibility to guests. The absence of any mention of the mirror hall inside likely indicates its removal.[6] The incorporation of the other three new scenes aligns with literati aesthetics of water and Lake Tai rock. The scene previously known as 'Reside Phoenix Pavilion' was renamed 'Bamboo Valley' (Fig. 1.6a), implying it might no longer serve as guests' accommodation. It is likely that visitors now stayed in Ask Plum Pavilion and the courtyard of Holding Moon Peak, as the number of monks diminished in the early Ming.

Wang Yi's approach to appreciating the temple markedly differed from that of Gao Qi's group. He was not well-versed in Chan Buddhism nor familiar with the temple's origin from Mount Heaven Eye. He appreciated Lake Tai rocks for their unique, bizarre forms, rather than for any Buddhist connotation. This is exemplified in his criticism of the name of Erected Jade Peak rock. Have not recognised that the rock was associated with Erected Jade Pavilion at Mount Heaven Eye, Wang Yi criticised the name as unreasonable because the rock bore no resemblance to jade. He thus suggested to rename it 'Earth Lung' as more appropriate. This emphasis on form aligns with the literati's admiration of Lake Tai rocks, a tradition dating back to the Tang dynasty.

Besides applying the long-standing literati rock aesthetics to the lake Tai rocks in Shizi Lin Temple, Wang Yi's writing, *Record of Touring Shizi Lin* (see Appendix, Translation 1), presented a new spatial perception of the site. He first identified the mound as the primary topographical feature, then designated several rocks as landmarks: Lion Peak as the central peak, flanked by Holding Moon Peak to the west and Teeming with Rays Peak to the east. Utilising these rocks as spatial anchors, he further indexed other site features in relation to them: artefacts, the river encircling the mound and flowing into the urban canal, and the pathways connecting scenic spots. This is an early holistic garden record which mapped out the landscape's topography with clear orientation.

Wang Yi's record also reveals an interest in figuring out pathways through touring. He was captivated by the experience of navigating the intricate routes amidst Lake Tai rocks atop the mound, enjoying the discovery of a different path each time and gaining new impressions of the site.[7] He identified two main paths of the temple from the mound's summit: an eastern route leading the Dharma hall courtyard and further reaching Lying Clouds Chamber and Point to Cypress Hall; and a route on the western slope towards Ask Plum Pavilion and Icy Pottery Well. Wang Yi found great delight in wandering through the temple's topography, and his language rendered the touring experience as a journey amongst mountains and peaks.[8] He was never tired of touring the site and often invited his friends to join.

Wang Yi's view of Shizi Lin Temple, emphasising the form of rocks, topography, and movement, exemplified a literati's approach to the site. This contracts with Gao Qi's group, who emphasised Chan implications, integrating Buddhist and literati's modes of appreciation. While the twelve scenes eulogised by Gao Qi's group were not organised by spatial sequence, Wang Yi structured the scenes by the touring path and their spatial relations to the primary Lake Tai rocks. Both of their records and poems were inscribed on steles in the temple, promoting new ways of experiencing the landscape. The temple's cultural divergence from Chan philosophy resonated socio-political changes, where the social power of the gentry was rising while Buddhism was declining.

A Productive Garden

These literati's patronage of Shizi Lin was short-lived. Two years after their 1372's gathering in Shizi Lin, several members suffered political persecution. Gao Qi and Wang Yi were executed by the emperor in the spring of 1374, and Ni Zan passed away later the same year in winter. The temple's connections with implicated literati likely rendered it a taboo place, resulting in a decrease in visits and patronage. Meanwhile, the temple's assets became vulnerable as the Ming emperor Zhu Yuanzhang imposed policies to censor and restrain the political power of Buddhists.

During Jiajing's reign, Shizi Lin was transferred to the Lu family's possession.[9] Historical archives mentioned the family's name without further details, which implies that Lu family were affluent yet not actively engaged in the local culture. Instead of preserving the Buddhist and literati culture of the site, Lu family repurposed it for pragmatic uses such as housing servants, rearing poultry, and weaving silk.[10]

A notable shift observed during the High Ming period is that not only Buddhist estates were transferred to the gentry, but also estate representations. Shizi Lin exemplifies this change, as the temple's former paintings began to leak out and circulate among the gentry with increasing popularity. The three paintings of Shizi Lin, created by Zhu Derun, Xu Ben, and Ni Zan (Fig. 3.3), which used to be preserved within the temple, were now part of gentry collections and subjected to literati viewing.[11]

The circulation of Shizi Lin paintings incentivised lots of imitations and forgeries, as illustrated in Figure 3.4. Xu Ben's painting album leaves was first adapted by Du Qiong (1396–1474) into a hanging scroll in 1468 (Fig. 3.5).[12] Then Wen Zhengming reworked it, but into a handscroll format. Shen Zhou and Qian Gu also imitated and copied the album. Ni Zan's

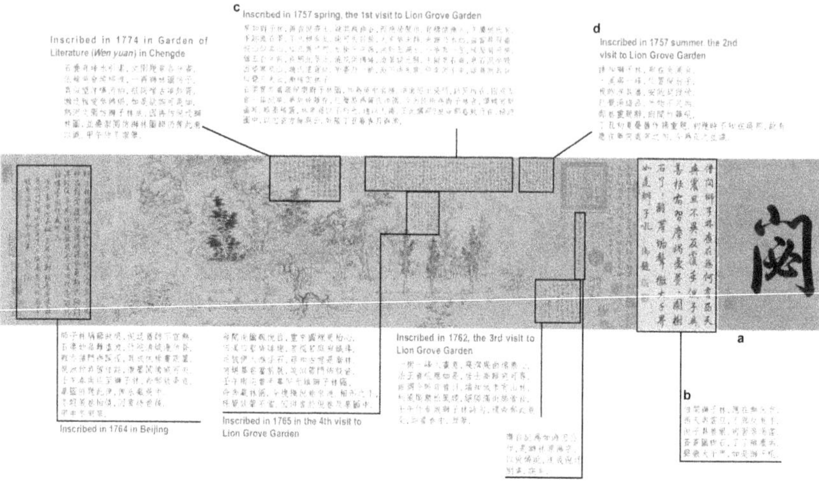

Figure 3.3 Ni Zan, *Painting of Shizi Lin,* handscroll, ink on paper, 28.3×39.28 cm. Beijing: The Palace Museum. Diagram and annotation by the author.

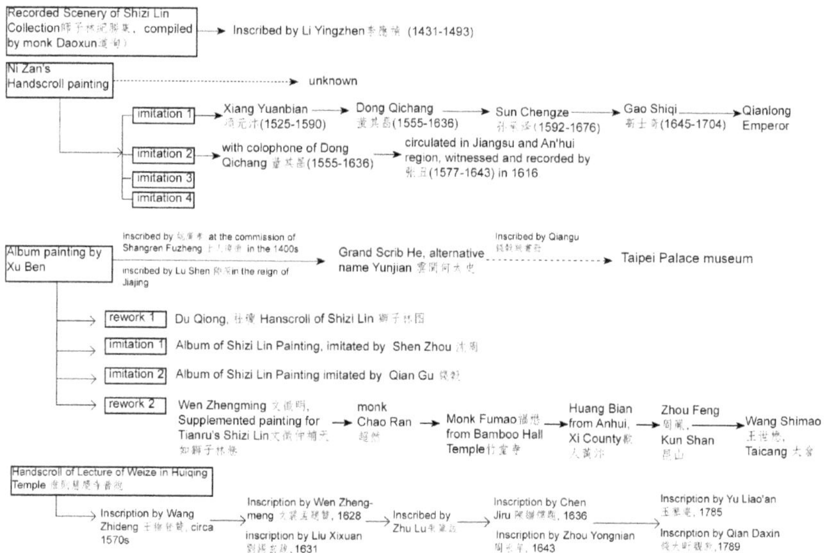

Figure 3.4 The circulation of artworks of Shizi Lin during the Middle Ming and High Qing dynasties. Source: Author.

Figure 3.5 Du Qiong (1396–1474), *Painting of Shizi Lin* (Shizi Lin tu), 1468. Colour ink on paper, hanging scroll, 88.8 cm × 27.9 cm. Taipei: Collection of the National Palace Museum.

handscroll, renowned for the artist's esteem, became the subject of several forgeries. One forgery eventually arrived at the treasure collection of Emperor Qianlong during the Qing dynasty (Fig. 3.3).[13]

The circulation of these derivative paintings, as diagrammed in Figure 3.4, suggests a feasible impact of Shizi Lin paintings on literati pictorial iconography. The iconic image of a pavilion perched atop the mountain, as depicted in Du Qiong's hanging scroll of Shizi Lin, became a popular motif in Ming gardens, hinting at the impact that Shizi Lin paintings had on garden aesthetics.[14]

Reconstruct the Temple

Shizi Lin was returned to Buddhists in the Late Ming after its four decades' ownership by the Lu family. Monk Qing'an Mingxing initiated a reconstruction of the temple in the 1590s, which was carried on by Guangyun Fazhi subsequently in the early 17th century. The reconstructed temple featured grand architecture, sublime Buddha statues, and an extensive collection of sutra documents. While the temple estate occupied the same area as it did during Tianru's time, religious artefacts were concentrated in the southern flat area, leaving the northern hill and river topography unattained and gong'an scenes unrecovered.

The mountain gate, Buddha hall, and sutra storage pavilion (*Cangjing ge*) defined the axis of the new temple, as simulated in Figure 3.6. A grand golden Buddha statue was enshrined inside the Buddha hall, 'which was solemn and compassionate,' as recorded by the magistrate Jiang Yingke.[15] Mingxing did not manage to complete the sutra storage pavilion, with only its foundation established. Thus sutras endowed from the imperium were temporarily stored and revered in the Buddha hall.[16]

Fazhi took over the reconstruction in 1642 and recommenced the construction of the sutra storage pavilion (Fig. 3.6(3)).[17] However, his plan was disrupted by the outbreak of wars during the Ming-Qing transition, which also caused damage to the Buddha hall.[18] Mingxing resumed promptly after the war. Under his supervision, the sutra storage pavilion was built in grandeur, using high-quality materials and elaborate decorations, as described by Xu Fang (1622–1694):

> The sutra storage pavilion was quickly completed, and soon after, Buddha was also finished. Both were constructed with fragrant cedar for beams and covered with bronze tiles. Their golden and jade-like brilliance surpassed the daylight and clouds. The craftsmanship was exquisite and sturdy, the design unparalleled. Not only did they surpass anything in the three Wus, but they were also unrivaled by the Five mountains and Ten monasteries.[19]

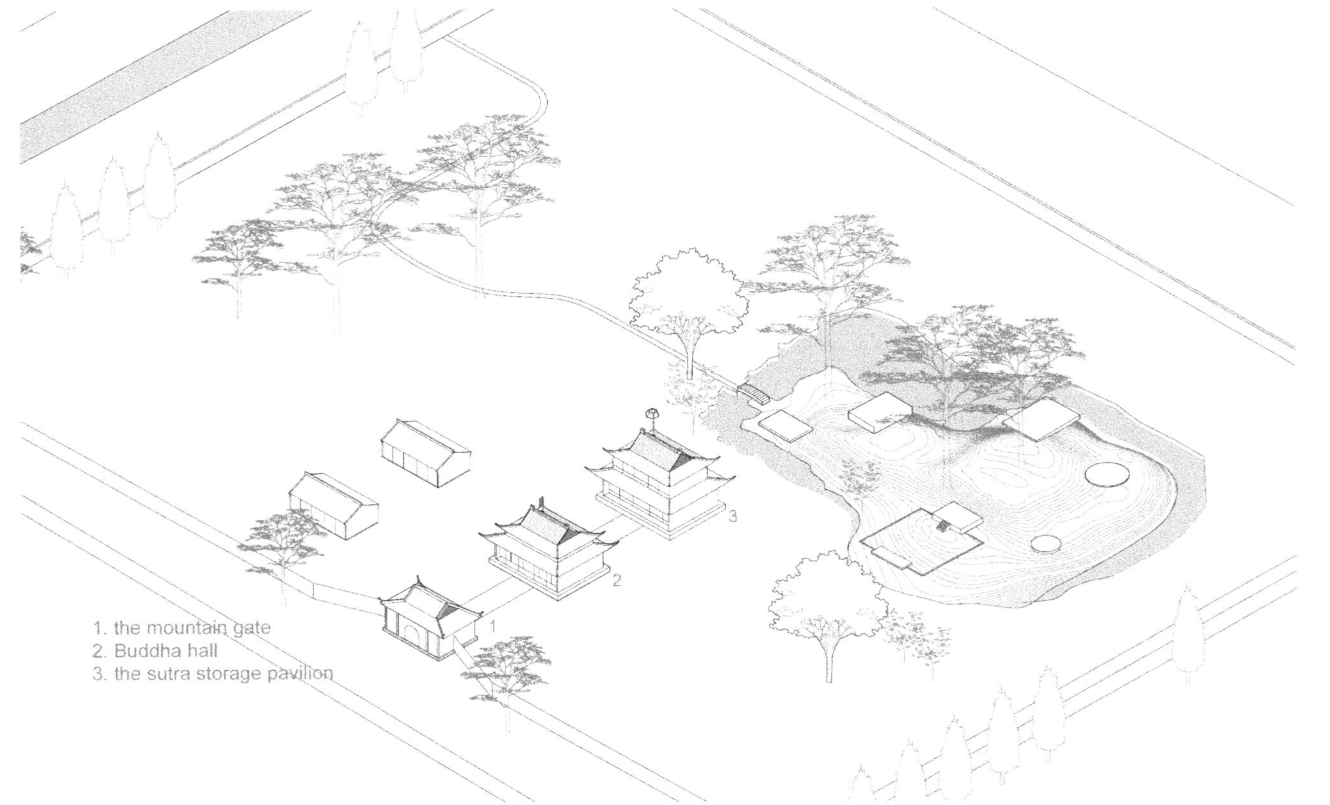

Figure 3.6 Shizi Lin Temple (Imperial Benevolence Temple) in the Late Ming. Source: Author's modelling and drawing. Only representative Lake Tai rocks are illustrated.

Comparing the 3D models of Shizi Lin in the Late Ming and Yuan dynasties in Figure 3.6 and 2.5 respectively, it is revealed that Mingxing and Fazhi's reconstruction did not follow the temple's dispersed layout in the Yuan but implemented a traditional seven-hall axial layout. The return to the cruciform layout resulted in notable changes in the temple's visuality and ritual programmes. In Tianru's temple, the sacred centre was the northern mound, where Lion Peak and other gong'an scenes were dispersed among the topography, facilitating Chan practices. In contrast, Mingxing and Fazhi's reconstruction shifted the ritual activities to the southern flat area, imposing visitors with the architectural prominence of monastic buildings. The absence of Buddha hall and the cruciform axial order in the Yuan was re-emphasised in the reconstruction. This change marked a late Ming monastic rupture from the scenic gong'an practices in the Yuan, reverting to the traditional ritual of viewing and worshipping Buddha. Consequently, the northern scenic area lost its appeal and became peripheral. The religious and the scenic started to split.

The temple's entrance in the Yuan dynasty was positioned on the east, with the river leading visitors to the mound and the Dharma hall. However, by the late Ming, the waterway had diminished and possibly disappeared. Adapting to the reduced water availability and the new axial layout, the new entrance was established at the southern street, which was aptly renamed 'Shizi Lin alley (*Shizi Lin xiang*),' as recorded in *Map of Gusu City*.[20] Upon entering through the grand mountain gate, visitors were impressed by the magnificence of Buddhist architecture. The northern scenic topography was relegated to the rear in neglect.

Why did Mingxing and Fazhi choose not to recover scenes in the north? It was not due to the lack of reference or awareness. *Anthology of the Recorded Scenery of Shizi Lin* was widely accessible in the late Ming prints, and Mingxing certainly had one.[21] Instead, I would argue the decision was rooted in the belief about what should and should not be seen. Mingxing and Fazhi considered Buddha images and architecture should be the focus of Buddhist visuality, instead of sceneries. Fazhi even regarded the northern scenery as distracting his Buddhist practice and deliberately avoided seeing it.[22] The local gentry Xu Fang's praise of Fazhi's avoidance of viewing the garden indicates that garden practices were no longer tied to Buddhism in the Ming period.[23]

Mingxing and Fazhi's reconstruction of the temple demonstrated resistance to the decline of Buddhism in the late Ming. In contrast to Tianru's time, when Buddhism was widespread, with its images and followers ubiquitous, Mingxing and Fazhi faced a society where monastic institutions had significantly diminished, leading to a less presence of Buddhist portrayal. Fazhi expressed concerns about the waning influence of imaged-based Buddhist teachings and criticised the concurrent decline in public morality.[24] While Tianru's non-iconic approach addressed the pedagogical crisis within the pervasive Buddhist practices, Mingxing and Fazhi were challenged by how to revitalise faith in Buddhism. Fazhi thus advocated, 'Without relying on sight and hearing, it is inadequate to cultivate true faith. Without illuminating the ears and eyes, it is inadequate to guide one's body and mind.'[25] As such, they endeavoured to build the magnificent Buddha hall and sutra storage pavilion, hoping that the splendour of Buddhist statues and architecture could captivate and draw people back to Buddhism.[26]

Reconstruct the Garden

The temple's northern section, losing its significance in Buddhist practice, was sold to Zhang Shijun, an affluent gentry in the early Qing. This sale likely relieved monks' financial strain by freeing them from the burden of retaining a garden. For a gentryman like Zhang, the site's growing cultural stature, bolstered by the circulation of its artworks and literature, made it an ideal venue for gathering. The gentry, rather than the Buddhists, were attuned to the site's artistic significance. An expansive tall wall was built to separate Zhang's garden in the north from the temple in the south (Fig. 3.7).[27] This wall not only marked the property boundary but also separated garden art from Buddhists.

Upon its transference to be Zhang Shijun's estate, the site's twelve Buddhist scenes were unrecognisable, a result of accumulated neglect and poor management by the Lu family, Mingxing, and Fazhi. The existing mound, as the geo-centre of Zhang's garden, presented potential for new landscaping. It was possible that some remnants of previous structures still existed, such as the stone platform basis of Ask Plum Pavilion, the Dharma hall, and Stand in Snow Hall (see Fig. 3.6).

Meanwhile, Shizi Lin's hydrological environment underwent significant changes, parallel with the general shrinkage of urban canals in Suzhou over the past three decades.[28] The waterway section which used to connect the temple's Dharma courtyard with the external public canal became increasingly sedimented and eventually vanished. This transformation is evidenced in *Map of Gusu City*.[29] The river encircling the central hill also shrunk, leaving only its western portion resembling a pond. The bamboo population on the mound decreased, concurrent with the condensation of the earth layer of the mound and its less water retention.

Zhang Shijun renovated the site as his Lion Grove Garden. His reconstruction was likely informed by extant historical traces of previous structures. The site's pictorial and textual representations might also play a crucial role in guiding Zhang's reconstruction efforts. The frequent references made by Zhang Shijun's guests to Xu Ben's album and poems in

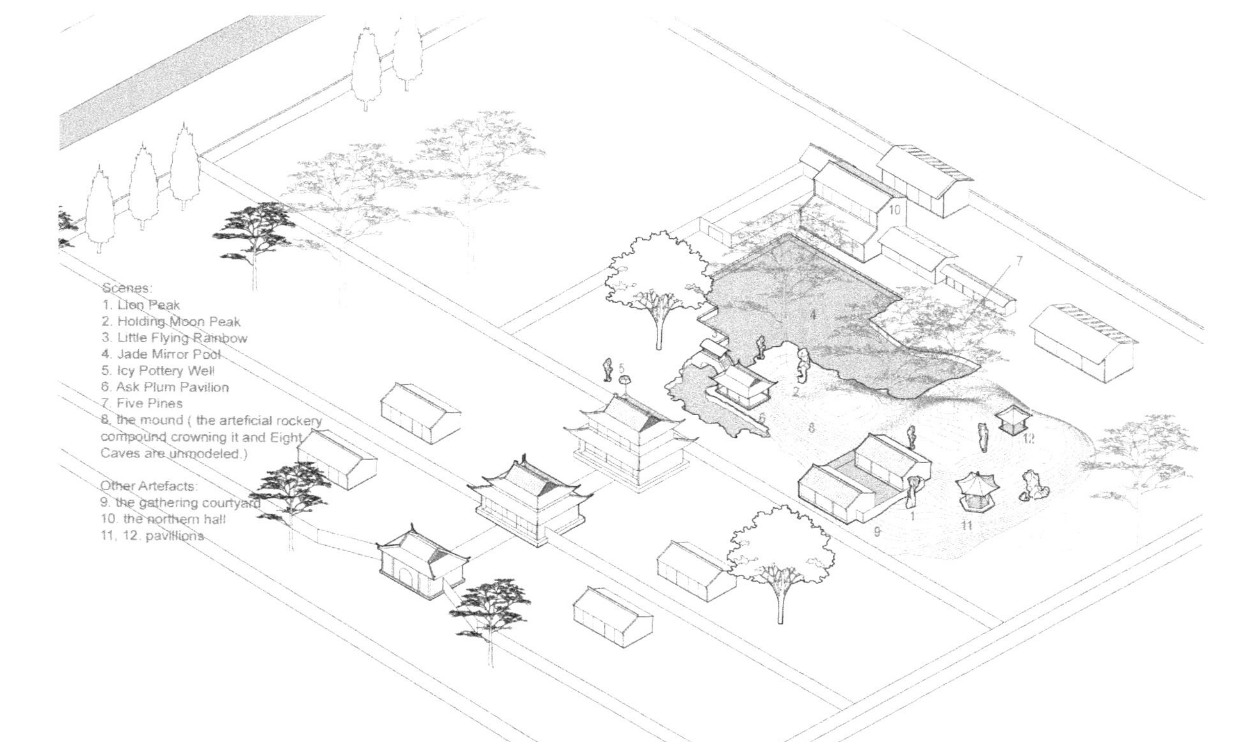

Figure 3.7 Shizi Lin Temple and Lion Grove Garden in the Early Qing. The locations of Lion Peak and Holding Moon Peak are speculated upon due to lack of historical records. Only representative Lake Tai rocks are illustrated. Scene no. 8, Eight Caves, referring to the artificial rockery compound atop the original earthen mound, is not modelled. Source: Author's modelling and drawing.

Anthology of the Recorded Scenery of Shizi Lin strongly suggest that the twelve scenes of Shizi Lin in the Early Ming served as models for their restorations.[30]

Zhang Shijun's Lion Grove Garden and Shizi Lin Temple in the Early Qing are visualised in Figure 3.7.[31] A salient renovation by Zhang was the construction of an artificial rockery complex atop the original earthen mound. The Lake Tai rockery complex, a technique that developed in the Late Ming and matured in the Early Qing, was a sought-after feature among wealthy garden owners.[32] The rockery complex used Lake Tai rocks to structure interconnected cavern spaces, differing from the earlier practices before the Late Ming when Lake Tai rocks were scattered on the earthen mound or displayed in flower pots.

Zhang Shijun's garden featured eight scenes, identified by his guests as Lion Peak, Holding Moon Peak, Little Flying Rainbow, Jade Mirror Pond, Icy Pottery Well, Ask Plum Pavilion, Five Pines, and Eight Caves (*Ba dong*) (Fig. 3.7, Fig. 3.8).[33] They were irrelevant of Buddhist connotations, with meanings differed from the twelve scenes eulogised by Gao Qi's group in the Early Ming.

The two new scenes, Five Pines and Eight Caves, were particularly remarkable in the garden. The Five Pines in the garden were identified by guests as being inherited from the Yuan dynasty. Rooted in the earthen mound, they stretched out through the rockery composite towards the sky.[34] Eight Caves were nestled inside the artificial rockery compound. Guest Cao Kai (fl. 17th c.) portrayed it vividly:

> Ridges and peaks intersect and span, within them lie eight celestial caves. The voids intertwine and crisscross, with each cave leading into the next in a spiral configuration. Visitors wander in and out, lost in their paths, feeling as if they were immortals in Wuling.[35]

While six scenes in Lion Grove Garden retained their names from Shizi Lin Temple in the Early Ming, their locations, styles, and interpretations underwent significant changes. 'Lion Peak' and 'Holding Moon Peak' referred to the Lake Tai rocks protruding from the rockery composite. Being eulogised for their form, they were no longer Buddhist metaphors. Ask Plum Pavilion and Little Flying Rainbow Bridge, rebuilt at their original Yuan dynasty locations, shifted from a simple homespun style to an ornate design. The central pond in the garden was newly designated as Jade Mirror Pond, with its handrail crafted to reflect the depiction in Xu Ben's painting.

Four gong'an scenes depicted in Xu Ben's painting were not recovered, including Point to Cypress Hall, Lying Clouds Chamber, Stand in Snow Hall, and Chan Hut (Fig. 3.8). The first two scenes were unavailable for

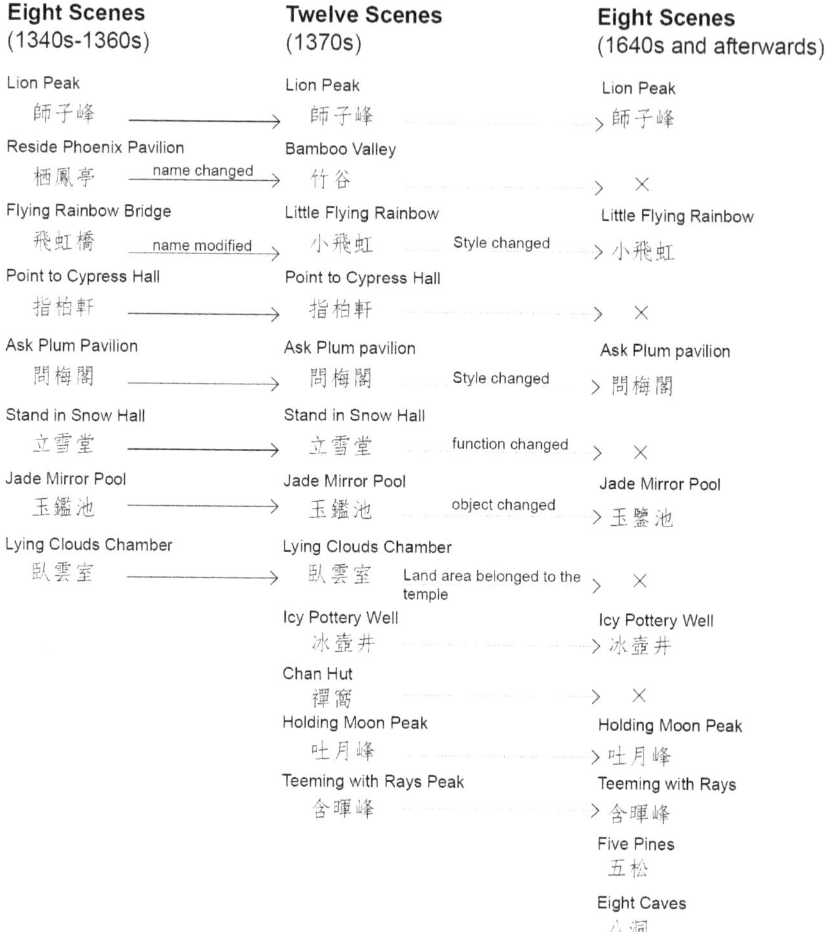

Figure 3.8 Transformation of Scenes of Shizi Lin from the Yuan to the Early Qing. Source: Author.

restoration as their located areas belonged to the temple after the site divided. The Dharma courtyard of Stand in Snow Hall was repurposed to host a gathering for Zhang's friends. The stony foundation of Chan Hut and Reading Sutra Platform to its southeast likely served as the bases for two new pavilions adorning the artificial hill (Fig. 3.7)

Zhang's deliberate choice of not restoring Buddhist scenes informs a broader trend among the gentry of his time: a preference to distance the culture of their garden estates from Buddhism, despite the estates' Buddhist origin. The new configured eight scenes reflect literati traditions and tastes, detached from gong'an practices and Chan interpretations. The allure of Xu Ben and Gao Qi's group's gathering in the temple overshadowed the legacy of the temple's founding master, monk Tianru. With its Buddhist history being neglected, the site was transformed into an aesthetic literati garden.

Upon completing the renovation, Zhang Shijun frequently hosted banquets and poetry sessions. Echoing the linked verse composition as initiated by the 14th-century Yuan masters, Zhang's 16 friends composed new linked verses in Shizi Lin. They also analogised theirs and the former masters' gatherings in the garden. Contributed to the poems included Zhu Yizun (1629–1709) and Wang Ying (1651–1726), prominent scholars and poets of the Early Qing dynasty.[36] They began by introducing the site's urban location, portraying the arrival of guests from near and far. The poets then likened the garden's topography to a real mountain, expressing their amazement at ascending and descending the rockery composite.[37] They vividly imagined Lake Tai rocks as myriad lions frolicking on a lavish blanket. Given the vivid view of 'lions approaching towards,' it could be deduced that these verses might have been composed at Ask Plum Pavilion (Fig. 3.7(6)) or in the gathering courtyard (Fig. 3.7(9)).[38]

The modelled drawing in Figure 3.7 revealed that the Ask Plum Pavilion, the gathering courtyard, and the northern hall were ideally suited for large gatherings. Nestled under a majestic pine, Ask Plum Pavilion opened its four sides, offering panoramic views of the surroundings, with the water at the front and rocks at the rear. The courtyard encouraged an upward view towards the sky, fostering a sense of vertical expansiveness. The northern hall, facing the water, provided a sweeping horizontal view that incorporated the nearby Little Flying Rainbow bridge and the temple's Buddha hall in the distance. Additionally, the wall separating the garden from the temple was adorned with Lake Tai rocks and plants, adding new picturesque views within the garden.[39]

Two tiny pavilions adorned the rockery complex, one perched upon the hill's summit and the other nestled at the eastern slope (Fig. 3.7(11,12)). The pavilions were seamlessly integrated into the artificial rockery, creating an engaging experience of being in the mountains. Their intimate scale fostered spaces conducive to individual contemplation. Zhao Zhixin's (1662–1774) poem vividly depicted his sentiments in the upper pavilion: 'a high pavilion commands a mound, with grotesque rocks encircling on all sides. Sitting here, I feel as if summer clouds are rising; turning around, the autumn mountains appear in disarray.'[40] It was during Zhang Shijun's possession that Shizi Lin began to assume a layout close to the current, which has later become iconic in Chinese garden design.[41]

The garden was a vibrant social space for Zhang Shijun to entertain his friends.[42] Likely with the intention to separate from the Zhang family's residential quarters, the garden had a designated corridor and entrance on the northwest. Some of Zhang's guests lodged in the garden for extended periods.[43] Most of the poems of the garden were penned by guests, with Zhang himself contributing only a quatrain during one gathering. Zhang's Lion Grove Garden highlights the garden's role in forging social connections among elites in the Early Qing.[44]

The Emperor's Garden

Around the 1730s, Garden of Five Pines was transitioned to Huang Xuan's (fl. 18th c.) family estate.[45] The Huang family refurbished the garden to accommodate Emperor Qianlong's six visits between the 1750s to 1780s.[46] These visits by Qianlong introduced new perspectives on appreciating the garden, identifying it as Ni Zan's literati garden. Furthermore, the emperor encapsulated the garden's typology, replicated it in two imperial palaces, and elevated its status as an emblematic literati garden.

Qianlong's initial interest in Lion Grove garden was ignited by a replica of Ni Zan's painting (Fig. 3.3), which entered his imperial collection around 1739. The artwork was held in high regard by Qianlong and referred to as 'the deified work (*shen ping*).'[47] While reading the painting, Qianlong mistakenly took the garden depicted as Ni Zan's Pure Secluded Pavilion (*Qingbige*). He thus inscribed 'Yunlin Qingbi' as the frontispiece of the handscroll (Fig. 3.3(a)). Left to the frontispiece, the emperor wrote several verses expressing curiosity about the location of this garden (Fig. 3.3(b)).[48]

During his second inspection tour to Suzhou in the spring of 1757, Qianlong was informed about the existence of Lion Grove Garden and made an immediate visit.[49] His poem expressed his delight in discovering Ni Zan's garden. He praised the garden's artificial hill for its resemblance to a real mountain, which echoes the sentiments of Zhang Shijun's guests in the Early Qing. With the belief that Ni Zan owned and crafted the garden, Qianlong expressed admiration for Ni Zan's gardening techniques. He made another inscription lauding the garden for composing various elements—the hill, the lake, the pines, and the artefacts—in harmony (Fig. 3.3(c)).[50] Imagining Ni Zan entertaining his guests in this garden five centuries ago, Qianlong perceived the garden as a spiritual platform to connect him with the immortal and sublime.[51]

The emperor's fascination of the artificial hill's intricate spaces was visualised in Qian Weicheng's painting *Complete View of the Lion Grove* (Fig. 3.9, *Shilin quanjing tu*).[52] The painting represents the artificial hill as a natural mountain. Progressing through the scroll, a viewer is transported into a grand mountainous landscape: soaring peaks pierced into the sky, clouds floating amidst the rugged terrain, and dispersed pavilions invite a virtual climb. The journey concludes at a serene lake adorned with tiny

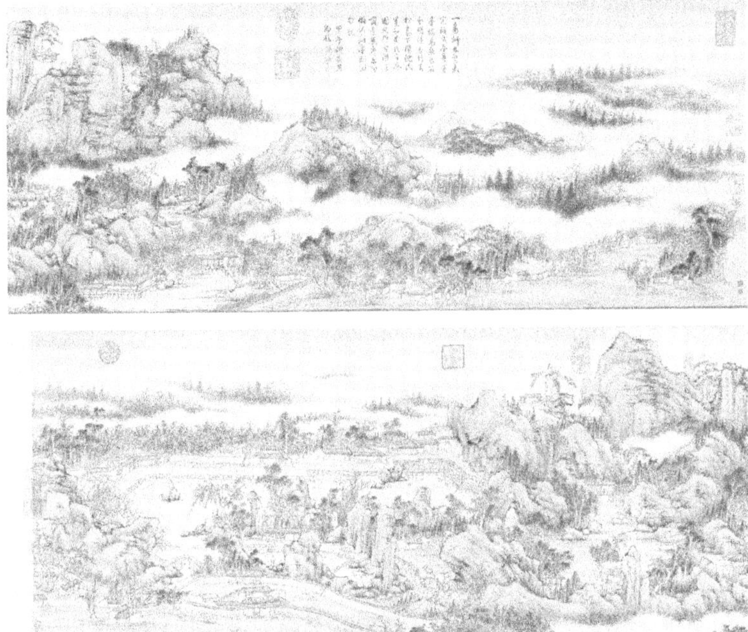

Figure 3.9 Qian Weicheng, *Detail of Complete View of the Lion Grove*. Ink and colour on paper. Mactaggart Art Collection (2004.19.60). University of Alberta Museums. Gift of Sandy and Cécile Mactaggart.

houses. Qian Weicheng stated in the postscript that he was depicting the garden, and he meticulously drew its wall boundaries at the beginning and the end of the scroll. This artistic representation conveyed the emperor's perception of the garden as a mountain.

Qianlong emperor advocated a new way of viewing the garden through Ni Zan's painting. He emphasised the garden as a historical site of literati culture. Believing that Ni Zan's painting held greater authenticity than the site, the emperor argued that the most fitting way to appreciate the garden was through Ni Zan's art. Therefore, he ordered the painting, which was stored in the Purple Palace in Beijing, to be swiftly transported to Suzhou.[53]

In the early summer of 1757, Qianlong revisited the garden, Ni Zan's painting in hand, for a detailed comparison between the representation and the represented.[54] He penned another poem on Ni Zan's painting (Fig. 3.3(d)), reflecting on how he resonated with Ni Zan's pure and lofty character when wandering in the garden. Some scholars interpreted Qianlong's profound reverence for Ni Zan as a political gesture—by aligning with Ni Zan, a paragon of literati-ness, Qianlong symbolically asserted his authority over the essence of Chinese culture.[55] Qianlong's engagement transformed the garden into a historical site, whose current significance was validated by its past.[56]

To accommodate and attract more imperial visits, the Huang family modified their garden estate, much like the local officials' refurbishment of Surging Wave Pavilion around the same period. The modifications were identified by the inscribed characters in the painting of *Grand Ceremony of the Southern Inspections* (Fig. 1.5b). The initial east street entrance to the garden was deemed insufficiently imposing, leading to the construction of a new entrance avenue stretching from the temple's main gate to Huang family's garden (Fig. 1.5b and Fig. 3.10).

The garden's redesign also conformed to imperial rituals. The courtyard to the south of the mound (Fig. 3.10(7)), the imperial stele pavilion (Fig. 3.10(6)), and the northern hall facing the pond (Fig. 3.10(3)) formed a triangular spatial relation, each offering a unique view of the site. The courtyard was named 'Landing Place (*luozuo*),' designated for the emperor's carriage to stop, where he would disembark and commerce the garden tour. The imperial stele pavilion (*Yubei ting*) housed a stele bearing the poem composed by the emperor during his first visit to Lion Grove Garden.[57] Although the emperor visited only once every five years, the presence of the imperial stele as a simulacrum of imperial authority continuously commanded respect from visitors, asserting the emperor's interpretation of the garden.[58] In the centre of the rear hall, a chair marked as 'Imperial Seat (*yu zuo*)' further signified the imperial presence (Fig. 3.11a(4)).

A new structure added to the garden was the two-storied Pine Breeze Tower (*Songfeng ge*) to the east of the hill (Fig. 3.10(10)). Its upper floor commanded a panoramic view of the garden's hill and lake, and its surrounding loggia offered the emperor an extensive view of the entire city. This building, along with the three aforementioned structures, is depicted in the painting *The Famous Lion Grove Garden in Suzhou* (Gusu Mingyuan Shizi Lin) (Fig. 3.11b(3)), illustrating the garden's adaptation to its newfound imperial significance.

Emperor Qianlong transformed the layout of Lion Grove garden into a replicable model. Qian Weicheng's 1757 artwork captured the garden's configuration (Fig. 3.9): the western part was characterised by a rockery mound, veteran pines, and adorned pavilions. Intertwined with this rockery mound is the pond occupying the eastern part, juxtaposed by axial buildings and overarched by a bridge. To replicate with precision, Qianlong commanded local Suzhou official Shu Wen (fl. 18th c.) to submit measured architectural drawings and a detailed 1:20 scale ironed paper model (*tang yang*) of Lion Grove Garden to the court in 1771.[59] Referencing the painting, drawing, and model, Qianlong constructed Everlasting Spring Garden (*Changchun yuan*) in Garden of Perfect Brightness (*Yuanming yuan*) in

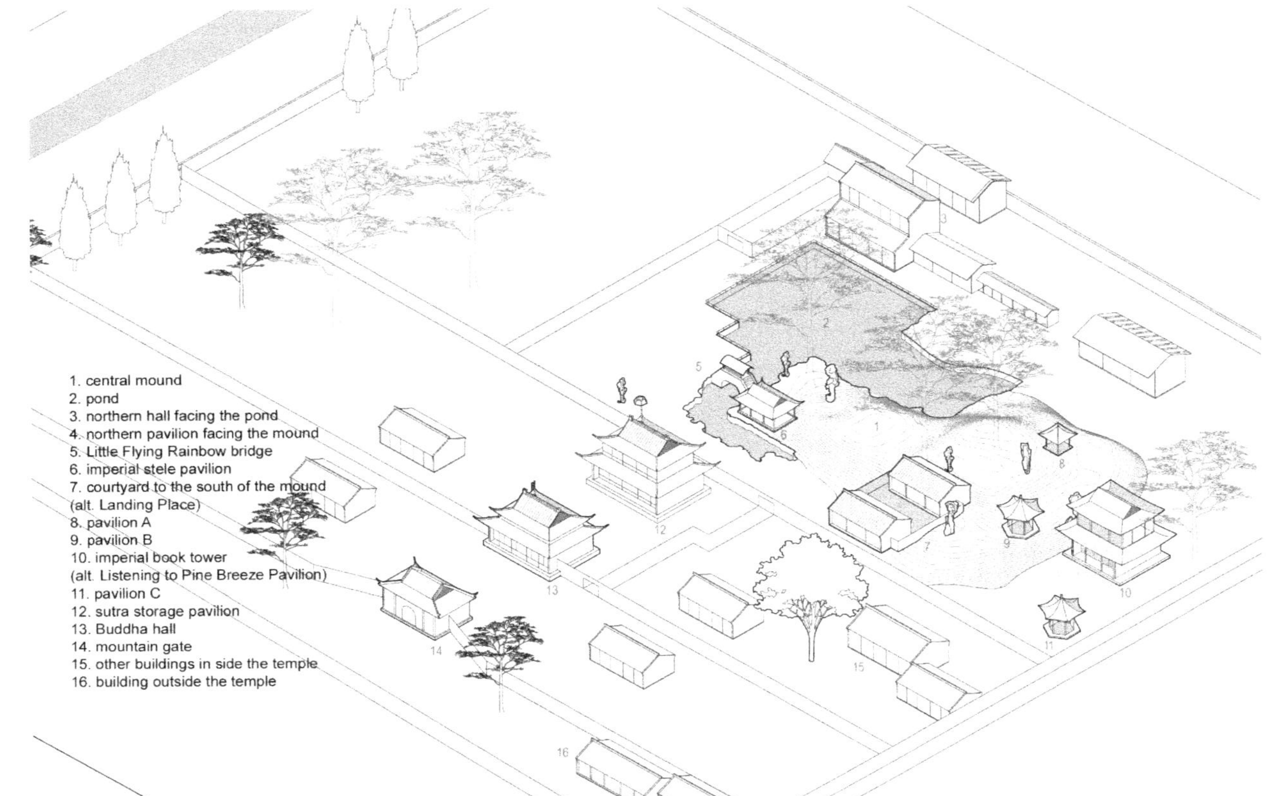

Figure 3.10 Shizi Lin Temple in the High Qing period. Source: Author's modelling and drawing.

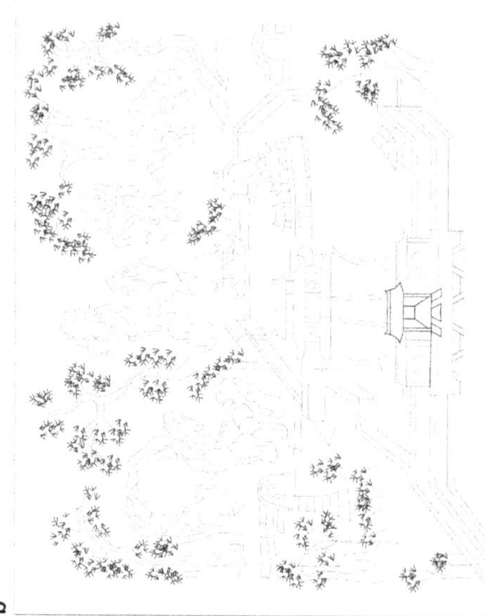

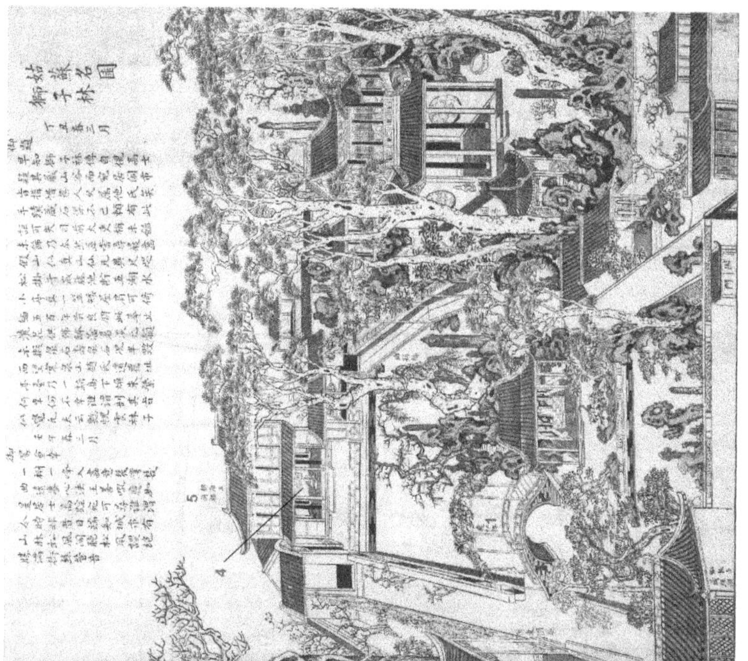

Inscriptions of Artefacts in the left painting:
1. 飛虹僑 Flying Rainbow Bridge; 2. 御書亭 imperial inscription pavilion; 3. 松風閣 Pine Breeze Pavilion; 4. 御座 the imperial seat; 5. 朝南五間樓 five bays tower facing the south.

Figure 3.11 Two paintings of Lion Grove Garden in the High and Late Qing. (a) Anonymous, *The Famous Lion Grove Garden in Suzhou* (Gusu mingyuan Shizi Lin). Tenri Central Library. (b) Guo Zhongheng, *Lion Grove Temple*, 1759, wood block print. Source: Author's drawing following Guo.

Beijing and Garden of Literature (*Wen yuan*) in Mountain Estate to Escape the Heat in Chengde (*Chengde bishu shanzhuang*).[60]

Qianlong's six visits to Lion Grove Garden incentivised the reproduction of paintings of Lion Grove Garden in the court. On his second visit, the emperor created an imitation of Ni Zan's painting of Shizi Lin and decreed it to be kept in the Lion Grove Garden in Suzhou. Later upon the completion of Everlasting Spring Garden and Literature Garden, the emperor replicated Ni Zan's painting for these new locations, ordering that these replicas also be housed there. Qianlong's artisans further recreated paintings reworking Ni Zan's painting and depicting the two imperial gardens.[61] The compositions and components of Ni Zan's painting and Lion Grove Garden became endurable garden iconography, consistently reproduced in paintings and on-site.[62]

After Qianlong's visits, Lion Grove Garden was transformed from an exclusive elite garden to a public place, obligating citizens to pay homage to the emperor's inscriptions.[63] Visitors' pictorial representations revealed at least three perspectives on viewing the garden. The first is a literati appreciation for rockeries, pines, and water, as revealed by Jiang Yun's portrayal of the garden as a sphere of lion-like rockeries (Fig. 3.1a). This way of seeing originated in the Early Qing, when Zhang Shijun's guests appreciated the garden as a rockery labyrinth. The garden typology, centred around an artificial rockery mountain, was further developed by Emperor Qianlong during the High Qing period.

For the public, the garden was perceived as 'a grove of lions,' or 'lions in the grove.' Lions were the beast-like rockeries, and groves were the contorted pines, as depicted in the woodblock print in Figure 3.11b.[64] A third, an imperial interpretation of the garden is found in *Famous Garden in Suzhou: Lion Grove Garden*, a painting acquired by Japanese tourist Ozawa Kei, on his visit to Suzhou in 1882 (Fig. 3.11a). The painting depicts the buildings as refurbished for the imperial, with the emperor's poems inscribed atop. It demonstrated the emperor's authority over the garden while enhancing the Huang family's local influence, associating their garden with the imperial.

As Lion Grove Garden gained popularity, some elites expressed their contempt. Shen Fu (1763–1832), for instance, disdained the garden's artificial rockery during his early 1800's visit: 'The most famous place in Suzhou, Lion Grove Garden, was supposedly created in the style of Yunlin with splendid rocks and many old trees. But to me, it looks more like a pile of coal dust covered with moss and ant hills, without the least suggestion of the atmosphere of mountains and forests. In my humble opinion, there is nothing particularly wonderful about it.'[65] Official Liang Zhangju (1775–1849) showed similar apathy to the garden. When invited to revisit the garden, he declined, remarking: 'amongst all the gardens in Suzhou,

I dislike Lion Grove the worst of all because its space was cramped and made me feel squeezed.'[66]

Summary

The five stages of Shizi Lin's evolution demonstrate the transference of garden practice from Buddhists to the gentry. The wall dividing Lion Grove Garden and the southern temple in the Early Qing (Fig. 3.7) could be read as a spatial metaphor, evidencing and clarifying that the transference of garden culture was not due to the close exchanges between the two groups, but to their contentions over estates.

Garden practices were closely imbricated in the formation of their owners' social networks during the Ming and Qing Periods. In the Early Ming, the scenic temple attracted visits from the gentry and secured the temple both political and literature patronage despite the postwar depression. In the Early Qing, the garden served as a platform for merchant Zhang Shijun to reinforce his social network with prominent gentrymen, which made his book printing business prosper. In the Middle Qing, the garden assisted Huang family in gaining frequent visits and favour from the emperor, elevating them to higher political privilege. The modifications of gardens facilitated social events.

The garden's urban hydrological function began to collapse since the High Ming period. The hydro-topography established by Tianru in the 14th provided foundational infrastructure for the Lu family's agricultural production but likely decayed due to poor management. In the Early Ming, as the estate's water became disconnected from urban canals, Zhang Shijun constructed an artificial rockery mountain and a central pond. The garden water became as decorative as its rockeries. The prototype of a Chinese garden, featuring an artificial mountain accompanied by a pond, could be seen as a landscaping response to the shrinkage of Suzhou's waterways since the 17th century.[67]

The representations of gardens were shifted from reflecting reality to becoming a simulacrum that precedes the actual. In the Early Ming, the physical garden inspired paintings and representations. Since the Middle Ming, the paintings of the garden scenes, circulating outside the Buddhist realm, began to take on significance of their own. In the Qing dynasty, these paintings inspired the garden's reconstruction meanwhile surpassed the garden in terms of perceived authenticity.[68] As the gentry demarcated their garden culture from the Buddhists, they also erased the garden's Buddhist connotations. This erasure of Buddhist traces in estates was more violent in the conversion of Great Propagation Temple into Unsuccessful Politician's Garden, as will be discussed in the next section.

From Great Propagation Temple to Unsuccessful Politician's Garden

Imperial Censor Wang Destroyed the Temple

Great Propagation Temple was located in the northeast corner of the city, next to Qi Gate. Its imperial title board was obtained in the Yanyou's reign of the Northern Song dynasty. It used to be the ancient Celebrate Longevity Temple (*Qingshou si*) in the past. The temple approximated the imperial censor Wang Xianchen's family estate. Imperial Censor Wang was taken with hills and valleys and befriended with gentry scholar such as Wen Zhengming.

[Wang] did not believe in Buddhism all his life. He demolished this temple to expand his garden. He commanded young thugs to pull down statues of Buddha, Bodhisattva, and the Kings of devas. They used knives to scrape golden plating from deities' faces. People nearby strongly admonished Wang not to do so, but he did not stop. In a moment, monastic buildings were demolished. Monks fled, steles torn down, and the temple became desolate.

The garden that was expanded from [the temple] was named 'Unsuccessful Politician's Garden'. Inside the garden, there were thousands of towering trees. These trees used to be the temple's old things. They were supreme in the Wu region.

Qian Xiyan (1562–1638), 'Retrogression'[69]

The above account outlines the aggressive and violent transformation of Great Propagation Temple into the Unsuccessful Politician's Garden by Wang Xianchen in the 1500s. The rapacious invader intruding on Buddhist space stands in stark contrast to 'the literati who retreated from the world,' as depicted in Wen Zhengming's painting (Fig. 3.12a). Wang destroyed Buddhist statues, eliminated Buddhist traces, and even scraped off the golden plating from statues. As an artist commissioned and remunerated by Wang, Wen chose to sidestep the site's temple history, a monastic origin often acknowledged by the estate's subsequent owners.

Wen Zhengming's portrayal of the garden as a lofty literati retreat, along with his exclusion of its Buddhist traces, has oriented scholarly interpretation of the site, leading to an overlook of the contentious and even dark aspects of the Late Ming gentry's garden origins. The construction process of Unsuccessful Politician's Garden remains enigmatic, as it was not elaborated in garden records and works by monks from Great Propagation Temple. Neglecting the garden's Buddhist history, scholars have often assumed that Wang Xianchen crafted all the scenic features of the garden, with Wen Zhengming serving as the design consultant.[70]

However, a site's physical construction should not be confounded with its cultural representation. Although monks in the Yuan invested massive labour in establishing the site's hydraulic foundation, the scant representation of their landscaping would lead to an overlook of contribution.

This section challenges the premise that Wang Xianchen crafted all the scenic features of the garden during the Ming. It argues that monks of Great Propagation Temple had already laid out the garden's hydraulic infrastructure during the Yuan. This hydro-topography, crucial for a garden's inhabitation and aesthetic appeal, likely made temples coveted by gentrymen such as Wang.

Architectural models are used to reconstruct the site's four transformative stages, spanning from the Yuan to the Qing dynasties. Different owners' contributions to shaping the garden are re-assessed in aspects of topography, plantation, architecture, and cultural constructs. The process of Wang Xianchen's conversion of Great Propagation Temple is unveiled, clarifying the site features which Wang chose to retain, exploit, or erase. The findings reinforced the argument presented in Chapter 1: monks' hydraulic landscaping in the Yuan not only laid the foundation for the prosperity of garden-scape in the Ming dynasty but also exerted a lasting influence on garden design in subsequent periods onto today.

The Current Site's Thick Biography

The current site of Unsuccessful Politician's Garden consists of western, middle, and eastern portions (Fig. 3.12b). Each part comprises mounds and ponds which were unconnected to each other and the urban canal. To model the current site, surveying data are collected from three sources. The first is the survey carried out by Liu Dunzhen in the 1950s, which provides the plan of the garden and elevations of each mound.[71] The second is the plan of the garden's urban surroundings, which is sourced from the plan of the Suzhou Museum design project and Google map.[72] The comprehensive surveying data of artefacts inside the garden, published by Suzhou Garden and Greenery Bureau in 2018, informed the architectural modelling.[73] By synthesising discrete data of plans and sections from these three resources, a model of the existing site was developed (Fig. 3.13). The five mounds north to the central pond are sequentially labelled from N1 to N5, the two southern mounds are designated as N6 and N7, and the three eastern mounds are identified as N8 to N10.

The current garden exhibits a harmonious interplay of water, topography, and artefacts. In the western section, Mound N2 is surrounded by a meandering waterway extending towards the north and south. Floating Emerald Pavilion (*Fucui ge*), positioned atop Mound N1, offers an expansive view of the entire garden. Perching on the protruded eastern crag of Mound

N2 sits the fan-shaped pavilion 'Who Sits Together with Me' (*Yushui tong-zuo xuan*). Opposite Mound N1 and N2, on the northwestern shore of Mound N6, lies Thirty-Six Pairs of Mandarin Ducks Hall (*Sanshiliu yuan-yang guan*) and Hall of Eighteen Mandala Flowers (*Shiba Mantuoluo hua guan*). The eastern part of the garden featured mounds N8 to N10. Mound N8 is encircled by a serpentine river, with Open Vista Pavilion (*Fangyan*

Figure 3.12 Unsuccessful Politician's Garden in painting and in aerial photograph. (a) Wen Zhengming, scene no. 23, 'Xiang Bamboo Valley,' in album *Thirty-One Scenes of Unsuccessful Politician's Garden*, 1533, ink on paper. (b) Aerial Photo. Source: Author.

ting) perched on its top. Mound N10 featured a concave shape, surrounding Heaven Spring Pavilion (*Tianquan ting*) at its front (Fig. 3.13).

The central part features mounds N3 to N7 and an expanse of water. Mounds N4 and N5 resemble islands floating above the water's surface. Fragrant Snow and Splendid Clouds Pavilion (*Yunxiang xuewei ting*) graces the peak of Mound N4, while Northern Mountain Pavilion (*Beishan ting*) crowns Mound N5, and Embroidered Damask Pavilion (*Xiuqi ting*) adorns the summit of Mound N7. Each pavilion constitutes a unique scene within the landscape while also offering views of other scenes. The southwestern part of Mound N4 protrudes into a triangular peninsula close to water. At its centre stands the hexagonal Lotus Breeze at Four Sides Pavilion, aptly named to evoke the summer experience of breezes wafting through blooming lotuses. The pavilion also acts as a transitional hub, liking mounds N3, N4, and N7 with bridges. The tallest structure is Seeing Mountain Tower (*Jianshan lou*), located at the northeast corner of mound N3. Its second story offers visitors a comprehensive overview of the garden after touring through all the scenes.

This central part has been widely acclaimed by scholars for representing the open, relaxed, and antique character of the Ming style.[74] Its layout, featuring islets floating above expansive waters, was lauded by Maggie Keswick as a 'water labyrinth.'[75] Liu Dunzhen, Chen Congzhou, and Pan Guxi all praised the central part for its unique blend of vast open water space and skilful spatial division.[76]

Spatial techniques represented by the central part have been also extensively studied in the modern Chinese garden discourse. First, the adept plotting of visual connection between artefacts among mounds exemplified the design techniques of 'facing the scene' and 'framing the view.'[77] By manipulating the locations, orientations, and openings of architecture, intricate and complex vistas have been crafted. Second, the central part maintains a harmonious blend between spaciousness and density. This spatial contrast is achieved by configuring courtyards on the southern bank. An exemplar is the water courtyard of Lesser Surging Wave. Layers of scenes were arranged along the elongated river branch, 'yielding an illusory sense of depth.'[78] Other praised features include the extensive verandas separating the middle, east, and west portions, and the borrowed view of the pagoda in Repay Kindness Temple to the east of the garden.[79]

As diagrammed in Table 3.2, the ownership of Unsuccessful Politician's Garden underwent several changes. Established by monks in 1297, it remained as Great Propagation Temple for nearly two centuries despite an intermittent occupation by Pan Yuanshao (fl. 14th c.). In 1508, Wang Xianchen forcibly acquired the site but only held it for three decades. Following Wang's death, his son lost the garden to Xu Shaoquan (fl. 16th

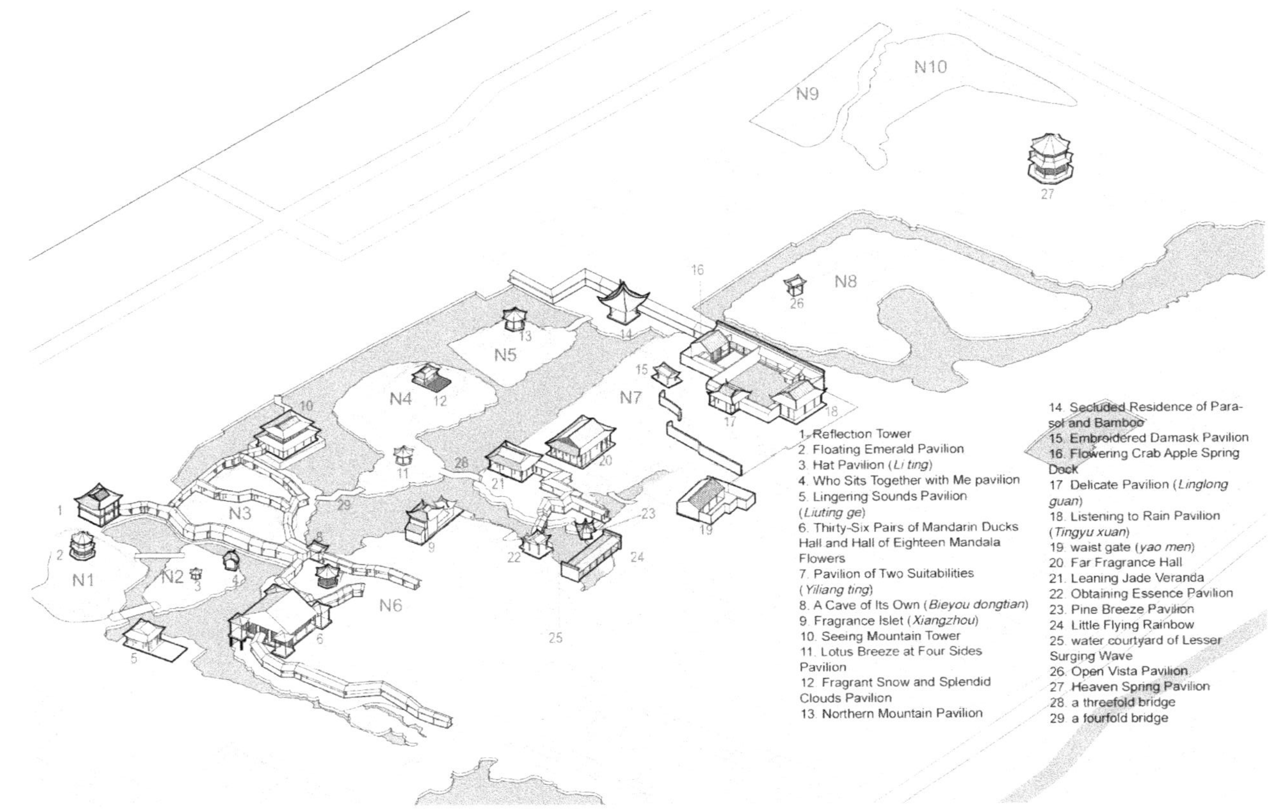

Figure 3.13 Model of the current site of Unsuccessful Politician's Garden. Source: Author.

c., alt. Xu Jia) in a gambling bet. Xu family then maintained ownership for about one century, from the 1560s to 1630s.

The estate experienced divisions in the late Ming. Around the 1640s, Wang Xinyi (1572–1645) acquired and renovated the eastern part of the site, naming it 'Returning to Farm Dwelling (*Gui tianyuan ju*).' The western and middle parts were later owned by Chen Zhilin (1605–1666) in the 1650s, and by Wang Yongning (d.1672) in the 1660s, before its conversion into Bureau of the Circuit of Suzhou, Songjiang, and Changzhou (*Su Song Chang dao xinshu*) decade later.

In the 1680s, the western and middle parts came under the ownership of the Wang and Gu families, respectively. The middle part then became

Table 3.2 Site Evolution Biography of Unsuccessful Politician's Garden

Decade	Western Part	Middle Part	Eastern Part
Temple Stage, Late Yuan to Early Ming, Figure 3.22			
1290s-1340s	Great Propagation Temple, constructed by monk Panquan Youlan in 1297.		
1350s- 1360s	The Palace of Pan Yuanshao		
1370s-1500s	Great Propagation Temple		
Transformative Stage 1, High Ming to Late Ming, Figure 3.20			
1510s-1540s	Wang Xianchen's Unsuccessful Politican's Garden		
1550s-1630s	Transferred to Xu Family		
1640s	Qian Qianyi		Wang Xinyi, Returning to Farm Dwelling
1650s	Chen Zhilin, Garrison General's Mansion (*Zhufang jiangjun fu*)		
Transformative Stage 2, Early Qing, Figure 3.24			
1660s	Wang Yongning reconfigured the garden as islands above water.		
1670s	Bureau of the Circuit of Suzhou, Songjiang, and Changzhou		
1680s-1720s	Wang family	Gu family	
1730s-1790s		Jiang Qi, Recovery Garden	
1800s- 1810s		Zha Shitan's garden	
1820s-1850s		Wu Garden, Wu Family	
1860s	Li Xiucheng, Mansion of the Loyal Prince		
Transformative Stage 3, Late Qing, Figure 3.25			
1870	Zhang Lüqian, Additional Garden, 1877	Guild of Eight-Banner Members from Fengtian and Metropolitan Provinces	Ye Shikuan, Book Garden

Jiang Qi's (fl. 18th c.) Recovery Garden (*Fu yuan*) and later Wu Jin's (fl. 18th c.) Wu Garden (*Wu yuan*). In the 1860s, the whole site served as Li Xiucheng's (1823–1864) Mansion of the loyal Prince (*Zhongwang fu*). In the 1870s, the eastern part was refurbished by Ye Shikuan (1689–1755) into Book Garden (*Shu yuan*), the middle part became Guild of Eight-Banner Members from Fengtian and Metropolitan Provinces (*Baqi fengzhi huiguan*), and the western part was renovated by Zhang Lüqian (1838–1915) as Additional Garden (*Bu yuan*). In the 1950s, the three parts of the garden were transferred to the government entirely and renovated into a public garden.[80]

Table 3.2 reveals that owners possessed the site over diverse timespans and in different proportions. As presented by the scholarship, Great Propagation Temple held the longest ownership of the site for two centuries, from the 1300s to 1508. Following it, Xu family's ownership spanned the second-longest period. The middle part of the site, under Jiang Qi's possession, ranked third in duration, lasting about seven decades from the 1730s to the 1800s. Contrary to its prominent recognition, Wang Xianchen's actual ownership of the site was relatively brief, spanning only three decades.

Modelling Unsuccessful Politician's Garden and Great Propagation Temple

Utilising the 3D model of the present site and analysing historical records, architectural models representing the site's significant renovations across different stages could be reconstructed. Records about Great Propagation Temple, though scarce and discrete, provided some basic information. The temple was established in the early Yuan by Panquan Youlan, and later inaugurated by Master Jing (fl. 14th c.). The legal imperial recognition received by the temple indicates its comprehensive Buddhist architecture and substantial monk community. Wang Xing's underscore of the temple's prestigious status in the Yuan also implies a thriving monastic community.[81] Lu Xiong and Wang Ao's (1450–1524) gazetteer maps (Fig. 3.14) show the temple was situated adjacent to the main street in the area of the current Unsuccessful Politician's Garden.

There were also recorded landscape features of the temple. The aforementioned Tiantai master Tianquan Yuze spent his later years in the temple's eastern section. He constructed the Eastern Chamber (*Dong zhai*) and dug a well named after him, which remains in Heaven Spring Pavilion of the current site (Fig. 3.13(27)). The fact that Pan Yuanshao, a relative of Zhang Shicheng (1321–1367), took up this temple as his residence further indicates the temple's well-established landscape architecture.

In contrast to the monks' sparse documentation about their temples, the gentry produced abundant representations of gardens. Wen Zhengming's paintings of Unsuccessful Politician's Garden secured the estate's fame. Between 1508 to 1513, Wang Xianchen forcefully acquired the temple, evicting monks to a small plot in the eastern part.[82] Two decades later, Wang invited Wen Zhengming to depict his garden.[83] Wen's first painting in 1528 was a hanging scroll offering a panoramic view of the entire garden

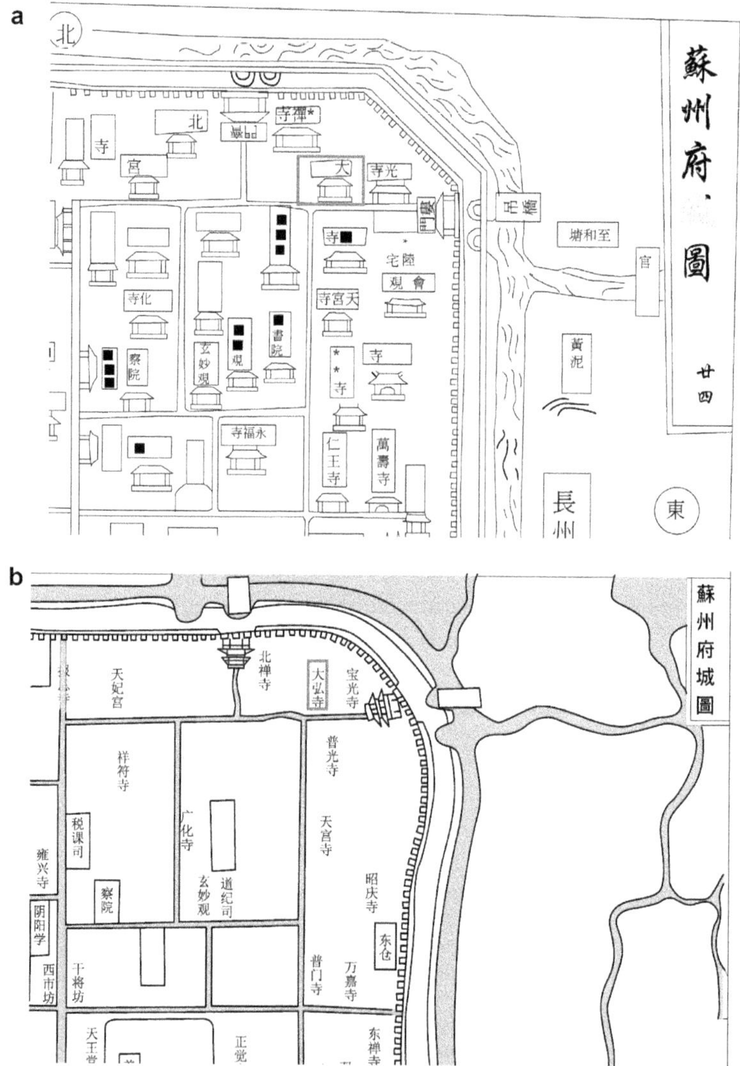

Figure 3.14 Great Propagation Temple in gazetteer maps of the Early and Middle Ming. (a) Early Ming map of Suzhou in *Gazetteer of Suzhou in the reign of Hongwu*. Author's traced drawing following Lu Xiong. (b) Mid-Ming map of Suzhou in *GSZ*. Author's traced drawing following Wang Ao.

(Fig. 3.15a). Later, in 1533, he created *Thirty-One Scenes of Unsuccessful Politician's Garden*, an album where each scene was accompanied with a note and a poem on the opposite page (see Fig. 3.12a). Wen also wrote a garden record at the end of the album. Wen's artwork rendered the garden as serene and reclusive, an ideal realm for its lofty owner.

Wen's 1528 hanging scroll, *Garden and Pavilions for Mr. Sophora Rain* (*Huaiyu xiansheng yuanting tuzhou*) captures the garden's topographical, architectural, and vegetational features. It presents a comprehensive view of the garden from a series of hovering views, as if the artist's gaze sweeps across the garden in mid-air, moving along the central elongated pond from east to west (Fig. 3.15a). The composition of the painting aligns with the site's layout (Fig. 3.15b). The curving waterway zigzagging on the canvas corresponds to the linear pond centred on the garden.

Analogous to the on-site waterbody's partition of the garden into zones, the meandering river in the painting divided the pictorial space. Key artefacts included in the painting are a hall, a veranda, a tower, a bridge, and a pavilion (Fig. 3.15). They are oriented towards the river and are strategically positioned at the edge of the scroll. The two men conversing beneath a tree looked at and oriented towards the river. Similarly, in the lower-left hall (Fig. 3.15(3)), a man in a yellow colour robe, likely Wang Xianchen, is depicted as gazing towards the river in the distance. The lush groves in the scroll's upper section and the bridge arching across the river in the lower section both reflect the garden's actual features.

The waterbody's on-site facilitation of touring is visually conveyed in the painting by the movement of figures, further emphasised by leveraging the hanging scroll's vertical format. When examining the hanging scroll upward, the viewer's attention is first drawn to two servant boys crossing a bridge in the lower-left corner, carrying tea trays, presumably to serve gentlemen under the tree. Another servant boy traversing the covered veranda draws the viewer's gaze upward, leading to a grey-robe gentleman who strolled along the riverbank towards a hall at the right side of the scroll (Fig. 3.15(2)). As the eye moves up, it catches a servant with a hoe entering the dense grove. The visual progression in the painting indicates how a visitor would experience and navigate the actual garden, being guided by the water.

The third verifiable geographical detail in the hanging scroll is its portrayal of the central river as flowing and navigable. Wen Zhengming meticulously illustrated waves on the river surface, indicating the flow. The presence of a boat docked at the shore further hints at the navigability. In Wang's garden, this inner waterway was linked to the external canal at the south, flowing from west to east (Fig. 1.10a). This direction is reflected in Wen's hanging scroll, where the water flows downwards. It remains in speculation whether this congruity is coincidental or intentional.

Figure 3.15 View Simulation of Wen's 1528 Hanging Scroll. (a) Wen Zhengming, *Garden and Pavilions for Mr. Sophora Rain* (Huaiyu xiansheng yuanting tuzhou), 1528, hanging scroll, ink on paper, 122×48.5cm. Beijing: The Palace Museum. (b) Simulated View, the author's modelling and drawing.

Wen's 1533 album of the garden provides more verifiable geographical features to inform the modelling process. The album blends four artistic and literary forms: painting, scenery notes, poetry, and records. Each leaf in the album represents one scene of the garden. The scenery note and poems on the opposite leaf were inscribed in four styles, including regular, running, cursory, and seal script. The notes identified the location of the scene, while the poems enrich scenes with metaphorical meanings from literati culture.[84] Despite the individual leaves disassembling the garden space into discrete realms, the record at the end of the album weaves a connecting route through these scenes.

Wen's record identified 31 scenes sequentially labelled from no. 1 to no. 31 in Appendix, Translation 2 and in the model drawing in Figure 3.16. The sequence of scenes follows the chronological order of Wen's tour through the garden.[85] The first half of the route was circular, from scene no.1 to no.15, as Wen climbed up the northern slope of mound N4. The mounds connected him with and also distanced him from the water.

Lu Andong has discovered the typological relations between locales by closely reading Wen's record and notes. Projecting Lu Andong's typological diagram onto the model of the garden, Wen's ascension and descension around mounds are revealed (Fig. 3.16). It could be inferred that Wen began his tour from Hall Like a Villa (*Ruosu tang*), as suggested by the gate depicted in leaf no. 2.[86] From this hall, he could see mound N4 across the river. He then crossed Little Flying Rainbow bridge to approach the mound. Following the shoreline bending westward and passing Hibiscus Bend, Wen stopped at Lesser Surging Wave gazebo. Here, he saw mound N6 on the opposite shore adorned with bamboo, which he identified as the eighth scene, Purifying Will Place (*Zhiqing chu*). Climbing the protruded crag known as 'Thoughts Afar Terrace (*Yiyuan tai*),' he turned northwards to the Water Flower Pond (*Shuihua chi*) and Deep Tranquil Pavilion (*Shenjing ting*), entering a realm more calm and secluded. After a brief stay, he turned eastwards and ascended Waiting for Frost Pavilion (*Daishuang ting*), perching atop mound N3. Beginning from Scene no.16, Flushed with Pleasure Place (*Yiyan chu*), located between mounds N4 and N5, Wen embarked on the second half tour to explore the eastern part of the garden, an area abundant with multiple fruit trees.

Four artefacts in Wen's painting leave no. 5 could be matched with actual locations on the current sites: Hall Like a Villa, Dream of Recluse Tower (*Mengyin lou*), Leaning Jade Veranda (*Yiyu xuan*), and Little Flying Rainbow bridge (Fig. 3.17a, Fig. 3.17d). The three buildings, interconnected by the bridge, create a triangular spatial relation. Little Flying Rainbow bridge corresponds to the location of the present triple bend bridge (Fig. 3.13(28)). Hall Like a Villa aligns with today's Far Fragrance Hall (*Yuanxiang tang*) (Fig. 3.13(20)), while Leaning Jade Veranda

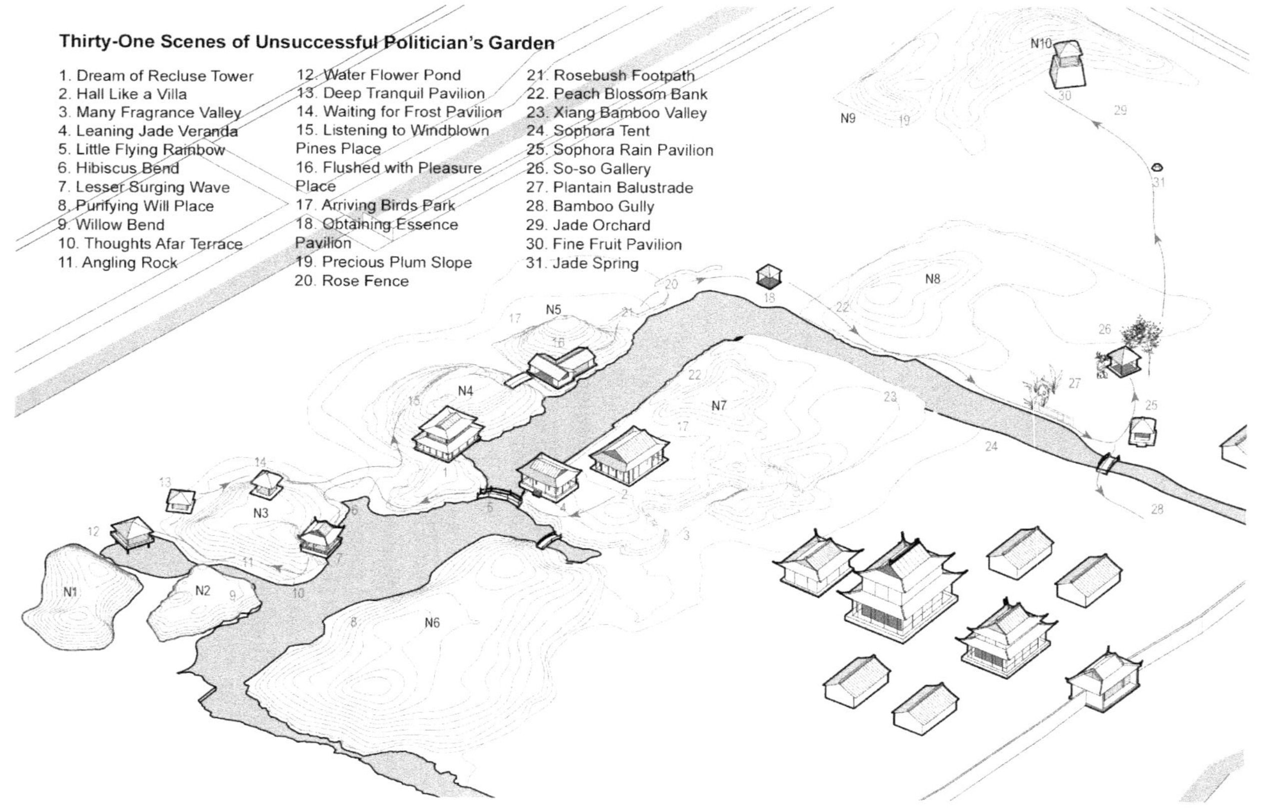

Figure 3.16 Wen's Touring Route in Unsuccessful Politician's Garden. Author's modelling and drawing.

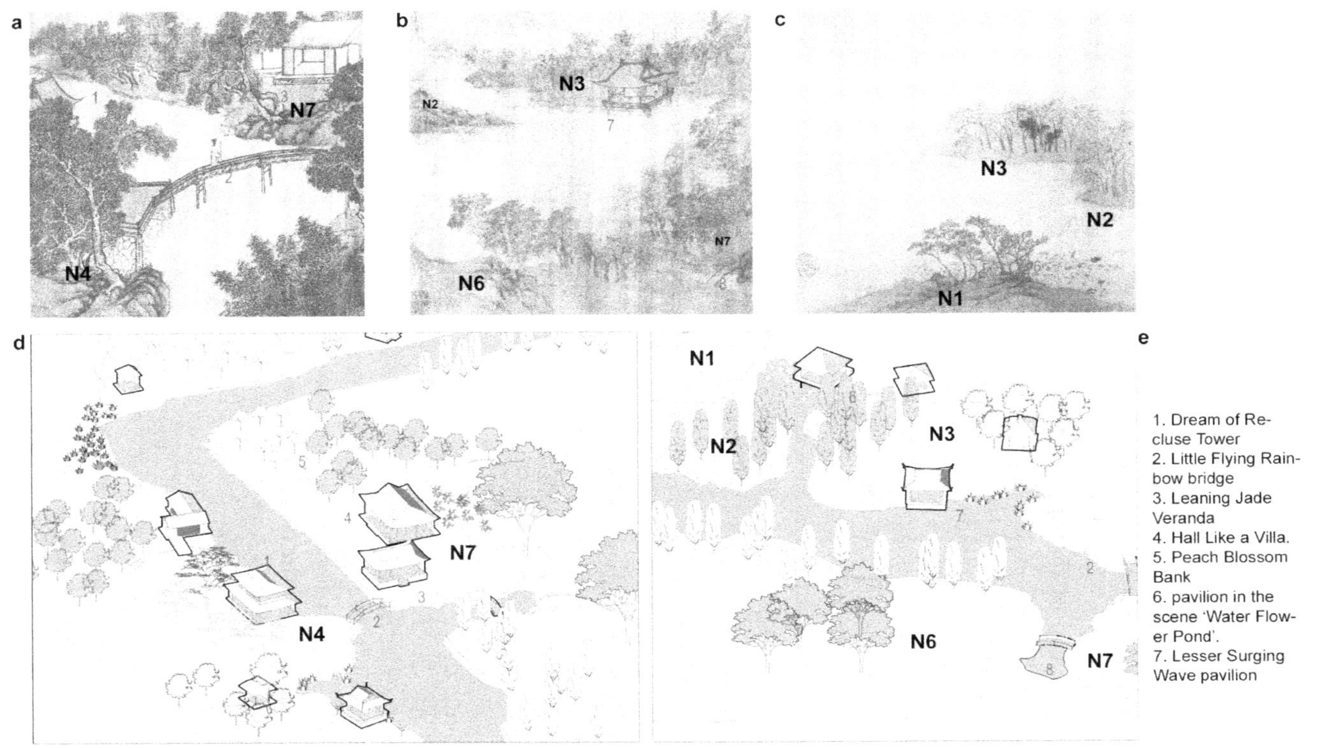

Figure 3.17 Translating artefacts and mounds in Wen Zhengming' painting into the model of Wang Xianchen's Garden. (a) Leaf no. 5, 'Little Flying Rainbow.' (b) Leaf no. 7, 'Lesser Surging Wave.' (c) Leaf no. 12, 'Water Flower Pond.' Author's modelling and drawing.

corresponds to the present artefact with the same name (Fig. 3.13(21)). This correlation was initially noted by Liu Dunzhen during his renovation of the garden in the 1950s and has since been agreed upon by other scholars who attempted to find archaeological clues of Wang Xianchen's garden.[87]

The current location of two other pavilions at the western and eastern ends of the river in Wang's garden could be deduced. The un-named pavilion in Wen's leaf no.12 *Water Flower Pond* (Fig. 3.17c) corresponds to Reflection Tower (*Daoying lou*, Fig. 3.13(1)) of the current site. Wen wrote: 'At the end of the stream, another little pond has been dredged, with lotus planted inside, which was named as "Water Flower Pond".'[88] Obtaining Essence Pavilion (*Dezhen ting*) in leaf no. 18 likely corresponds to the area of today's Secluded Residence of Parasol and Bamboo pavilion (*Wuzhu youju*, Fig. 3.13(14)).[89] Wen wrote: 'at the end of the park, four juniper trees are bound together to form a tent, which is named.'[90]

The ten mounds present at the current site can be identified in Wen's album leaves, verifying their existence in the 1500s. Figure 3.17 shows this topographical correspondence as identified by the simulation of Wen's painting leaves no. 5, no. 7, and no. 12. Mounds N4 and N7 are found in leaf no. 5 (Fig. 3.17a), and they are opposed diagonally in the painting and are connected by Little Flying Rainbow bridge. As simulated in Figure 3.17d, this painting actually embedded a very long vista, which extends towards the far east of Peach Blossom Bank (*Taohua pan*). In leaf no. 7 (Fig. 3.17b), the gazebo Lesser Surging Wave was backgrounded by mound N3, commanding the view of mounds N2, N6, and N7 (Fig. 3.17e). In leaf no. 12 (Fig. 3.17c), the view was composed of the pavilion at Water Flower Pond, which was in front of mound N1 and looking towards mounds N3 and N2. Wen left the background of this painting empty, giving the impression of a vast water expanse.

In leaf no. 16, *Flushed with Pleasure Place* (Fig. 3.18a), mounds N4 and N5 were placed horizontally in the painting and connected by a flat tiny bridge. Mounds N8, N9, and N10 could be found in leaves no. 22 (Fig. 3.18b), no. 29 (Fig. 3.18c), and no. 30 (Fig. 3.19d). Wen manipulated these mounds on the pictorial frame to create different spatial effects. In leaf no. 22, mound N7, Hall Like a Villa, and Leaning Jade Veranda formed the foreground. Wen collaged Dream of Recluse Tower to behind these two buildings, forming a cluster of artefacts. The literati sitting along the shore guide our attention to mound N8 afar, where numerous peach flowers began budding. In leaf no. 29 (Fig. 3.18c), mounds N9 and N10 were horizontally positioned. A servant boy shouldering a hoe seems to walk into the fence to approach the garden estate, where thousands of plums are in blossom.

Apart from mounds and slopes, three other topographical features correlate the current site with those painted by Wen: platforms, crags, and

Figure 3.18 Scenes painted by Wen Zhengming. (a) Leaf no. 16, Flushed with Pleasure Place. (b) Leaf no. 22, Peach Blossom Bank. (c) Leaf no. 29, Jade Orchard. (d) Leaf no. 14, Waiting for Frost Pavilion.

valleys. These landforms rendered the portrayed literati as lofty. Two platforms were depicted in leaf no. 14, Waiting for Frost Pavilion (Fig. 3.18d) and leaf no. 15, Listening to Windblown Pines Place (*Tingsongfeng chu*, Fig. 3.19a). To convey the view from the higher grounds, Wen left the background in these two paintings empty, contrasting the ground directly to the sky.

The crag protruding from mound N3 is represented in scene no. 10, Thoughts Afar Terrace (Fig. 3.19b). Wen depicted a gentleman standing at the tip of the crag, with pine trees canopying a space designated for his solitude. Wen raised the horizon to the upper part of the painting, creating an illusion that the water merged with the sky. The gentleman was gazing towards this horizon as if his thoughts were drifting far from the

Figure 3.19 Scenes painted by Wen Zhengming. (a) Leaf no. 15, Listening to Windblown Pines. (b) Leaf no. 10, Thoughts Afar Terrace. (c) Leaf no. 25, Sophora Rain Pavilion. (d) Leaf no. 30, Fine Fruit Pavilion.

mundane world. While crags provided expansive views symbolising the owner's grand vision, the valleys formed enclosed spaces representing the owner's retreat. These deep valleys were depicted in leaf no. 23 (Fig. 3.12a) and leaf no. 28.[91]

A model of the Unsuccessful Politician's Garden of the 1500s is constructed in Figure 3.20 by extracting the touring route from Wen's record (see Appendix, Translation 2), establishing correlations of architectural and topographical features between the current site and Wen's painting leaves. A pertinent question arises: were these mounds, river, and artefacts constructed by monks of Great Propagation Temple during the 13th–15th

Thirty-One Scenes of Unsuccessful Politician's Garden

1. Dream of Recluse Tower
2. Hall Like a Villa
3. Many Fragrance Valley
4. Leaning Jade Veranda
5. Little Flying Rainbow
6. Hibiscus Bend
7. Lesser Surging Wave
8. Purifying Will Place
9. Willow Bend
10. Thoughts Afar Terrace
11. Angling Rock
12. Water Flower Pond
13. Deep Tranquil Pavilion
14. Waiting for Frost Pavilion
15. Listening to Windblown Pines Place
16. Flushed with Pleasure Place
17. Arriving Birds Park
18. Obtaining Essence Pavilion
19. Precious Plum Slope
20. Rose Fence
21. Rosebush Footpath
22. Peach Blossom Bank
23. Xiang Bamboo Valley
24. Sophora Tent
25. Sophora Rain Pavilion
26. So-so Gallery
27. Plantain Balustrade
28. Bamboo Gully
29. Jade Orchard
30. Fine Fruit Pavilion
31. Jade Spring

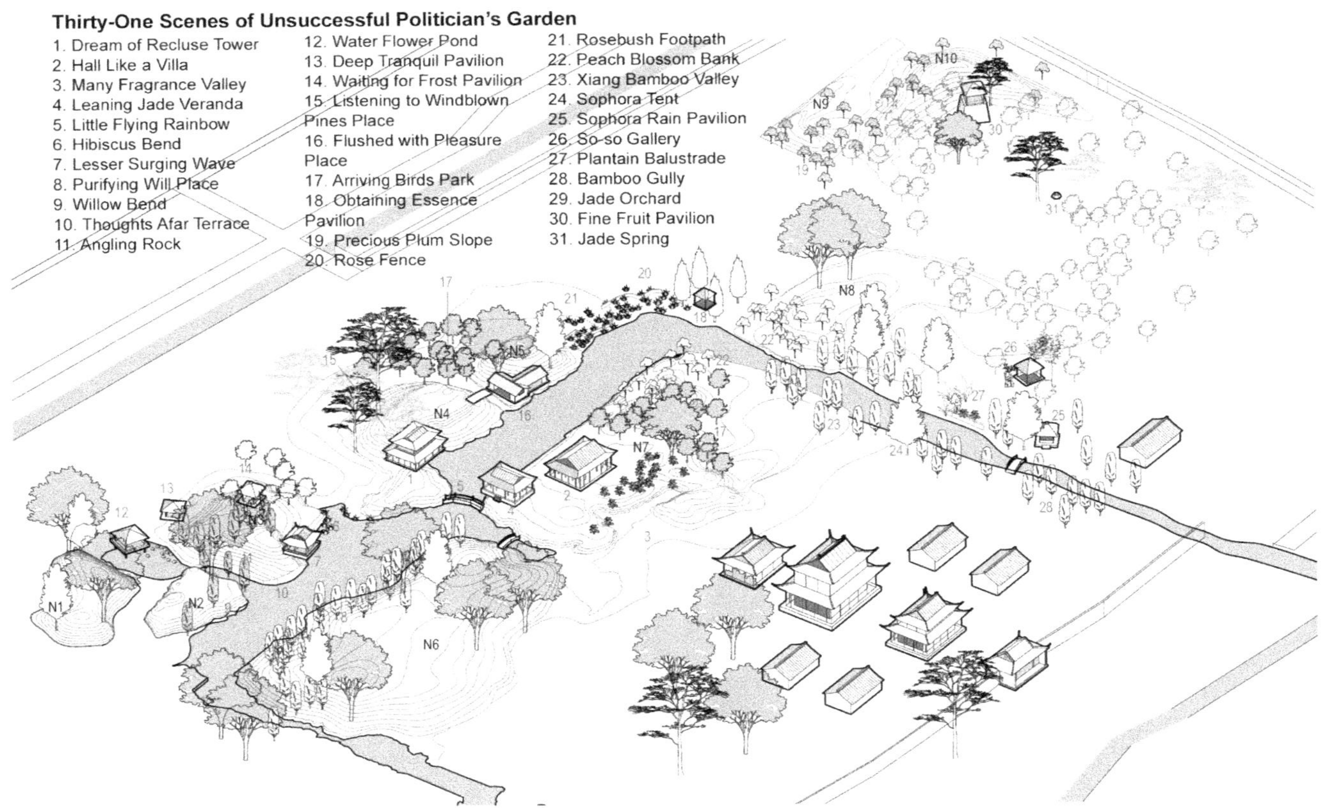

Figure 3.20 Model of Wang Xianchen's Unsuccessful Politician's Garden in the High Ming. Author's modelling and drawing.

centuries, or by Wang Xianchen in the early 16th century? I propose that the hydraulic landform was laid out by monks, not Wang.

My first evidence comes from Wen and Wang's records regarding the construction process. Wen succinctly depicted Wang's construction at the beginning of the record:

> Mr. Sophora Rain Wang Jingzhi lives in the north-eastern part of the city of Suzhou. His residence lies between Lou and Qi Gates and consists of a large tract of residual land with a pond at the centre. He slightly dredged the land and planted groves of trees. On the southern side [of the mound], he built a storied house named as 'Dreaming of Recluse Tower.' The hall on the northern side [of the mound] is called 'Hall Like a Villa.'[92]

The second sentence articulates that a central pond was present upon Wang's acquisition of the estate. All Wang had to do was to 'slightly' uncluttered silts disposed in the watercourse and planted new trees. The pre-existing topography of the estate meant that Wang's efforts were geared more towards restoring its former functionality rather than crafting it from scratch. The modifier 'slightly' implies that Wang did not undertake substantial earth moving, otherwise the work would be 'laborious.'

The inscription by Wang Xianchen in Wen Zhengming's 1533 album affirms that Wang was not the one who established the hydraulic infrastructure of the site:

> After retiring from office, I started teaching my servants daily, removing filth, planting trees, feeding cows, pouring milk, wielding spade, and carrying pots. the uncleaned, planted fences. We served cows, got the milk. We shouldered tools and hold containers in the arm. We planted for daily supplies. Day by day in accumulation the garden is accomplished. Inside the garden, the houses, rooms, terraces and pavilions were hastily built. Names of the garden sceneries are inspired by the classics and recent matters.[93]

Wang's inscription also informed us that his on-site work could be classified into four kinds. The first is to clear messy and uncleaned areas. The second is to erect fences, to firmly establish the boundary of his newly acquired estate. The third involved engaging in animal husbandry, agriculture, and horticulture. The fourth was crafting artefacts. Notably absent from this inventory is any earthwork, such as dredging waterways and piling up hills. Wang's description of the construction of artefacts as 'hastily accomplished' implies an efficiency, feasible because of the adaptation of former structures or re-used of their foundations. These revealing details lead to the conclusion that the expansive pond and the ten mounds were constructed by monks and subsequently inherited by Wang. Based on this hydro-topography, Wang built his 'one hall, one tower, six pavilions,' and crafted 23 scenes 'in the categories of gallery, balustrade, pond, terrace, valley, and gully,' making a total of 31 scenes.

After clarifying that it was monks who established the hydro-topography, let us proceed to examine who planted the site's vegetation, which is featured in each of Wen's painting leaves and also contributes to the garden's current aura as 'natural ancientness (*cang gu*).' The site's lush vegetation could be classified into four groups, with their location and covered area diagrammed in Figure 3.21.

The first sort was the evergreen trees, comprising pine, sophora, cypress, and juniper trees. Some scenes were named after these evergreen trees, and meanings were deduced from their characteristics. For instance, scene no. 15, Listening to Windblown Pines Place (Fig. 3.19a), captures a moment when a person sat under pines, hearing the wind tilting their branches. Wen further depicted the shadows cast by tree branches in his poem: 'sparse pines absorb cold spring, the wind from the mountain fills the hall. Floating clouds pass the empty valley, gently they leave delicate shades.'[94] The 'hall' Wen referred to was likely the second story of Dream of Recluse Tower. In the model drawing of Figure 3.20, it can be seen that these pines provide views and sounds for the tower's upper floor. In the last four lines, Wen eulogised the garden scenery as purifying secular concerns. He further glorified the person amid these pines as the famous hermit Tao Hongjing (456–536).

Other examples of how Wen assigned literati allusions to ancient trees include scenes no. 25 Sophora Rain Pavilion (*Huaiyu ting*, Fig. 3.19c) and no. 24 Sophora Tent (*Huai wo*).[95] Sophora was a symbol of the garden since Wang Xianchen's literati name was 'Mr. Sophora Rain.' Wen depicted Wang sitting inside the pavilion and beneath the tree in these paintings. Scene 'Sophora Tent' was so named because the canopies of sophora trees expanded like a tent. In his poem, Wen associated the sophora in the scene with the sophora tree under which Chunyu Fen (8th c.) fell asleep and dreamed of his prosperous life as the governor in a place called Nanke. The metaphor cunningly implied Wang's previous political career as successful yet illusory, and the current garden Wang retreated to was more real. Veteran trees served as crucial components in Wen's every leaf. They continued to be the major attractions of the garden in subsequent dynasties and remain so today.

Were those evergreen trees planted by Wang Xianchen, or existed since the era of the Great Propagation Temple? Considering that woody trees require centuries to mature and became gnarled, it would be impossible for them to reach an aged state within merely three decades of Wang's

Figure 3.21 Plantation plan in Wang Xianchen's Unsuccessful Politician's Garden. Source: Author.

management. Therefore a more viable scenario is that Wang Xianchen inherited the veteran trees from the temple. This deduction is confirmed by Qian Xiyan's writing, quoted at the beginning of this section: 'Inside the garden there were thousands of towering trees. They used to be the old things of the temple, which were supreme in the Wu region (Suzhou).' Wen Zhengming's language also implies these trees' pre-existence. When describing pines in leaf no. 15 (Fig. 3.19a), he wrote 'there are many tall pines in the place,' and in describing the sophora of leaf no. 24, he wrote 'there was an old sophora.'[96]

Taking into account the growth circle of plants and textual evidence, it could be ascertained that evergreen trees such as pines and sophoras already existed, rather than being planted by Wang Xianchen. As discussed in Chapter 1, woody trees were typically planted on the tops and slopes of mounds to reinforce the overlaid earthwork. Therefore, it's more plausible to deduce these trees were planted during monks' construction of the mounds.

The second category of plants in Wang's garden was fruit trees, serving as the garden's primary productive species. These consisted of oranges, apples, peaches, plums, and apricots. Clunas has provided a comprehensive analysis of the economic value of each fruit tree.[97] The locations of these fruit trees are identified in the plantation plan of Figure 3.21, which also corresponds with the trees in the model drawing of Figure 3.20. Apple, peach, and orange trees, all of which require substantial water supply to generate juicy fruits, were planted on the mound slopes close to the river. Orange trees were the primary character of scene no.14 Waiting for Frost Pavilion (Fig. 3.18d). Although Wen mentioned in the text that many orange trees surrounded this pavilion, he only suggestively painted seven of them, with oranges on branches. The owner Wang Xianchen, dressed in a Daoist robe, was meditating inside the pavilion, attended by a servant boy nearby.

The two kinds of plums, the common plum (Chinese: *li*, Latin: *prunus domestica*) and flowering plum (Chinese: *mei*, Latin: *prunus mume*), which required less water, were planted on Mounds N9 and N10 (Fig. 3.21), farther away from the river. In leaf no. 19, *Precious Plum Slope* (Zhenli ban), Wen depicted common plums suggestively; even though there were many, Wen only drew several, clustered on the slope of Mound N10.[98] For the flowering plum, Wen first represented in large quantity and then focused on details of tree branches. In leaf no. 29, *Jade Orchard* (*Yao pu*, Fig. 3.18c), Wen depicted the plum tree flowers in blossoms resembling clouds, praising their visual resemblance to the other world. In the subsequent leaf no. 30, *Fine Fruit Pavilion* (Jiashi ting, Fig. 3.19d), two gnarled plum trees extended branches to canopy a pavilion on the terrace. A man was climbing along the path to reach the plums.

I prefer to consider these fruitful trees were planted by Wang Xianchen. This is because the verb frequently employed by Wen in the scenery notes was 'plant (*zhi*).'[99] Wang also emphasised that one of his tasks was 'planting for daily supply.'[100] Despite Wang being the one to plant fruit trees, the monks' construction of the mounds provided fruit trees with a good topographical foundation. The plantation map in Figure 3.21 illustrates that Wang planted fruit trees on the upper slopes of mounds. We could further deduce that the river previously dredged by monks provided steady irrigation for Wang's plantation. Given that most temples were self-sufficient agricultural estates during the Yuan dynasty, it is also worth considering what crops the monks planted when they laid out the hilly landform. Is it feasible that some fruit trees have been planted by the monks? What were previously planted on the slopes now covered by Wang's fruit trees?

The third type of vegetation were willows and bamboo, being planted on the embankment of valleys between mounds N1, N2, and N3, and the gully between mounds N7 and N8. They were the major characters in leaves no. 8 *Purifying Will Place*, no. 9, *Willow Bend*, and no. 23, *Xiang Bamboo Valley* (Fig. 3.12a).[101] I propose that these plants were originally planted by monks when they constructed the hydraulic infrastructure, as willows and bamboo are typically used to stabilise the embankment after dredging and piling up.

The fourth kind of vegetation included various flowers, such as rosebush, rose, hibiscus, and lotus. Given their high water demand, they were typically planted along the shore where water flows less rapidly, or around the small pond at the rear of Hall Like a Villa (Fig. 3.21). Considering that these plants require annual replanting, it can be inferred their plantation is part of Wang's work. Wen also used the term 'planting (*zhi*)' when referring to these flowers. For instance, he noted: 'In front of Hall like a Villa, a variety of flowers, such as peony, Chinese peony (Paeonia lactiflora), osmanthus, Chinese flowering crab apple, Bauhinia were planted in a mixed fashion.'[102]

After distinguishing monks' site works from those carried out by Wang Xianchen, the model of the Great Propagation Temple is presented in Figure 3.22. This deduced model argues that the temple already featured a beautiful landscape which fits into the acclaimed garden feature of islands, circulating waters, old trees, and dispersed artefacts. Its topography resembles Shizi Lin Temple in the Yuan facilitating a similar hydrological process.

From the analysis provided, it could be seen that the primary scenic features of Unsuccessful Politician's Garden—such as mounds, rivers, evergreen trees, bamboos, and willows—which Wen leveraged to construct the thirty-one scenes, were in fact established by monks of Great Propagation Temple in the Yuan dynasty. In other words, Wen created a literati cultural narrative that originated from sceneries crafted by monks. Although site documentation from the monks is not extent, and they might have been less active in representing, their efforts in laying out the hydraulic landscape, preserving evergreen trees, and planting willows and bamboos should be acknowledged.[103]

While the gentry erased Buddhist symbols when converting temples, they inherited the most valuable parts of a Buddhist estate: the hydraulic infrastructure, lush vegetation, solid architectural foundations, and a topography conducive to agricultural production and scenic beauty. Based on this landscape, the gentry constructed artefacts, planted productive species, and conceptualised the literati garden. The garden culture of the Ming dynasty emerged while the gentry distinguished their literati culture from Buddhism, as they occupied temples and removed former monastic traces. The secluded literati garden of the Ming dynasty came into being amidst the active and violent estate conversion.

Evolution of the Hydraulic Infrastructure

The landscape of the estate, originally shaped by monks during the Yuan dynasty, experienced two major hydraulic transformations afterwards. In the Late Ming and Early Qing period, the site was partitioned, with Wang Xinyi acquiring the eastern part for his estate and Wang Yongning the western and middle sections. Both owners undertook extensive dredging efforts, creating new waterbodies and mounds, in response to the challenges posed by diminishing urban canals. In the Late Qing period, the western and middle parts were further divided into two estates. These new owners extended the waterways southward and introduced various artifacts and courtyards, enhancing the visual and experiential qualities of their respective garden sections.

Scholars widely concur that from the Early Ming to Late Ming period, the focus of garden culture gravitated from 'the productive landscape of moral good' towards an aesthetic landscape for strolling and displaying luxurious rocks and artefacts.[104] The split between productive and aesthetic is evident in Wang Xinyi's garden. Wang purchased the eastern part of the site of Unsuccessful Politician's Garden and renovated it into a garden named 'Returning to Farm Dwelling.' The 1966's estate map showed that Wang Xinyi's Garden attributed the eastern part for productive yields, including sorghum fields and lotus pond, and the western part as aesthetic, featuring Lake Tai rocks and artefacts for strolling and appreciation.[105] Wang Xinyi detailed his construction process as follows:

> Where the land allows for a pond, create a pond. Excavate soil from the pond, pile it up to form a mound. Where a mountain can be made,

make a mountain. Above the ponds and between the mountains, where a house can be built, build a house.[106]

While Wang Xinyi crafted the landscape according to circumstances, the resulting landscape was highly decorative and artificial. Wang dredged a meandering waterway that wrapped around the base of the existing mound. Lake Tai rocks and artefacts were strategically positioned at the bends of the river and along paths. These focal points served to guide visitors as they meandered through the garden. Contrary to Wang Xianchen's garden, where the main attractions were topographic, Wang Xinyi's garden featured abundant use of Lake Tai rocks, a vogue in the late Ming.[107]

Dredging waterways and constructing mounds were not only labour-intensive but also financially demanding. Acquiring Lake Tai rocks was also transacted with high prices in the late Ming. Wang Xinyi invested four years and a significant sum of money in these endeavours. His descendants diligently maintained the estate, preserving its beauty and integrity. Sixty years after its creation, Wang's grandson commissioned Liu Yu (fl. 18th c.) for a painting of the garden's western part.[108] Liu employed two techniques to present garden's exquisite rocks and architecture. The first was the archaic blue-and-green landscape style, which effectively rendered Lake Tai rocks as majestic mountains meanwhile subtly referenced Tao Yuanming's (365–427) idyllic Peach Blossom Garden (*Taohua yuan*).[109] Secondly, he applied the meticulous ruled-line painting (*jie hua*) technique, ideal for presenting dedicated architectural details.

Unfolding this handscroll, one is virtually transported on a stroll through the garden from east to west. The visual journey begins with Rinsing Water Pavilion (*Shushi ting*), seemingly afloat above the pond. Behind this pavilion, rocks amalgamated resembling a wall, aptly named 'Allying Walls Peak (*Lianbi feng*).' Inside this rockery complex there was a grotto named 'Little Peach Origin (*Xiao taoyuan*)' one could enter.

As one progresses through the handscroll, the primary building, Orchid Snow Hall (*Lanxue tang*), comes into sight. The estate founder, Wang Xinyi, who passed away forty years prior to the painting's creation, was depicted as seated composedly in the hall, looking over his creation. In front of him lies Containing Blueness Pond (*Hanqing chi*), a water body he had meticulously dredged. Positioned slightly to the left, his gaze leads views towards the rockery named 'Linked Clouds Peak (*Zhuiyun feng*)' to the left of the hall, resembling clusters of clouds. These rocks conglomerate, with its posture leaning towards the left, guides the observer's eye to further explore the garden's western end, where water pavilions, bridges, a lake, and bamboo groves are revealed. The presence of Wang Xinyi in the painting links the current landscape with the founder's original vision.

Figure 3.22 Model of Great Propagation Temple in the Yuan. Author's modelling and drawing.

The monks remained in the cloister to the east corner of the site. In Liu Yu's painting, unlike Wen Zhengming's deliberate exclusion of Buddhist elements in the depiction of the Unsuccessful Politician's garden, the monks' cloister was depicted as the background of Wang Xinyi's garden. Since Wang Xinyi was not the one who expelled the monks, he displayed a welcoming attitude to Buddhism in his garden experience. He composed a poem reflecting on the subtle experience of living near the temple:

Beyond the plum trees, there are bamboos, neighbouring a monk's abode.

At dawn and dusk, the chant of Buddhist scriptures occasionally drifts from within the bamboo.[110]

Wang Yongning was the third person who undertook a significant renovation of the estate's hydraulic infrastructure. As a relative of the influential King Wu Sangui (1612–1678), Wang Yongning led an opulent lifestyle in Suzhou, acquiring the estate from Chen Zhilin for a substantial sum. His passion as a wealthy collector and aficionado of Kun Opera transformed the site into a cultural hub for gatherings and theatrical performances.[111] With ample financial resources, Wang Yongning dredged a waterway behind mounds N4 and N5, configuring them into islands surrounded by a circular river.[112] Similarly fascinated with Lake Tai rocks, Wang Yongning adorned these islands with Lake Tai rocks. The transformation of the linear waterway into circular shape was likely a tactical move to counter the reduced connection of the inner waters with external canals, enhancing flow and aeration while preventing stagnation. These islands possibly served as stages for opera performances.

Wang Yongning also renovated the artefacts extensively. Records highlight his construction of residential halls crafted from Phoebe Zhennan (*nanmu*), a dense and esteemed hardwood traditionally used by the imperial.[113] The artefacts he renovated included Profound Lustre Loft (*Zhanhua lou*) at the northeast end of Mound N3 and Bright Snow Pavilion (*Yanxue ting*) on mound N4. The garden after Wang Yongning's renovation was depicted by Yun Shouping (1633–1690) in 1682, later imitated by Fang Shishu (1693–1751) (Fig. 3.23a).[114] The inscription on the hanging scroll reads:

In the autumn of the year of Renzi (1682), while visiting Wu Gate, I stayed at the old Unsuccessful Politician's Garden. At the end of summer and beginning of autumn, the air was clear and refreshing after the rain. Sitting alone in the Southern Veranda (*Nan xuan*, Fig. 3.23(1)), I looked across to the opposite bank (Fig. 3.23(N4)) where steep, piled rocks towered above a clear pond. The winding paths were encircled by tall trees of sophora, cypress, willow, and pine. Their branches soar distinctively above the forest. Surrounding the embankment were lotus flowers, their red and green hues intertwining. Looking down into the clear waters, I could see the fish swimming, evoking a sense of serene leisure reminiscent of Hao and Pu rivers. From Southern Veranda, passing through Bright Snow Pavilion (Fig. 3.23(3)) and crossing Red Bridge (*Hong qiao*, Fig. 3.23(4)) to the north, following the path along the horizontal ridge, at the foot of the mountain (Fig. 3.23(N3)) where the path ends, there is a dike leading to a small mound. The area is densely wooded, and upon the pond is Profound Lustre Loft (Fig. 3.23(5)), facing the opposite corridor (Fig. 3.23(7)) across the water. This is indeed the most exquisite spot in the garden.[115]

Chen Congzhou identified that Southern Veranda in the record corresponds to Leaning Jade Veranda (Fig. 3.13(21)), Red Bridge corresponds to the current fourfold bridge (Fig. 3.13(29)) that connects mounds N4 and N3, and the newly built Profound Lustre Loft corresponds to today's Seeing Mountain Tower (Fig. 3.13(10)).[116] These structures are marked in the hanging scroll in Figure 3.23a with a view simulated in Figure 3.23b.

The scroll's inscription and imagery indicate a shift in how the garden was viewed and toured. Mound N4 is central in the pictorial composition, with mound N7 foregrounded at the lower right corner and mound N3 backgrounded to the left, which vividly conveys Yun's view when he is seated in Southern Veranda. This south-to-north vista contrasts with Wen Zhengming's west-to-east tour, as demonstrated in the composition of Wen's 1528 hanging scroll (Fig. 3.15a). Yun's depiction implies a tour starting at Red Bridge and proceeding northward towards Profound Lustre Loft, resulting in a hanging scroll shorter than Wen's 1528 scroll. Additionally, it is noted that Hall Like a Villa, the two-story pavilion landmarked Wen Zhengming's scenes, disappeared in the late 17th century.

Yun Shouping's touring experience was picturesquely visual driven: seated in Southern Veranda, he first glimpsed the Bright Snow Pavilion, nestled among dense foliage on mound N4, and likely caught sight of Profound Lustre Loft in the distance, drawing him northward. Reaching the loft, he turned back to take in the scenes he had seen, identifying a vista between the loft and Southern Veranda. The hanging scroll also created a diagonal relation between Profound Lustre Loft and mound N7 to suggest the un-depicted Southern Veranda. Lu Andong posits that Wen Zhengming's 16th-century experience of touring the Unsuccessful Politician's garden resembled a mountainous journey, with 31 scenes visually independent of each other.[117] Contrastingly, Wang Yongming's 17th-century renovations crafted visual connections between places, plotting artefacts in a way encouraged exploratory movement. This suggests that the garden technique of 'opposite the view,' a concept critical to Chinese

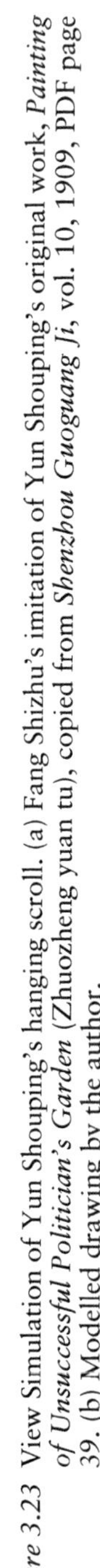
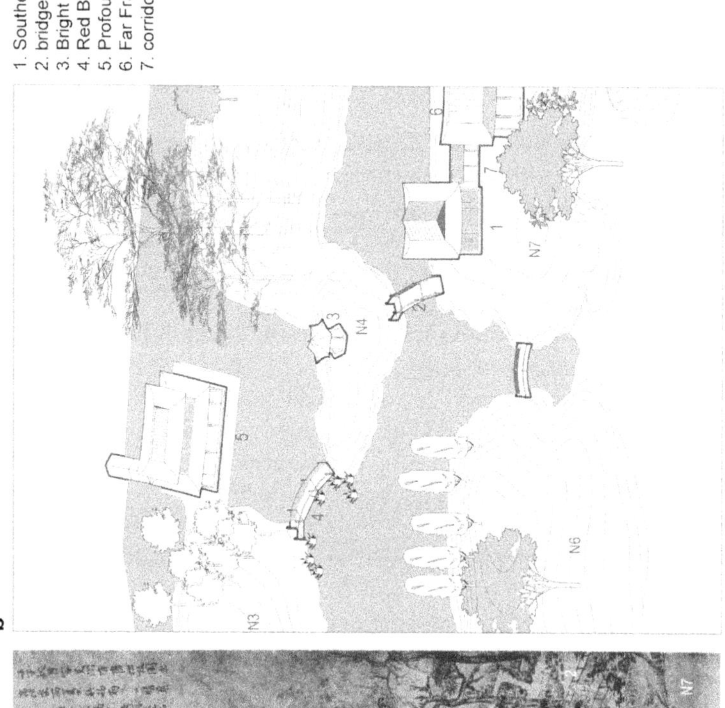

Figure 3.23 View Simulation of Yun Shouping's hanging scroll. (a) Fang Shizhu's imitation of Yun Shouping's original work, *Painting of Unsuccessful Politician's Garden* (Zhuozheng yuan tu), copied from *Shenzhou Guoguang Ji*, vol. 10, 1909, PDF page 39. (b) Modelled drawing by the author.

garden studies, emerged as early as the 17th century, as seen in Wang Yongning's garden.

During the early High Qing around the 1680s, Wang Yongning's estate was partitioned into western and middle estates (see Table 3.2). The western estate featured a north-south lengthy river. Wang family utilised several strategies to transform this liner river into a circular one. They excavated a branching river to encircle mound N2, configuring it into an island as the garden's topographical centre. Artefacts were tactically placed to oppose each other around mound N2, and a two-story tower was constructed at the north end, providing a panoramic view of the garden (Fig. 3.13). These spatial techniques developed by Wang Yongning witnessed their application by subsequent garden owners.

Meanwhile several courtyards were developed in the middle part, including Flowering Crab Apple Spring Dock (*Haitang chunwu*, Fig. 3.24(1)), Loquat Courtyard (*Pipa yuan*, Fig. 3.24(2)), and the water courtyard of Lesser Surging Wave (*Xiao canglang shuiyuan*, Fig. 3.24(3)). These courtyards subdivided spaces and created layers of scenic views, as captured in the painting *Guild of Eight-Banner Members from Fengtian and Metropolitan Provinces* in Figure 3.24. The painting also identified two corridors as boundaries separating the estate in the middle from the estates in the west and east.

The water courtyard of Lesser Surging Wave showcased the technique of creating layered scenes. A slender river branching southwards from the central pond was dredged in the late Qing. Viewing from Lotus Breeze at Four Sides Pavilion (*Hefeng simian ting*, Fig. 3.24(4)), the scene unfolded in layers: the newly constructed boat pavilion Fragrance Islet (*Xiangzhou*) in the foreground (Fig. 3.24(5)), the covered veranda Lesser Flying Rainbow (*Xiao feihong*) as the middle layer (Fig. 3.24(6)), Pine Breeze Pavilion (*Songfeng ting*) and Lesser Surging Wave Water Pavilion (*Xiao Canglang shuige*) forming the background (Fig. 3.24(7, 8)). The end of the river was cleverly obscured behind the hall, creating an illusion of endlessness. Chen Congzhou praised this water courtyard for its 'gradation in depth' and ability to 'generate the illusory visual effect in a narrow space.' Scenic courtyards emerged as popular subjects in garden paintings during the period, as evidenced by Wu Jun's (fl. 19th c.) painting album of the garden, where nine out of twelve leaves depicted these semi-enclosed courtyard spaces (Fig. 3.25).

Temples' Influence on Gardens

After tracing the transformation of Unsuccessful Politician's Garden from a renowned temple into a typical Chinese garden, different owners' contribution to the formation of the current site is elucidated. The mounds and

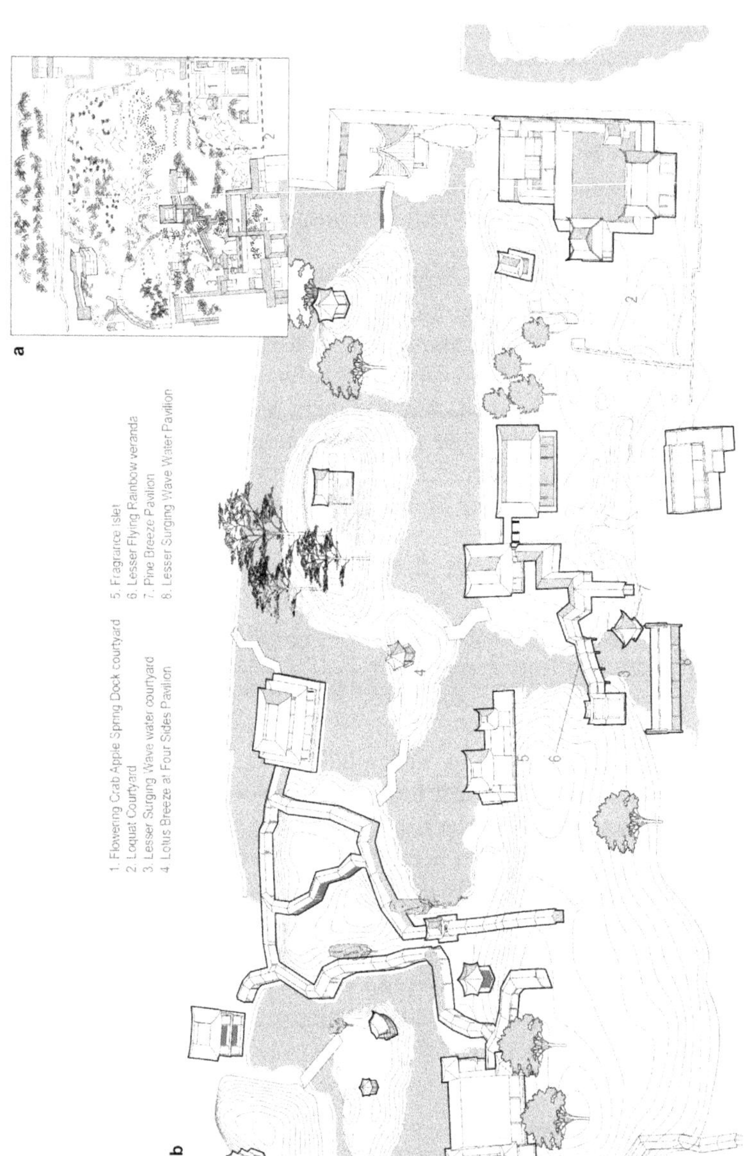

Figure 3.24 Island topography after Wang Yongning's renovation and the courtyards in the Late Qing. (a) Anonymous. *Painting of the Guild of Eight-Banner Members from Fengtian and Metropolitan Provinces*, ca.1870s. Author's line-traced drawing based on the painting from Liu, *Classical Gardens of Suzhou*, 302. (b) View Simulation. Author's modelling and drawing.

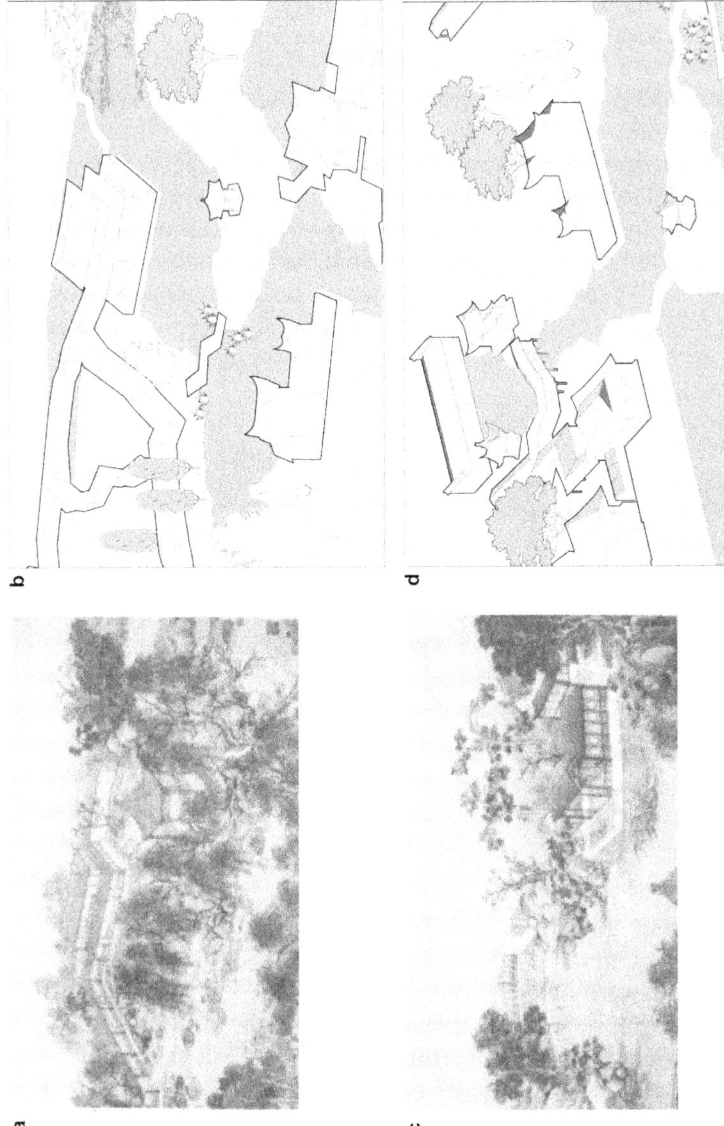

Figure 3.25 Wu Jun's painting of Unsuccessful Politician's Garden. (a) (c) Wu Jun, *Painting of Unsuccessful Politician's Garden* (Zhuozheng yuan tu), ca.1870s. Copy from Dunzhen Liu, *Classical Gardens of Suzhou*, 301. (b) (d) Author's modelling and drawing.

waterbody of the current site were initially laid out by monks of Great Propagation Temple. Then Wang Yongning dredged the waterway behind the three mounds in the Early Qing, creating a circular river around them and forming the layout of three islets above water. Later estate owners further dredged rivers southwards from the primary waterbody. The hydro-topography, veteran trees, and architectural foundations established by monks in the Yuan endured throughout centuries and evolved to be garden features lauded by modern scholars as 'topographical interplay' and 'natural ancientness.'

The garden techniques canonised in modern scholarship, such as 'opposite the view' and 'borrow the view,' emerged in the Late Ming and evolved in the Qing. To promote the circular touring experience, owners plotted interlocking vistas between scenes. The spatial manipulations, such as 'layers of scenes' and 'depth of scenes,' developed in the late Qing as owners configured courtyards within the garden to enrich the viewing experience. Corridors were then introduced as a spatial enhancement to the boundaries after estate divisions.

The hydraulic landscaping by monks set the foundational hydro-topography for Wang Xianchen's garden during the Ming dynasty. This hydraulic infrastructure consistently influenced the craft of scenes in the Late Ming and Qing dynasties. It is ironic that the literati gardens, so appreciated by modern scholarship and deemed as irrelevant to Buddhism, were deeply rooted in a monastic origin. This non-representation of monastic garden practices originated in Wang Xianchen's forceful acquisition of the temple, as he scraped off the golden plating from Buddhist sculptures and expelled the Buddhists.

Formation of Literati Garden-scape

Gazetteers and local records, usually compiled by officials and the gentry, tended to downplay the less commendable aspects of gentry history. The perspectives and voices of Buddhists are often less represented or even omitted, leading to a partial and skewed portrayal of historical events. The dominated gentry's documentation leads to the availability of only brief and sanitised accounts of temple conversions. For example, *Gazetteer of Changzhou*, compiled in the late Ming period, barely touched upon the conversion of Shizi Lin temple, summarising the event in a mere sentence: 'During the reign of Jiajing, a powerful family occupied Shizi Lin and made it into their garden.'[118] The gazetteer gave a similar cursory mention in the conversion of Great Propagation Temple by Wang Xianchen: 'In the reign of Jiajing, Imperial Censor Wang Xianchen constructed his residence on the abandoned land of Great Propagation Temple.'[119]

The third documented temple conversion was Illuminous Celebration Temple, an imperially patronised temple prominent in the Yuan dynasty as discussed in Chapter 2. *Gazetteer of Suzhou Prefecture* briefly records: 'At the beginning of Hongwu's reign, it was incorporated into Northern Chan Temple. Later it decayed and became a residence.'[120] *Gazetteer of Wu County*, in a similar concise way, identified that the temple became the garden of Wang family.[121]

The fourth recorded conversion was Return to the Origin Chan Temple into Xu Taishi's (1540–1598) Western Garden (*Xi yuan*). As examined in the last chapter, the temple was initially constructed by Cao Ruli in the late 13th century. Its conversion was recorded by the local gazetteer as:

Jiechuang Lü Yuan Temple was located in the ninth *du* of Zhifang Bang. It was originally the Western Garden of Xu Taishi, the Minister of Imperial Stud during the Ming dynasty. His son, Xu Rong of the Ministry of Works, later donated it to be transformed into Return to the Origin Temple.

According to previous records, Return to the Origin Temple was located the east of Changdang area. During Zhiyuan's reign (1264–1294), local Cao family constructed it for monks in Tiger Hill … it was later renovated into Xu Rong's Western Garden, before being donated again to serve as a temple. Therefore, it retained its old name but was prefixed with the words 'Fugu' (Restoration of the Antiquity).[122]

These succinct records indeed hint at the gentry's massive occupation of temples during the Ming. Huang Xingzeng's (1490–1540) petition to officials reveals this dramatic transformation:

Though the prosperity and magnificence of gentry are bequeathed by hundreds of lands, the gentry show no gratitude. The temples become deserted, and Buddhists live a meagre life. If they could get one inch of land to settle upon, the Buddhists could still manage to live. The innocent flush [monastic] lands were converted into areas for the wealthy and noble's touring and entertainment.'[123]

Huang's phrase 'The areas for the wealthy and noble's touring and entertainment' very much refers to the literati gardens of the Late Ming era, primarily serving as spaces for touring, social gatherings, and aesthetic enjoyment. Huang Xingzeng highlighted that these prosperous gardens were converted from monastic lands covered with plantations.

Peng Nian (1505–1566) vividly depicted the wealthy and noble's encroachment of monastic lands during the High Ming period:

Recently, spiritual and grand temples, which declined due to burdensome taxation and corvee labour, are snatched by the wealthy usurpers. Beautiful Jade Palaces [of these temples] were converted into trade areas. Precious timber materials [of these temples] were broken into firewood. There are even those who have gone excessive to scrape off the golds plated on the full-moon faces of the Buddha statues, dug out treasures stored in the Buddha statues (Buddha's indestructible bodies).[124]

The people referred to by Peng Nian as 'the powerful usurpers (*hao qian*)' not only seized control of temples but also ruthlessly exploited their valuable materials. This portrayal of the greedy invaders by Peng Nian accords with Qian Xiyan's account of Wang Xianchen's takeover of the Great Propagation Temple, illustrating the widespread and unscrupulous exploitation of religious sites during this era.

Fan Jinmin's research indicates that during the reign of Jiajing, gentry families emerged as the largest landholders by incorporating lands from Buddhists and farmers.[125] Other research shows that the gentry and local government owned 50–90 percentage of a country's land.[126] For instance, the Xu family, who incorporated Return to the Origin Temple as their Western Garden, possessed extensive estates during the High Ming, which was inventoried by Gu Zhentao:

Xu lüxiang's (ca. 16th c.) estates include Jiangxi Guild, Taojia Pond, Flower Port, Shifang estate, Liufang estate, Peach Blossom Mound along Xiatang River near Chang gate. Xu was the wealthiest in the region of Jiangsu (Three *Wu*s). Changchuan Bang River is where he anchored his boats of collecting debts. He has three grave lands. One is in Clouds Mountain has thousands of *mu*. The second is in Eastern Dragon Pond, of three hundred *mu*. The third is in Yaofeng Mountain, of ten hundred *mu*. These are the largest graveyards in Suzhou.[127]

Within these estates, Gu Zhentao specified Eastern Garden (*Dong yuan*), the present-day Lingering Garden (*Liu yuan*), and Western Garden, now known as Western Garden Temple, as prominent garden estates. It was when the gentry became the largest landholders in the High Ming period of the 16th century, their garden estates flourished and were recognised by the local gazetteer as a distinct estate category. Drawing upon Wei Jiazan's inventory and Mei Jing's mapping of these recorded garden estates, I developed a GIS map of the distribution of these gentry garden estates in Figure 0.2.[128]

Contrary to the significant transfer of monastic lands to gentry families as observed by gentry scholars such as Huang Xingzeng and Peng Nian, detailed documentation on the gentry's conversion of temples is notably lacking. This scarcity, when considered alongside Derrida and Foucault's notion of archives as hegemonic ways of recording, points to the gentry's deliberate omission of conversion records. Such omission further suggests the inappropriate or controversial nature of converting monastic lands.[129] Nevertheless, the elitist archives can still be read subversively.[130] I have endeavoured in this chapter to use 3D modelling to track the landscape transformation of Shizi Lin Temple and Great Propagation Temple to exemplify how this subversive reading is feasible. The two conversions serve as in-depth references to reflect upon other unrecorded transformations of temples and their potential contributions to the garden-scape in the Late Ming period.

Notes

1 '師字如海, 高昌人. 有禪學, 又能喜文詞, 所謂地因人而勝者也.' in Wang Yi, 'Shizi Lin Shisi yong 師子林十四詠 (Forty Odes on Shizi Lin),' in *JSJ*, juan xia, 11. Yuan patrons of Shizi Lin seemed to disappear during the Early Ming, possibly because lots of Uyghur migrants returned to the north.

  In the Early Ming, some local Suzhou residents were forced to migrate to other regions under Zhu Yuanzhang's decree. For more on Zhu Yuanzhang's migration policy in Suzhou, see Chen, 'The Scattering of the 'Red Flies' and the Northern Jiangsu Migrants'; Wang et al., *Outlined History of Suzhou*, 189–192. Also see Marmé, *Suzhou: Where the Goods of All the Provinces Converge*, 75–76.

2 These intellectuals represented a generation of Suzhou's artistic and literature talent. The poems of Shizi Lin they created are considered supreme works in garden literature, as noted by Chang, 'Literature of the Early Ming to Mid-Ming (1375–1572),' 3–5.

3 These literati were part of the renowned group 'Ten Friends of the North Wall (Beiguo shiyou).' For more of their gardens and gatherings, see Sensabaugh, 'Life at Jade Mountain.'

4 See Chang, 'Literature of the Early Ming to Mid-Ming (1375–1572).' Gao Qi and Xu Ben migrated to Suzhou in 1370, and Zhang Shi returned to Suzhou in 1371. Other sites for their gatherings in Suzhou included Zhang Shi's estate, Pleasure Patch 樂圃 (*Le pu*).

5 Gao Qi, 'Preface to the Twelve Odes of Shizi Lin.'

6 The removal of the mirror installation in Chan Hut is evident from comparing Yuan and Early Ming depictions. During the Yuan, descriptions primarily used an outsider's perspective, noting the mirrors but lacking reports of interior experiences. Tianru's writings indicate he practiced Chan Buddhism there, suggesting the hut was used for his personal practice rather than for visitors in the late Yuan. In contrast, Ming visitors documented interior experiences, describing the interior as containing only a couch bed and walls for meditation, without reference to the mirrors.

7 Wang Yi, 'Record of Touring Shizi Lin.'

8 Ibid.

9 '聞十餘年前, 獅子林尚在. 而所謂十二景者, 亦半可指數. 今已轉授民家. 陸氏縱織作畜牧其中. 而佛像峰石, 老梅奇樹之類, 無一存者. I've heard that a decade ago, Shizi Lin Temple still existed. Half of its twelve scenes could still be recog-

nised. Now, it has been sold to the Lu family. They used the land for weaving and raising livestock. The Buddha images, rocks, old plums, and trees were no longer exist.' In Wang Shizhen, 'Writings on Wen Zhengzhong complementing Tianru's scroll of Lion Grove 書文徵仲補天如獅子林卷 (Shu Wen Zhengzhong Bu Tianru Shizi Lin Juan),' in *Continued Manuscripts of Yanzhou*, 16–17. Another stele record from Jiang Yingke 江盈科 (1553–1605) also noted that during the High Ming, all the scenes created by Tianru were non-recognisable: '自天如涅槃, 弟子散去. 庵之水石花竹, 日就荒蕪. 至國朝并庵亦废. 昔所稱含暉, 吐月, 立雪, 昂霄, 栖鳳亭, 小飛虹, 指柏, 問梅諸境, 一切淪沒于荒烟野草, 殘霞落照間. 久之, 折入豪门, 構为市居. 傭保雜作, 錯處其上. 如是者數十年, 而獅林之额幾不可識.' Jiang Yingke, 'Record of Imperial Endowment to Reconstruct Lion Grove Imperial Benevolence Temple 敕賜重建獅子林聖恩寺記 (Chi Ci Chongjian Shizi Lin Sheng En Si Ji),' in *Anthology of Jiang Yingke*, vol. 1, 382–384.

10 Many garden estates, such as Wu Kuan's Eastern Estate in the Early Ming and Wang Xianchen's Unsuccessful Politician's Garden in the High Ming were productive agricultural sites. See Clunas, *Fruitful Sites*; Guo, *The Garden History of Suzhou in the Ming Dynasty*. For the economy of Suzhou during the Early and Mid-Ming, see Marmé, *Suzhou: Where the Goods of All the Provinces Converge*, 40–127.

11 The collectors' seals, inscriptions, and records of these paintings allow a trace of their changing owners and locations, which Clunas referred to as 'the biography of an object.' See Clunas, *Art in China*, 139. This concept 'the biography of things' is borrowed from Kopytoff, 'The Cultural Biography of Things: Commoditization as Process.'

12 See Fu, 'Ming Dynasty Suzhou Paintings and the Literati's Monastery Tours.' Du Qiong's inscription wrote: '徐幼文嘗為如海師作師林圖. 几十二段, 各係以五言詩, 詞意簡捷. 後有少師榮公跋. 按是詩, 當是榮公手蹟也. 余於月舟上人山寮見之, 屬擬其意, 回撮其大概作小幀, 並書原詠詩. 成化四年仲春鹿冠道士杜瓊.' Also see the painting in Jiang, *A Panorama of Paintings in the Collection of The National Palace Museum*, vol. VII, 26.

13 Xu, *Study on the Authentication of Ancient Books and Paintings*, vol. 3, 167–172, and 197–200.

14 Gu Kai noted that in the Late Ming, the image of a pavilion standing at the top of an artificial hill became popular, as witnessed in Yanshan Garden 弇山園, Zhi Garden 止園, and Eastern Garden 東園. See Gu, 'Rethinking on Rockery-Top Pavilion in Traditional Chinese Gardens.' For the paintings of Zhi Garden and Eastern Garden, see Cahill, Huang, and Liu, *Imperishable Groves and Spring*, 13–58, 82–88.

15 'Mingxing donated his personal savings and solicited every possible patronage to get the construction funding. The Buddha hall, temple gate, and Sutra hall were newly constructed. Inside the Buddha hall there was a golden Buddha statue, solemn and compassionate. Sutras were revered inside the Buddha Hall.' in Jiang, 'Record of Imperial Endowment to Reconstruct Lion Grove Imperial Benevolence Temple.'

16 Li Mo 李模 (1594–1674), 'Record of the Reconstruction of the pavilion in Imperial endowed Ancient Lion Grove Imperial Benevolence Temple 敕賜聖恩古師林寺重建殿閣碑記 (Chi Ci Sheng'en Gushilin Si Chongjian Diange Beiji),' in *JSXJ*, juan shang, 1–3. Jiang Yingke's stele reveals that Mingxing stored the endowed sutras in Buddha hall since the construction of the sutra storage pavilion was not completed. See Jiang, 'Record of Imperial Endowment to Reconstruct Lion Grove Imperial Benevolence Temple.'

17 '迄崇禎十五年, 日新自會稽來. 閱藏三載. 矢願建閣貯經. 會遭兵燹遂寢. 戊子, 復來吳門, 銳意經始. 竭蹶星霜, 鳩工庀材, 悉本心匠. 方五六年, 經閣既成, 大殿並峙, 翬飛霞起, 堅好殊特. 遠近來觀, 詫為神斤鬼斧.' in Li Mo, 'Record of the Reconstruction of the pavilion in Imperial endowed Ancient Lion Grove Imperial Benevolence Temple.'

18 Ibid.

19 Xu Fang, 'Inscription on the pagoda of Master Zhi, the abbot of Guangyun in the Lion Grove 獅林廣運大師智公塔銘 (Shilin Guangyu Dashi Zhi Gong Ta Ming),' in *Anthology of Yiju Hall*, vol. 15, 1–4.

20 Shen and Gu, *Gazetteer of Yuanhe County in Qianlong's reign*, vol. 2, 101.

21 Jiang Yingke wrote '既至, 而寺之故跡了不可覓. 不佞按舊志漸為稽復. Upon his arrival, the historical remains of the temple became non-recognizable. [He] recovered the site according to the old gazetteer.' in Jiang, 'Record of Imperial Endowment to Reconstruct Lion Grove Imperial Benevolence Temple.'

22 '寺雖廢, 中有林石陂池之勝. 而師一心求道, 盡屏外緣, 足不窺園, 目不交睫, 以披誦如來修多羅藏及毘尼藏阿毘曇藏. 始終三年, 無不週遍. 淪肌砭骨, 以寢食周旋於古佛聖僧之中. Although the temple was in a desolated status, it featured groves, rocks, and ponds. The master [Fazhi], with his heart set on seeking the way (*dao*), completely shut out external encounters. He never walked to have a glimpse of the garden. He almost did not sleep a wink, read and chanted sutras of the Tathagata, the Vinaya, and the Abhidharma collections. Over three years, he had read through all the sutras. His skin shrivelled, and his bones ached. Even while sleeping and eating, he was as if amongst ancient Buddhas and sacred monks.' in Xu Fang, 'Inscription on the pagoda of Master Zhi, the abbot of Guangyun in the Lion Grove.'

23 Xu Fang was a gentry well-versed in gardening. See Fu, 'New Examination on Real Landscape and Xu Fang's Painting of Thatched Cottage on the Gully.' Pan and Zhao, 'Study of Xu Fang's Jian Shang Garden in Early Qing Dynasty.'

24 '既而嘆曰, 世風日替, 象教晚秋. He sighed that the public morals declined day by day, the teaching of images (Buddhism) declined.' In Xu, 'Inscription on the pagoda of Master Zhi, the abbot of Guangyun in the Lion Grove 獅林廣運大師智公塔銘 (Shilin Guangyu Dashi Zhi Gong Ta Ming).'

25 '非假見聞不足以起正信, 非耀耳目不足以攝身心.' In ibid.

26 '吾既精心於藏海, 不當構傑閣以奉之乎. 乃兵燹充斥之後, 瘡痍未復之秋而師鳩工經始, 毅然不回.' In ibid.

27 See Han Qi 韓琪 (fl. 18th c.)'s poems in *JSJXJ*, juan zhong, 5.

28 Zhang Guangwei's research shows that Suzhou's urban canals dramatically declined during the Late Ming to Early Qing. See Zhang, 'The Documentation of Historic Maps of World Heritage Site City Suzhou.'

29 See Zhang, *The Atlas of Ancient Suzhou*.

30 For example, in the collaborative work *Linked Verses of Shizi lin*, Zhu Yizun wrote: '北郭徐賁, 繪圖以畕.' 1–4. In guest Mao Jinfeng 毛今鳳's poem made upon one of the social gatherings in the garden in 1702, he wrote: '徐考十二景, 記載入稗宮. 質疑叩鐘鼓, 蕾昧釋疑團.' Another guest Li Fu 李紱 (1675–1750) wrote: '安得徐賁手, 衣冠填青丹. 點入畫圖中, 千秋想遺顏.' Xu Ben's painting album was also mentioned and discussed by guest Qian Chenqun 錢陳群 (1686–1774): '遊師子林: 分圖十二段, 不隔尺與咫. 晨起汲井華, 手添養魚水 … 十二圖傳面面峰, 粉牆橫列碧芙蓉.' In *JXJSJ*, juan zhong.

31 Most of the artefacts in Gao Jin's painting were constructed by Zhang Shijun in the Early Qing.

32 Clunas, *Fruitful Sites*, 157.

33 '師林八景: 師子峰, 吐月峰, 小飛虹, 玉鑑池, 冰壺井, 問梅閣, 五松, 八洞.' in Cao Kai, 'Eight Views of Shilin 師林八景' in *JSJXJ*, juan zhong, 9–10.

34 See '師林八景, 五松: 高峰有五松, 羅列互爲友. 根蟠自宋元, 鐵幹終不朽. 餐之可長年, 何必還丹壽.' in ibid.

35 '師林八景, 八洞: 岡巒互經亘, 中有八洞天. 嵌空勢參錯, 洞洞相回旋. 遊人迷出入, 渾疑武陵仙.' In Cao Kai, 'Eight Views of Shilin 師林八景' in *JSJXJ*, juan zhong, 10.

36 *SZLZ*, 202–206.

37 See verses such as '有峰有岫, 有碉有湫.' in *JSJXJ*, juan zhong.

38 Many poems suggest that these gentrymen knew Ni Zan's painting of Shizi Lin, though they had not seen it.

39 The Lake Tai rocks and plants arranged along Zhang Shijun's garden's southern wall could be seen in Figure 1.5b, 'Shizi Lin' in *Grand Ceremony of the Southern Inspections*.

40 '高亭擅一邱, 怪石擁四面. 坐疑夏雲起, 顧覺秋山亂.' In Zhao Zhixin's poem in 1705 titled '獅子林贈主人張籥三 Dedicated to the Owner of Shizi Lin, Zhang Yusan (Zhang Shijun),' in *JSJXJ*, juan zhong, 5.

41 Ibid.

42 See 'Linked Verses of Shizi Lin 師子林聯句' in *JSJXJ*, juan zhong, 4–5.

43 The garden was first resided by Zhao Zhixin and subsequently by Wang Ying. See *SZLZ*.

44 Finnane has discussed that the garden constructions in Qing dynasty Yangzhou fostered merchants' social network. Finnane, *Speaking of Yangzhou*, 90–116.

45 Chen Zhong, '黃氏家族與獅子林 Huang Family and Lion Grove,' in *SZLZ*, 191–192.

46 The six visits were in years 1757, 1762 (2 visits), 1767, 1780, and 1784. See ibid., 7.

47 Zhao Yanzhe, 'Absorbing the Ancient, Embracing the Present,' 91–92. Before this painting was added to Qianlong's collection, the emperor had highly praised Ni Zan's works and amassed many. Sensabaugh noted that by the time the first part of 'Treasured Boxes of the Stone Moat (*Shiqu baoji*)' was accomplished, Qianlong already owned forty-one Ni Zan paintings. See Sensabaugh, 'The Lion Grove in Space and Time.'

48 Qianlong's first poem was: '借問獅子林, 應在無何有. 西天與震旦, 不異反覆手. 倪子具善根, 宿習摩竭受. 蒼蒼圖樹石, 了了離塵垢. 聲徹大千界, 如是獅子吼. 御題.'

49 See Zhao, 'Absorbing the Ancient, Embracing the Present,' 96.

50 See Qianlong's poem '假山似真山, 仙凡異尺咫. The artificial hill resembles a real mountain, the realms of immortals and mortals are seperated by mere inches.' in *JSJXJ*.

51 See '緬五百年前, 良朋此萃止. 澆花供佛缽, 淪茗談元髓.' in *JSJXJ*.

52 Qian Weicheng accompanied the emperor on his first visit.

53 Ni Zan's paintings were stored in Yangxin Dian 養心殿 in the Purple Palace. see Chiang, *Emperor Qianlong's Hidden Treasures*, 65.

54 The first poem made in the spring was transcribed in the centre of Ni Zan's painting, with another poem inscribed to its right.

55 Wu Hung argues that Qianlong's act of inscribing paintings depicting Jiangnan gardens could be read symbolically as a Manchu ruler's expression of his possession of Han culture. See Wu, *The Double Screen*, 200–236.

56 Although Zhang Shijun's friends in the Early Qing also recalled the history of Shizi Lin in their poems, they did not emphasise the past as predominant as Qianlong did.

57 The stele was inscribed with the poem made by Qianlong upon his first visit to Lion Grove Garden in the late spring of 1757. It reads: '早知獅子林, 傳自倪高士. 疑其藏幽谷, 而宛若鬧市. 肯構惜無人, 久屬他氏矣. 手跡藏石渠, 不亡賴有此. 詎可失目前, 大吏稱未飾. 未飾乃本然, 益當尋屐齒. 假山似真山, 仙凡異尺咫.' see *SZLZ*, 98.

58 The stele is still at the current site of Lion Grove Garden. It was moved into a pavilion attached to the southern wall. See *SZLZ*, 8–9, 139. The stele served as a substitute for the emperor. This finding is revealed by comparing the paintings of Shizi Lin with the paintings of Qianlong's imperial gardens which replicated Shizi Lin.

59 The original paper model and measured drawing of Lion Grove Garden are not extant. For paper model's utilisation in imperial projects in the Qing dynasty, see Valleriani and Schäfer, eds., 'Knowledge by Design.' See the measured drawing of Everlasting Spring Garden 長春園 in Liu, 'From Imperial Realistic Painting to Yangshi Lei Pattern.'

60 For researches on these two gardens, see Wang, 'The Garden-Making Art of Wen Garden'; Wang, 'Emperor Qianlong and the Lion Grove Garden'; Liu, 'Research on "Qianlong Ancient Costume Snow Scene Pleasure Trip Picture"'; Jia, 'Further Exploration into Shizilin Garden in Changchunyuan and the Picture of Emperor Qianlong in Snowscape'; Siu, *Gardens of a Chinese Emperor*, 51–106.

61 Typical works include Fang Cong 方琮's imitation of Ni Zan's painting, which was commissioned by Qianlong. Qianlong titled Fang's painting as 'The Dharma Escapes Its Essence 法逸其趣.' See Zhao, 'Artistic Conformity to the Secluded Path.'

62 Denis Cosgrove and Stephen Daniels advanced previous art theories on iconography by developing the concept of landscape as iconography. See Cosgrove and Daniels, eds., *The Iconography of Landscape*, 1–10.

63 Lion Grove Garden as an imperial site is also discussed in Li, 'Making the Qianlong Emperor's Private Garden.'

64 Guo Zhongheng, *Famous Places of Suzhou, with Pictures and Odes.*

65 See Fu, *Six Records of a Floating Life.*

66 In *SZLZ*, 144.

67 For the shrinkage of canals in Suzhou, see Qu, *Gazetteer of Rivers and Canals of Suzhou.*

68 The three orders of simulacra and simulation, see Baudrillard, *Simulacra and Simulation.*

69 '城東北齊門內大弘寺. 宋延祐中賜額, 即古慶壽寺也. 與王御史憲臣第宅隣近. 御史躭情丘壑, 與李長沙文待詔諸公交善, 而平生不信內典, 因拆毀此寺, 以廣園囿. 命惡少挽仆佛菩薩天生諸像于地, 用刀刮其面金. 左右疆諫不從. 須臾之間, 梵軸縱橫, 僧徒奔竄, 蘚碑剝落, 蓮社荒涼 … 所拓之園名拙政. 喬木千章, 皆寺中故物也. 為吳下之甲焉.' Qian Xiyan, 'Imperial Censor Wang Destroyed the Temple 王御史毀寺 (Wang Yushi hui si),' in *WDFC*, 883.

70 Gu Kai considered Wang designed the garden and Wen Zhengming was engaged in the process in *Gardens of the Jiangnan Area in the Ming Period*, 73–77. Guo Youming has noted the garden was previously Great Propagation Temple. He suggests that Wang Xianchen might be controversial among the network of Suzhou elites. See Guo, *The Garden History of Suzhou in the Ming Dynasty*, 154–164.

71 Liu, *Classical Gardens of Suzhou.*

72 Huang, 'A Building as a Biography,' 53–54.

73 Suzhou Shi Zhuozhengyuan Guanlichu, *Frozen Symphony, the Architecture of Unsuccessful Politician's Garden in Suzhou.*
74 '惟談園林之蒼古者, 咸推拙政. 今雖狐鼠穿屋, 蘚苔蔽路, 而山池天然, 丹青淡剝, 反覺逸趣橫生.' in Tong Jun, *Gazetteer of Jiangnan Gardens*, 28–29. The ancientness of Unsuccessful Politician's Garden was initially commented by Yuan Hongdao: '拙政園在齊門內, 余未及觀, 陶周望甚稱之. 喬木茂林, 澄川翠干, 周回里許. 方諸名園, 最為古矣. Unsuccessful Politician's Garden is within the Qi Gate, which I have not yet had the opportunity to visit. Tao Zhouwang has spoken highly of it, praising its towering trees and dense groves, tranquil streams, and verdant woods. The garden spans over a li in circumference and is the most ancient among all the renowned gardens.' In Yuan, 'A Brief Account of Gardens and Pavilions.'
75 Keswick, *The Chinese Garden*, 169–170.
76 Liu, *Classical Gardens of Suzhou*, 56–59; Chen, *On Gardens: Famous Chinese Gardens*, 224–233; Pan, *The Art of Crafting Scenes in Jiangnan*, 224–233.
77 Chen, *Chinese Gardens: Theory and Practice*, 121.
78 Chen, *On Gardens*, 1984.
79 Chen, *Chinese Gardens: Theory and Practice*, 144–146.
80 Shao and Li, eds., *Historical Records of Famous Gardens in Suzhou, Reconstruction Records of Suzhou Gardens*, 319–321. Xie Xiaosi also traced the site's multiple ownership since the Yuan dynasty. See ZZYZ, 243–246.
81 '且夫大弘之席, 昔云盛矣.' in Wang Xing, 'Inviting Lü monk Youzhang to abbot in Great Propagation Temple 請律有章住大弘寺致言 (Qing Lü Youzhang zhu Dahong Si zhiyan)' in *Supplementary Additions to the Banxuan Collection, Buddhism and Daoism*, 15–16.
82 Wang Xinyi recorded that when he owned the eastern part of the site, monks still resided in a small part of it. Wang Xinyi, 'Gui Tianyuan Ju Ji 歸田園居記 (Record of Return to Farming Dwelling),' in ZZYZ, 180–182.
83 Wen was about ten years younger than Wang Xianchen and got acquainted with Wang in his twenties. Clunas, *Fruitful Sites*, 23–24.
84 The format of Wen's 1533 album is discussed in Lu, 'Deciphering the Reclusive Landscape.'
85 Lu Andong's analysis further reveals that Wen employed both an external orientational system, using cardinal directions such as south, north, west, and east, and internal orientations, such as front, back, left, and right. See Lu, 'Deciphering the Reclusive Landscape.'
86 See the leaf in Kerby, *An Old Chinese Garden*, PDF page 25. The scanned pdf of the books is available at https://archive.org/details/oldchinesegarden 1922wenz.
87 Liu, *Classical Gardens of Suzhou*, 56; Gu, *Mingdai Jiangnan yuanlin yanjiu*, 73–77.
88 Wen, 'Record of Wang's Unsuccessful Politician's Garden 王氏拙政園記 (Wang shi Zhuozheng yuan ji),' in ZZYZG, 73–74.
89 For the painting 'Waiting for Frost Pavilion,' see Kerby, *An Old Chinese Garden*, PDF page 77.
90 Ibid.
91 See leaf no. 28, 'Bamboo Gully,' in Kerby, *An Old Chinese Garden*, PDF page 133.
92 '槐雨先生王君敬止, 所居在郡城東北, 界婁齊門之間. 居多隙地. 有積水亘其中. 稍加濬治, 環以林木. 爲重屋其陽, 曰夢隱樓. 爲堂其陰, 曰若墅堂.' In ibid.
93 '罷官归, 乃日课僮仆, 除秽植檞, 饭牛酤乳, 荷畚抱瓮. 业种艺以供朝夕, 竢伏腊. 积久而园始成. 其中室卢台榭, 草草苟完而已. 采古言即近事以为名. 献臣非往湖山，赴庆吊．虽寒著风雨，未尝一日去．屏气养拙几三十年．' In Liu, *Classical Gardens of Suzhou*, 56.

94 '疎松漱寒泉，山風滿清聽．空谷度飄雲，悠然落虛影．' in ZZYSBY, 32–33.

95 See leaf no. 24 in Kerby, *An Old Chinese Garden*, PDF page 117.

96 '聽松風處在夢隱樓北，地多長松．' and '槐喔：在槐雨亭西岸，古槐一株．蟠屈如翠蛟，陰覆數弓．' in ZZYSBY, 32–33, and 48–49. See leaf no. 24 in Kerby, *An Old Chinese Garden*, PDF pages 116–117.

97 See Clunas, *Fruitful Sites*, 15–56.

98 See leaf no.19, see Kerby, *An Old Chinese Garden*, PDF page 97.

99 For example, when referring to the common plum, Wen wrote: '在得真亭後，其地高阜，自燕移好李植其上．' in ZZYSBY, 38–39.

100 '业种业以供朝夕' in Liu, *Classical Gardens of Suzhou*, 56.

101 For leaves no. 8 and no. 9, see Kerby, *An Old Chinese Garden*, PDF pages 53 and 57, respectively.

102 '若墅堂之前，雜植牡丹，芍藥，丹桂，海棠，紫瑤諸花．' in *ZZYSBY*, 10–11.

103 Li Qi pointed out that in the Yuan, many monks, such as Tianru, did not place much emphasis on accumulating representation works of their temples: '昔之佛舍僧房，託名羣賢集中，以傳不朽者多矣．若師子林，則固無待於詩也．無待於詩而詩以美之者，當世之士大夫也．' Li Qi 李祁 (1200–1270), *JSJ*, 7–8.

104 See Clunas, *Fruitful Sites*, 58–100.

105 See the 1966's estate map of Returning to Farm Dwelling in ZZYZG, 171.

106 '地可池則池之．取土於池，積而成高．可山則山之．池之上，山之間，可屋則屋之．' in Wang Xinyi, 'Record of Returning to Farm Dwelling.'

107 See Pan, ed., *History of Ancient Chinese Architecture*, 398.

108 See Liu Yu's painting in Dong, *Selected Landscape Paintings of Suzhou Gardens*, 82–85.

109 See the discussion of blue-and-green painting style in Fong, *Beyond Representation*, 104–105.

110 '梅之外有竹．竹鄰僧舍．旦暮梵聲，時從竹中來．' in Wang, 'Record of Returning to Farm Dwelling.'

111 Wang Yongning's guests include famous Qing painters such as Wang Hui 王翬 (1632–1717), Wang Shimin 王時敏, Wang Yuanqi 王原祁. See Zhang and Bai, 'Early Qing Noble Collector Wang Yongning'; Qi, 'The Authenticity of Huang Zijiu's 'Autumn Mountain Painting.' Yu Huai 余懷 wrote two poems that vividly portrayed the Kun opera performance in Wang Yongning's Unsuccessful Politician's Garden, see ZZYZG, 116.

112 Wang Yongning's substantial renovation of the site's topography was recorded by Xu Qianxue as: '凡前次數人居之者，皆仍拙政之舊．自永寧始易．置邱壑，益以崇高雕鏤，蓋非復園記詩賦之云云矣． All those who previously resided here adhered to the old practices of Unsuccessful Politician's Garden. It was only with Yongning that changes were initiated. [Wang Yongning] crafted the hills and ravines, further enhanced by lofty buildings and elaborate carvings. It was no longer the garden what the previous records and poems described.' See Xu Qianxue 徐乾學 (1631–1694), 'Record of the New Bureau of the Circuit of Suzhou, Songjiang, and Changzhou 蘇鬆常道新署記 (Susong Changdao Xin Shu ji),' in ZZYZ, 182–183.

113 In ibid. Phoebe Zhennan material was conventionally used for imperial constructions and was also called 'imperial wood (*huangmu*),' as discussed by Campbell, *What the Emperor Built*, 53–54.

114 Yun Shouping's 1682 painting is not extant. Fang Shishu's 1732 work copied Yun Shouping's colophon verbatim except for changing the date to 1732. Considering that the garden was renamed Jiang Qi's 'Recovery Garden' in

1732, but Fang did not update the garden's name in his painting, it is reasonable to deduce that Fang's painting is a copy of Yun's original work. For the record of Yun Shouping's work, see *ZZYZG*, 145–146.

115 In Yun Shouping, *A Compendium of Fragrant Cup Studio*, 278. The recording of Yun's work *Hanging scroll of Unsuccessful Politician's Garden* 拙政園圖軸 was in Cai, *Yun Shouping*, 262. Translated by Ling Yuan in Chen, *Chinese Gardens: Theory and Practice*, 50. Note the corridor mentioned by Yun refers to the one connecting Southern Veranda and Far Fragrance Hall (Fig. 3.23(7)).

116 Chen, *Remaining Ink in the Carpenter's Studio*, 540–543.

117 Lu, 'Deciphering the Reclusive Landscape.'

118 '嘉靖間有勢家佔獅子林寺為之園.' in Li and Gu, *Gazetteer of Changzhou County*, vol. 18, 9.

119 Ibid., 8.

120 '昭慶寺在城東北隅 … 明洪武初歸併北禪寺, 後廢爲民居.' in Fu and Jueluo, *Gazetteer of Suzhou Prefecture*, vol. 32, 32.

121 '昭慶寺在城東北隅大儒巷 … 明洪武初歸併北禪寺, 後廢爲王氏園.' in Cao and Li, *Gazetteer of Wu County*, vol. 36.

122 '戒幢律院. 在九都冶坊浜. 舊為明太僕徐時泰西園. 子工部溶捨為復古歸原寺. 前志歸源寺, 在長蕩東. 元至元間, 里人曹氏為虎邱寺僧建. 其地與冶坊相近. 後改為徐溶西園, 復捨為寺. 故因其舊額, 而冠以復古二字.' In Li and Feng, *Gazetteer of Suzhou Prefecture*, vol. 42, 21–22.

123 Huang Xingzeng, 'Appeal to Magistrates Nie and Cai, Ministers Huang and Zhu to Stop Seizing Temples 與聶蔡二郡公黃朱二令止奪寺觀書 (Yu Nie Cai Er Jungong Huang Zhu Er Ling Zhi Duo Si Guan Shu),' in *Collection of the Five Sacred Mountains Hermit*, vol. 31, 12–14.

124 '近歲, 精藍古刹, 往往弊于賦役, 夺于豪僭. 溷琳宮为贾区, 摧琼木而薪櫋. 至有削金于满月之容, 凿寶于不壞之体者. 既而人随澌灭居成. 僧舍所得亦几何哉. 嗚呼.' in Peng Nian, 'Record of Reconstructing Auspicious Light Chan Temple 重修瑞光禪寺記 (Chongxiu Ruiguang Chansi Ji),' in *WDFC*, vol. 10, shang, 17–21.

125 Fan and Xia, *Social and Economical History of Suzhou Region*, 187–192.

126 See Dardess, *Confucianism and Autocracy*, 16–17; Chow, *The Rise of Confucian Ritualism in Late Imperial China*, 16–18.

127 Gu, *Wumen's Manifest and Hidden*, vol. 1, 10–11.

128 Wei, *History of Classical Gardens of Suzhou*; Mei, 'Study on the Changes in the Scale of Ming and Qing Dynasty Suzhou Gardens and Their Relationship with Urban Transformation.'

129 Zeitlyn, 'Anthropology in and of the Archives.'

130 Ibid.

4 Converting Temples into Confucian Institutions

During the 1520s to 1540s, anti-Buddhism swept Suzhou and lots of temples were repurposed into Confucian institutions by the government, including shrines, academies, and schools. Buddha halls were converted into Confucian shrines, and Buddha statues were replaced by Confucian ones. The monastic layout and landscape were redesigned to align with Confucian rituals and educational purpose. The architectural renovations reoriented the citizens' religious practices from worshipping Buddha to venerating Confucian patriarchs. The urban waterfront façade was also transformed with archways, steles, and artefacts advocating Confucianism.

This chapter examines the conversion of five temples. The first section traces Southern Chan Temple's evolution into the Confucian shrines and academies across centuries. The meaning of the Surging Wave shifted from representing the Buddhist philosophy of emptiness to embodying the Way of Confucianism. The second section focuses on the city's pivotal transitional period in the 1520s, when the government initiated anti-Buddhism campaigns, proactively converting four temples in order to orient the urban culture towards Confucianism. The final section delves into the peak of anti-Buddhism in the 1540s, exemplified by the conversion of Ten Thousand Longevity Chan Temple into Changzhou County School. The anti-Buddhism of the High Ming period dramatically transformed the temple-scape with far-reaching historical impact. By acquisition of Buddhist lands inside the city, the government gained urban control against the rising gentry's autonomy.

From Southern Chan Temple to Surging Wave Pavilion

In the 14th to 15th centuries, a stroll along the shore of Surging Wave (Fig. 4.1a) would reveal magnificent Buddha halls linking both banks, bustling with devoted monks and echoing with the chants of sutras.[1] The area was then Southern Chan Temple, an extensive abbey comprising three individual temples (Fig. 4.1b). However, by the 1800s, the scene had been transformed significantly. Visitors would encounter enshrined portraits

DOI: 10.4324/9781003387169-5

A. Garden of Suitability; B. the current site of Surging Wave Pavilion; C. Surging Wave pond. 1. Wash Tassel Spot; 2. Ladling Purity Hall (*Yiqing tang*); 3. Ladling Purity Pool. 4. Facing Water Gazebo (*Mianshui xuan*); 5. Place for Watching Fish; 6. Surging Wave Pavilion; 7. the mound; 8. Shrine of Five Hundred Worthies; 9, Illuminate the Way Hall; 10, 11. sutra pillars of Grand Clouds Cloister.

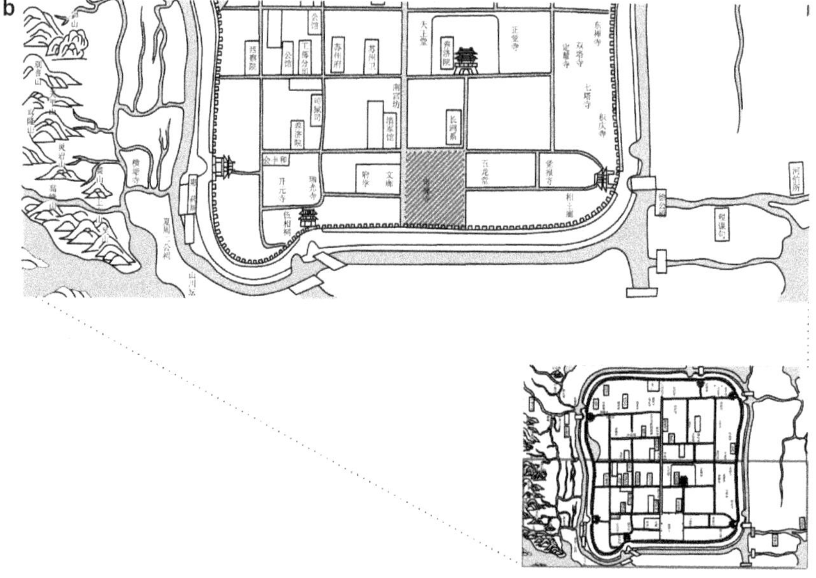

Figure 4.1 The site of Surging Wave Pavilion, current and during the Yuan to Middle Ming. (a) Author's aerial photo of the site taken in 2016.12. (b) Southern Chan Temple in *GSZ*. Author's traced drawing following Wang Ao.

of Confucian figures and the singing of the Six Classics. To the north of Surging Wave were two academies, Rectify Meanings Academy (*Zhengyi shuyuan*) and Learn the Classics Hall (*Xuegu tang*), which were renowned for evidential studies. South of the wave, Shrine of Five Hundred Worthies (*Wubai mingxian ci*) advocated the local activism, and Illuminate the Way Hall (*Mingdao tang*) promoted neo-Confucianism (Fig. 4.1a). The once dominant temples were now restrained to just the northeast and southwest corners. How did a Buddhist site of the 1300s evolve into a Confucian site by the 1800s?

This section periodises and models the transformation of Surging Wave Pavilion site into six stages, defined by shifts in land ownership, spatial configuration, and the evolving symbolism of Surging Wave (Table 4.1). Buddhists possessed the site from the 1300s to 1520s. The government's conversion of Buddha hall into Shrine of the King of Qi in 1523 altered the site's religious trajectory. During the early High Qing period, the third stage, the government purchased a key parcel of land south of Surging Wave, establishing axial shrines and scenic areas facing the water. In the fourth period of the late High Qing, the government further extended its hold over the southern region, centring around newly erected imperial steles. By the Middle and Late Qing period, the site had been transformed into a complex of shrines and academies encircling Surging Wave, advocating local activism and a Confucian social order. The architectural modelling integrates visual and textual evidence, illustrates estates' evolving boundaries, and maps out officials' spatial strategies.

Surging Wave within Temples

The name of the central pond (Fig. 4.1a(C)), 'Surging Wave,' dates back to the Northern Song dynasty and is attributed to Su Shunqin. After his dismissal from court, Su moved to Suzhou and bought the site.[2] The site featured a mound with water surrounding its three sides and filled with dense bamboo grove. Su constructed Surging Wave Pavilion at the mound's northern base adjacent to water (Fig. 4.2 (1)).[3] In this serene setting, Su composed lots of poems and writings. His *Record of Surging Wave Pavilion* received widespread acclaim and contributed to the site's popularity, particularly the lines 'My body at ease, my mind thus becomes untroubled; nothing perverse seen or heard, the Way thus becomes luminous.'[4] From then on, the water beneath the mound was known as 'Surging Wave.' Following Su Shunqin, ownership of the site passed to Zhang Dun (1035–1106), and later, Han Shizhong, the King of Qi, coerced Zhang to offer the garden.[5]

In the Yuan and early Ming periods, Surging Wave Pond was the water inside Southern Chan Temple. To the north was Grand Clouds Cloister and to the southwest was Gathering Clouds Temple (*Jiyun si*). Witty Retreat

Table 4.1 Site Evolution, Biography of Surging Wave Pavilion

Decade	Representative Spatial Practices
The temple stage, Middle Yuan to the Middle Ming	
1310s	Monk Zongjing (fl. 1314–1320) constructed Witty Retreat Cloister.
1340s	Monk Shanqing constructed Grand Clouds Cloister.
Transformative stage 1, High Ming	
1520s	1523, prefect Hu Zuanzong converted Buddha hall of Witty Retreat Cloister into Shrine of the King of Qi.
1550s	1546, monk Wenying (fl. 16th c.) constructed a Surging Wave Pavilion on the top of the mound of Witty Retreat Cloister and a gazebo on the waterfront.
1600s	1606, the government renovated Surging Wave Pond as part of the public canal.
Transformative stage 2, Early High Qing	
1690s	1695, Song Luo bought the architectural land of Witty Retreat Cloister, renovated its axial halls and waterfront, and established the site of Surging Wave Pavilion as a Confucian garden.
Transformative stage 3, Late High Qing	
1710s	1719, Wu Cunli (fl. 18th c.) installed imperial stele pavilion and renovated the site of Surging Wave Pavilion's waterfront.
1730s	Ying Jishan constructed Near Mountain and Grove (*Jin shanlin*) and Garden of Suitability (*Ke yuan*).
1740s	1747, An Ning (fl. 18th c.) established the second imperial stele pavilion. The entrance arch of the site was established.
1750s	1757, emperor Qianlong visited Surging Wave Pavilion.
Transformative stage 4, Middle Qing	
1800s	1805, Tiyeboo (1752–1824, alt. Tiebao) converted Near Mountain and Grove into Rectify Meanings Academy.
1820s	1827, Liang Zhangju and Tao Shu (1779–1839) converted the primary hall on the western axis of the site of Surging Wave Pavilion into Shrine of Five Hundred Worthies, and reconstructed Garden of Suitability as the teacher Zhu Jian (1769–1850)'s residence of Rectify Meanings Academy.
Transformative stage 5, Late Qing	
1850–1864	Taiping rebellion destroyed artefacts of Surging Wave Pavilion.
1870s	1872, Zhang Shusheng (1824–1884) reconstructed the site of Surging Wave Pavilion and Rectify Meanings Academy.
1880s	1888, the construction of Learn the Classics Hall academy.

Cloister occupied the mound overlooking the Surging Wave Pond, a place corresponding to the current site of Surging Wave Pavilion (Fig. 4.2).[6]

During the early Ming dynasty, Monk Baotan Shiying (fl. late 14th c.) petitioned Emperor Zhu Yuanzhang (1328–1398) to unify the three temples into a single abbey, arguing that they 'shared the same root.'[7] The architectural drawing in Figure 4.2 informs us the 'root' was likely referred to the Surging Wave Pond, the water resource serving the three temples.

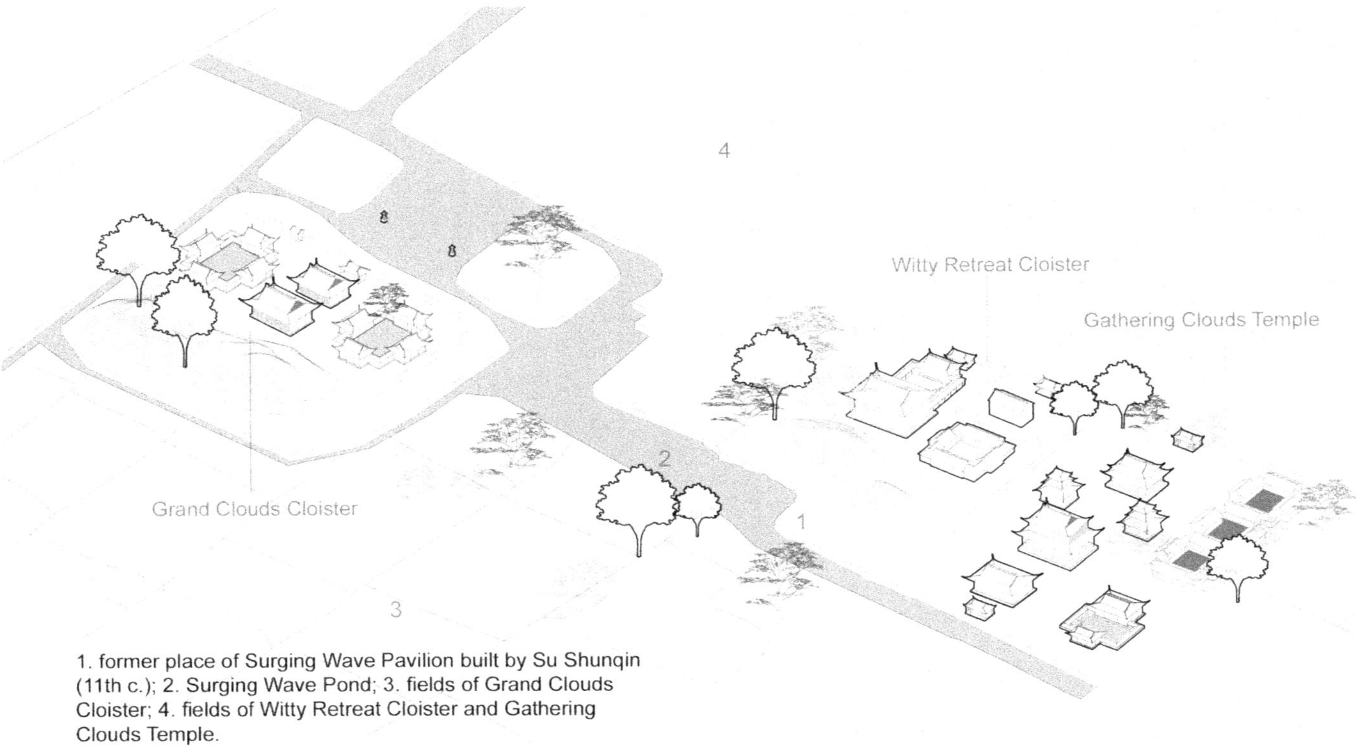

1. former place of Surging Wave Pavilion built by Su Shunqin (11th c.); 2. Surging Wave Pond; 3. fields of Grand Clouds Cloister; 4. fields of Witty Retreat Cloister and Gathering Clouds Temple.

Figure 4.2 Temples surrounding the Surging Wave during the Middle Yuan and Middle Ming, restored northwest isometric view. Author's modelling and drawing.

Despite the decline in monastic economies in Suzhou at that time, Baotan succeeded in revitalising Southern Chan Temple. Remarkably, it was the only temple in Suzhou gained imperial recognition in the early Ming and was also the largest in the city, boasting extensive fields to its north and south.[8] Figure 4.2 illustrates that each monastic compound was aligned axially towards Surging Wave Pond. Central to each axis, grand Buddha halls enshrining Buddha statues gazing over the water.

Grand Clouds Cloister enjoyed particular favour among the gentry. This cloister's model in Figure 4.2 was created by integrating elements from Shen Zhou's paintings (Fig. 1.9), historical records, and the archaeological remains. Notably, the location of the two stone sutra pillars in Shen Zhou's painting could be found at the site today (see Fig. 4.1a(10, 11)). The cloister was organised by three parallel axes. The grand Buddha hall dominated the central axis, serving as the focal point for worship, flanked by residential courtyards for monks and guests. In front of the monastic compounds was the Release Life Pond and at the rear was a ten-meter-high hill. A river, branching off from Surging Wave Pond, encircled the temple, supplying water to the northern paddy fields.

Gathering Clouds Temple, located southwest of Surging Wave Pond (Fig. 4.2), was held in high esteem by gentry elites, albeit not commonly visited by them. When the temple's artefacts were damaged in a fire, the abbot managed to secure restoration funds from the local wealthy and received a commemorative stele from the court official Wu Kuan.[9] The temple featured a central axis consisting of Buddha halls and a Bodhisattva hall, flanked by a drum tower on the west and a bell tower on the east.[10] The abbot's quarters and monks' residences were likely situated along the western axis.[11]

In comparison, Witty Retreat Cloister at the centre received the least patronage from the gentry in the Ming (Fig. 4.2). Referencing Zhou Chen's (1460–1535) painting and later records about the cloister's conversion, it can be inferred that the temple consisted of western and eastern axes, which corresponds to the axis of Shrine of Five Hundred Worthies and the axis of Illuminate the Way Hall of the current site (see Fig. 4.1a (8,9)). Zhou Chen's handscroll informed that the western axis of the cloister was residential quarters.[12] Additionally, Witty Retreat Cloister and Gathering Clouds Temple also owned vast paddy fields to their south.[13]

During this period, Surging Wave Pond belonged to the three temples. It marked the terminus of the public canal that originated from Feng Gate, as depicted in *Pingjiang Map* (Fig. 4.5a). The end water was described by Shen Zhou when he boated to Grand Clouds Cloister:

The water originates from the Feng River and flows westwards. It passes through Changzhou County Seat, branches towards the south, and bends towards the east. Then the river turns south and converges to the left of the cloister. It forms a belt around the temple and converges in the pond in front of it.[14]

During the two centuries of the Yuan and Early Ming, the area of Surging Wave functioned as a Buddhist space, and Su Shunqin's Surging Wave Pavilion did not exist. The religious presence was accepted by the gentry, and they revered the simple, pure, and serene Buddhist environment. This attitude was captured in Shen Zhou's poem *The Former Site of Surging Wave Pavilion Has been Resided by Buddhists*:

Do not ask who owns Surging Wave today. A hundred years of rise and fall are just like the waves.
Boil tea and ladle it into the Wu monk's bow. Wash the feet while nursery songs echo through.
It is only tranquil and serene here at the south of the city. Who has passed by, with the dust of carts and the tracks of horses?[15]

The gentry also inscribed the Surging Wave with Buddhist meanings. Zhou Lun (1463–1542) perceived the Surging Wave as a mirror of emptiness:

The water below Grand Clouds Cloister, is the old Surging Wave of Su Shunqin. With clear eyes to observe the mundane world, cleanse people's heart [the Buddha] opened the temple.
Fragrant lotus remains unsullied by filth. The water surface is like a mirror of wisdom, which receives reflections from all directions.
The empty and blue pond receives the rain. The sky contains free clouds and light.[16]

Zhou Lun likened the surface of Surging Wave Pond to a mirror, which reflects all things without holding onto any. His last stanza further delved into the Buddhist philosophy of 'non-attachment.'

Replacing Buddha with the King of Qi

Three incidents in the Ming dynasty reoriented the site from Buddhism towards Confucianism and the literati culture. The first is the government's intrusion of the Buddhist estate in the High Ming. Hu Zuanzong converted the Buddha hall on the eastern axis of Witty Retreat Cloister into Shrine of the King of Qi in 1523 (Fig. 4.3).[17] He ordered to replace the Buddha statue with that of the military hero, Han Shizhong (the King of Qi), criticising the Buddhist control of the estate as inappropriate.[18] Hu claimed to revert the site to its alleged original ownership by the King of

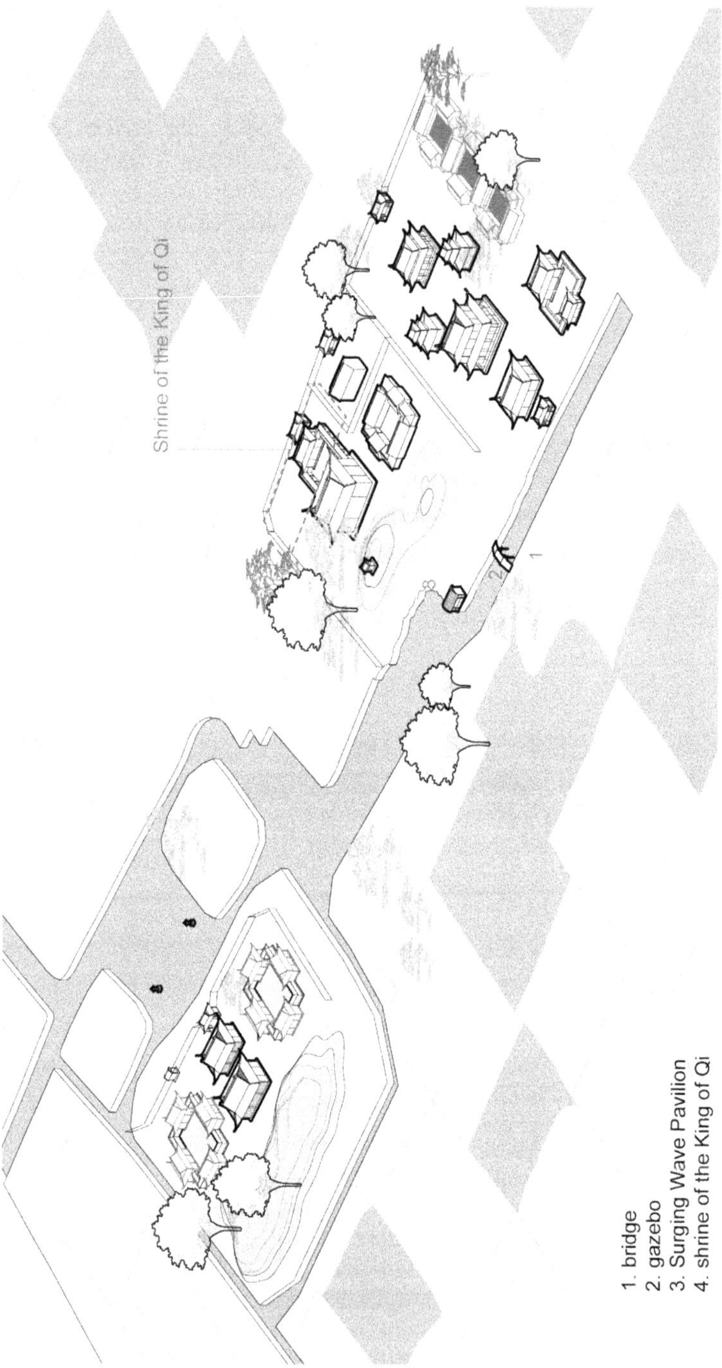

Figure 4.3 The site of the Surging Wave during the High Ming, restored northwest isometric view. The dashed line indicates the area occupied by the government. Author's modelling and drawing.

Qi.[19] He established annual worship ceremonies at the shrine and assigned its upkeep to monks.[20]

To prevent the site's potential reversion to Buddhist control, Hu erected two steles to document this conversion.[21] In the stele inscriptions, Hu and Huang Xingzeng portrayed that officials and local citizen all praised this conversion, with monks 'pleased to comply,' which exemplified the transformation of Buddhists.[22] It was quite dubious that monks would be pleased: they were coerced to shift their religious devotion from Buddhism to a local deity whom they might never be interested in. Furthermore, a significant portion of their land was seized by the government and opened to the public. The monks' suppressed indignation was revealed in their destruction of this shrine in the late Ming.

Concurrent with the government's encroachment on Witty Retreat Cloister, monks and the gentry elites collaborated to build two waterfront artefacts commemorating Su Shunqin's legacy. In 1546, Monk Wenying reconstructed a Surging Wave Pavilion to reminisce Su Shunqin. This new pavilion was depicted in Wen Boren's painting *Brisk Summer in Surging Wave* (Canglang qingxia) in 1554 (Fig. 4.4a(4)). Translating artefacts in Wen Boren's painting into the site's model reveals that Wenying did not adhere to Su Shunqin's original placement of the pavilion along the shore. Instead, he situated it atop a mound, offering expansive views. Additionally, a new gazebo was built along the shore (Fig. 4.4 (2)). In Wen Boren's painting, a literati was reading inside the gazebo, leaning against the handrail by the shore, suggesting the site's accommodation of gentry's leisure activities. Wenying's efforts in reviving literati culture received praise from gentrymen such as Gui Youguang (1506–1571).[23]

The government's integration of Surging Wave Pond into the public canal in the Late Ming further fragmented Southern Chan Temple. In 1606, the government connected Surging Wave Pond with the pond of the prefectural school to its west, transforming it into a transitional hub of public navigation (Fig. 4.5b).[24] The official in charge, Zhang Guowei, planned the water converging in Surging Wave Pond to continue its flow westwards into Washing Horse Pond (*Xima chi*) at the front of the prefectural school. He believed that this connection, providing the prefectural school with a substantial influx of water, would be auspicious for the success of its students in obtaining degrees.[25]

The incorporation of Surging Wave Pond as a transit of urban canal resulted in two spatial changes of the temple. First, the Surging Wave pond, as the inner part of Southern Chan Temple, was opened to the public. What was once a secluded and serene endpoint within the temple, as captured in *Pingjiang Map* (Fig. 4.5a) and paintings by Shen Zhou and Wen Boren (Fig. 1.9, Fig. 4.4a), had evolved into an active node of the public transportation, vividly depicted in Wang Hui's painting (Fig. 4.5c).

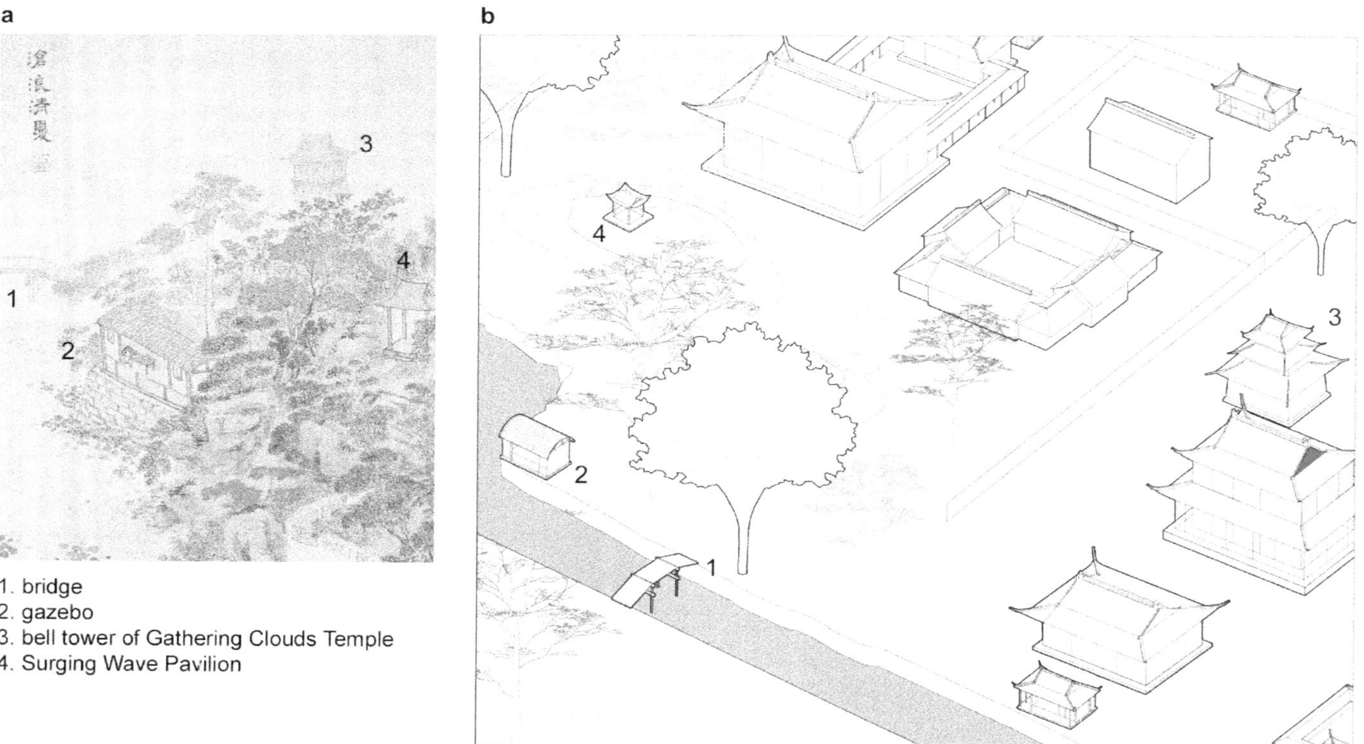

Figure 4.4 Translating Wen Boren's painting into architectural drawing. (a) Wen Boren, 'Brisk Summer in Surging Wave (Canglang qingxia),' leaf no. 3 in the album *Ten Scenes of Gusu* (Gusu shi jing tu), 1554, colour ink on paper, 32cm×25.6cm. Taipei: Collection of the National Palace Museum. (b) Author's modelled drawing simulating the scene.

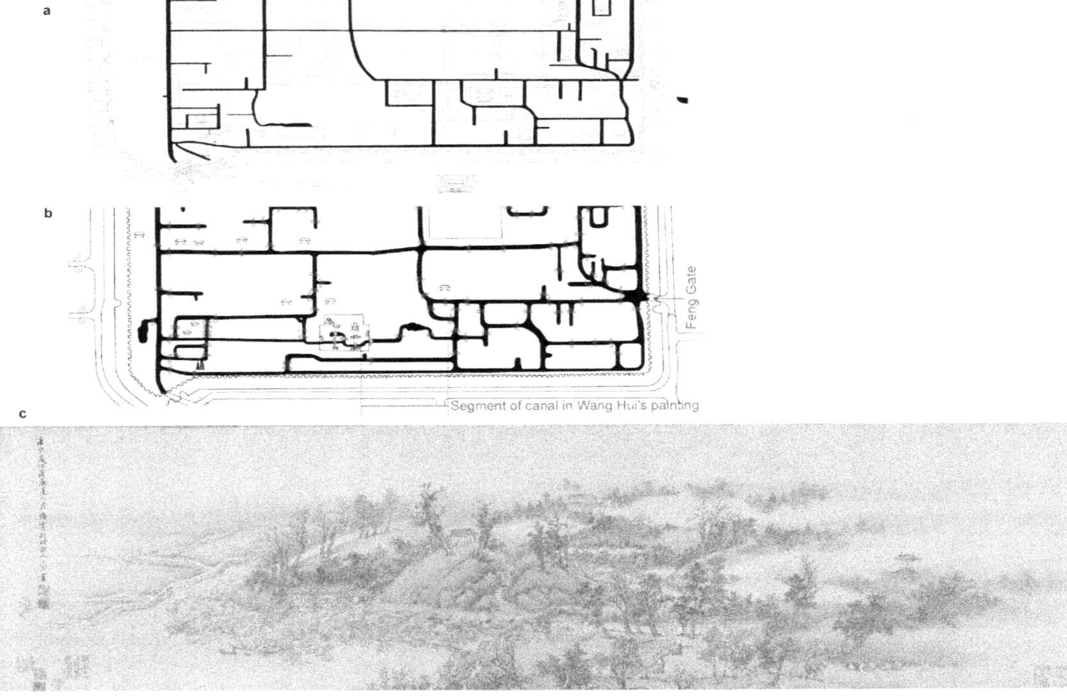

Figure 4.5 Canals connecting Surging Wave Pond, before and after the Late Ming renovation. (a) Urban canals before the 1606's renovation. Author's traced drawing following *Pingjiang Map*. (b) Urban canals after the 1606's renovation. Author's traced drawing following Zhang Guowei, *Map of the Waterways Management within Suzhou Prefecture City*. (c) Wang Hui, *Painting of Surging Wave Pavilion* (Canglang ting tu), 1700, handscroll, colour ink on paper. Nanjing Museum.

Secondly, the canal renovation brought the temples and the prefectural school into closer proximity. Before the 1630's hydraulic overhaul, these two estates drew water independently from west and east, a spatial relation evidenced in *Pingjiang Map* (see Fig. 4.5a). After the renovation, the temple and the school were connected by the waterway, as illustrated in Zhang Guowei's *Map of the Waterways Management within the Suzhou Prefecture City* in Figure 4.5b. These spatial shifts paved the way for the government's subsequent conversion of the site into Confucian establishments and the redesign of its waterfront façade, aligning it with the Confucian ethos.

Pavilions and Shrines Facing Surging Wave

In the early Qing period, government official Song Luo purchased the land of Witty Retreat Cloister, an area identified in Figure 4.6. To assert this new ownership, Song Luo erected a stele on the premise.[26] The front side of the stele inscribed his restoration work and new interpretation of Surging Wave. The reverse side officially declared the transaction of Witty Retreat Cloister into government's possession. Establishing a legal estate boundary retrospectively legitimised Hu Zuanzong's contentious encroachment on the temple one and half centuries earlier. To sustain the site, Song Luo bought several fields to generate income to finance the ongoing maintenance of artefacts.

Song Luo undertook a meticulous renovation of Witty Retreat Cloister, which was detailed in his *Record of Reconstructing Surging Wave Pavilion* (Chongxiu Canglang Ting ji). This record, together with other records and poems, was anthologised in *A Small Gazetteer of Surging Wave Pavilion* (Canglang xiaozhi). To visually document the new site, Song Luo commissioned Gao Jian (1634–1707) for a painting as the frontispiece of the gazetteer (Fig. 4.7a). He also invited Wang Hui to paint the new site's waterfront as a handscroll (Fig. 4.5c, Fig. 4.8). Integrating textual and pictorial evidence from these three sources, a 3D model is constructed in Figure 4.6.

Song Luo preserved the axial layout of the cloister meanwhile removed Buddhist elements. On the eastern axis, he refurbished Shrine of the King of Qi established by Hu Zuanzong (Fig. 4.7(10)). Buddha hall behind Shrine of the King of Qi was repurposed into Shrine of Su Shunqin (Fig. 4.7(11)).[27] A tablet honouring Su Shunqin was venerated inside and annual rituals was scheduled for his worship. Comparing the model of the site after Song Luo's construction with that of the original Witty Retreat Cloister (cf. Fig. 4.6 with Fig. 4.2), it reveals that Song Luo renovated monastic quarters on the western axis into a new social place for him and his friends' gathering.

Song reconstructed the pavilion atop the mound previously built by monk Wenying (Fig. 4.7(4)). He mounted a plaque on the pavilion's

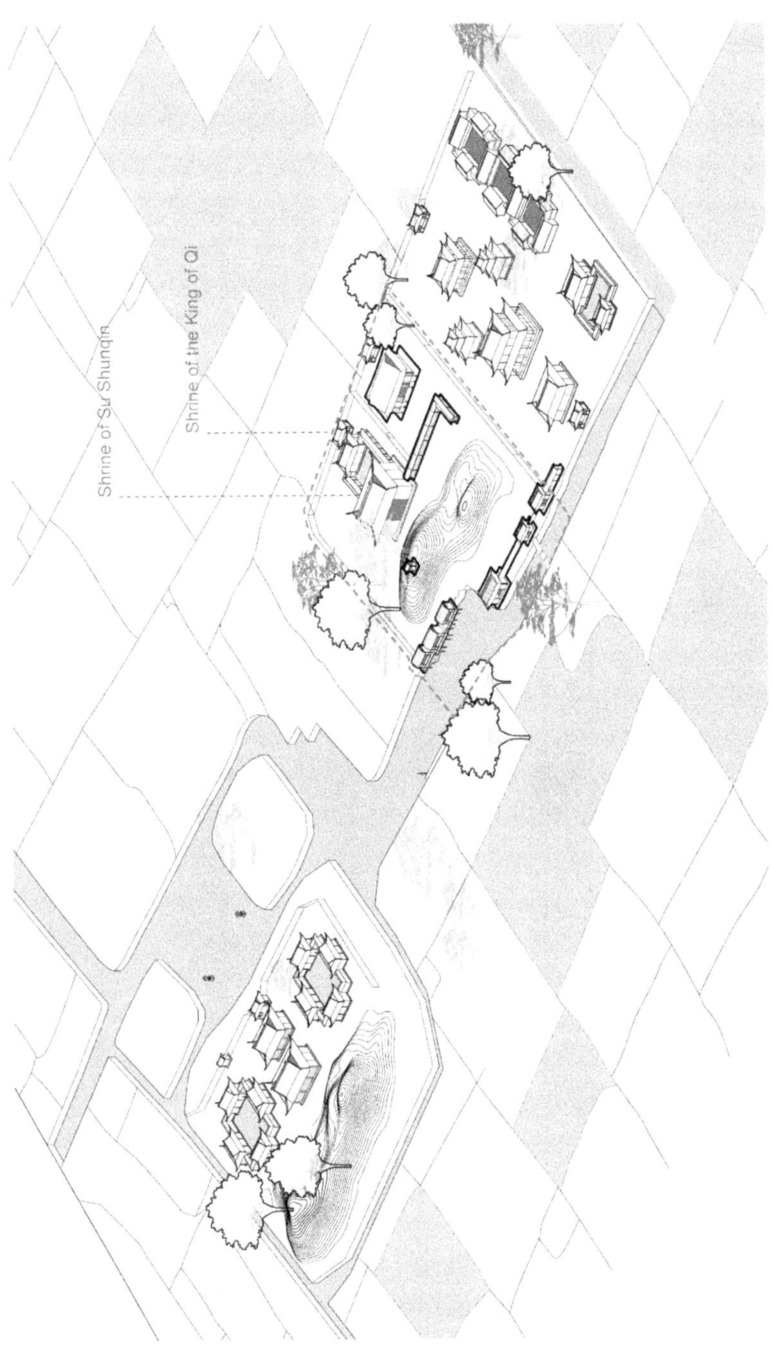

Figure 4.6 The site of the Surging Wave during the early High Qing period, restored northwest isometric view. The dashed line indicates the area occupied by the government. Author's modelling and drawing.

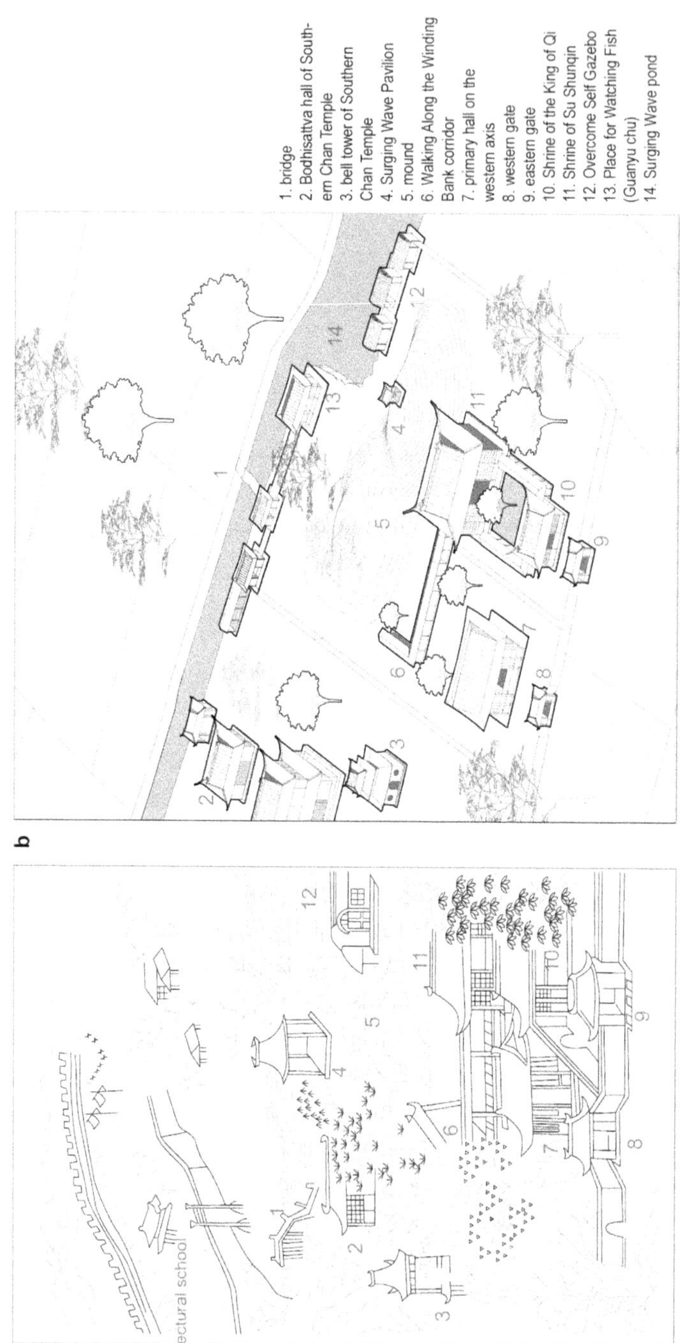

Figure 4.7 Translating Gao Jian's painting into architectural drawing. (a) Gao Jian (b. 1621), *Painting of Surging Wave Pavilion* (Canglang ting tu), 1702, stone block print, in *CLXZ*. Author's traced drawing following Gao Jian. (b) Simulating Gao Jian's view in 3D model. Author's modelling and drawing.

lintel, explicating its name as 'Surging Wave Pavilion.' By adapting Wen Zhengming's calligraphy for the name inscribed, Song Luo rooted Surging Wave Pavilion in Suzhou's literati tradition. The mound and the pavilion were connected to the axial shrines by a corridor named 'Walking Along the Winding Bank (*Bu qi*, Fig. 4.7(6)).'[28]

Reading Gao Jian's painting reveals Song Luo's first renovation strategy: to extend the inner axial courtyards towards the northern sceneries, and further reaching the waterfront. Visitors entering through the main gate were first directed to pay homage at Shrine of the King of Qi in the first hall (Fig. 4.7(10)), followed by Shrine of Su Shunqin (Fig. 4.7(11)). Song Luo conceptualised the site as a spiritual residence for Su Shunqin and the King of Qi, as he wrote: 'the spirit of general Han (the King of Qi) must come here, meeting and socializing with the Chief official (Su Shunqin).'[29] Proceeding through the corridor, the visitor ascended the hill to Surging Wave Pavilion at its summit. This vantage point offered views once enjoyed by Su Shunqin. As many added artefacts were named after phrases from Su Shunqin's poems, Song Luo created an environment to engage visitors with the poet's emotions. Gao Jian's painting captured the view commanded by the pavilion as described by Song Luo: 'beyond the stream, there are vegetable patches and houses intermingled like embroidery.' Gao Jian also included the prefectural school and the urban wall in the painting upper left corner, suggesting the site's public nature and integration into the broader urban fabric (Fig. 4.7(a)).

Song Luo's second landscaping strategy involved creating an architectural façade along the waterfront, as revealed by reading Wang Hui's handscroll. Unfolding the handscroll, the newly dredged canal comes into sight as well as the bell tower of Southern Chan Temple (Fig. 4.8d, also see Fig. 4.5c). The river was juxtaposed by the northern and southern banks, and three gentlemen were walking towards the triple bend bridge. The bridge led to the new entrance gate. It was connected to other structures by walls and corridors (Fig. 4.8a).

Proceeding with the handscroll, we encounter a hipped-gable pavilion on the stone platform (Fig. 4.8b). There are two literati depicted as near the opened window frame. The younger person seems to seek guidance from the elder one, who was seated facing the water, listening attentively. The shore adjacent to the pavilion was adorned with giant stone. The pavilion was named 'Overcome Self Gazebo (*Zisheng xuan*),' referencing to the famous phrase from Su Shunqin's *Record of Surging Wave Pavilion*. It was renovated by Song Luo from the previous tiny roll-shed water gazebo built by Wenying.

Continuing to unfold the scroll, a mound emerges, depicted by Wang Hui as resembling a large mountain (Fig. 4.8d, also see Fig. 4.5c). Atop this mound was the reconstructed Surging Wave Pavilion. To the left of

Figure 4.8 Wang Hui, *Painting of Surging Wave Pavilion* (Canglang ting tu). Nanjing Museum.

the mound was three newly constructed boat-shaped pavilions on stilts above water (Fig. 4.8c). Song Luo named them 'Place for Watching Fish (*Guanyu chu*),' inspired from Su Shunqin's poem *Watching Fish in Surging Wave* (Canglang guanyu). These three boat pavilions largely opened to the water and surrounded by lotus flowers (Fig. 4.8c). Fishermen in boats were picking up the lotus.[30]

Wang Hui's handscroll vividly captures the beauty and fluid continuity of the waterfront after Song Luo's renovation. The waterfront façade is rhythmically arranged, adorned with slanting veteran trees. Artefacts and paths along the shore enabled direct and frequent interaction with the water (Fig. 4.8). Wang Hui leveraged the handscroll's horizontal format and its unfolding motion to convey the experience of walking and boating towards the Surging Wave Pavilion from the prefectural school. Notably, it was the earliest Chinese painting to focus on the urban waterfront façade of a public garden.

Song Luo redefined the Surging Wave as capable of refreshing officials' mind. Thus they could be more astute and adaptable in managing the administrative mess. As a local official, Song Luo felt his sense dulling, being overwhelmed with a myriad of tasks.[31] But visits to the Surging Wave Pavilion site regained his vitality: 'Once I have distanced myself from such matters and rested awhile in a place of cool, fresh air, and vast openness, my faculties of sight and hearing become more open and clearer, and my aspirations and vigour are rekindled.'[32]

Xu Yinong has observed that, in the early Qing dynasty, the site of Surging Wave was no longer associated with political frustration, which was inscribed by Su Shunqin in the Northern Song dynasty. Instead, it evolved to symbolise success in officialdom. This view was resonated by Song Luo's guests. They furthered Song Luo's view that visiting the Surging Wave Pavilion could cultivate one's literature talents and augment success in the political realm. The surging wave thus took on a new connotation, blending the literati's poetic mastery with official aspirations.[33]

The relationship between officials and monks transitioned from being conflict to collaboration. The site was owned by the government, meanwhile its maintenance was managed by the temple, with monks receiving compensation from government for the work. Monks, seemingly accepting the transaction of their lands, endeavoured to foster positive relations with officials. They participated in Song Luo's community and presented poems. The construction of a Confucian garden on the former monastic land increased monks' chance in approaching officials and gaining support from the government, though not a reliable one. Furthermore, monks in Gathering Clouds Temple provided accommodations for some of Song Luo's guests.

'Like Myriad Stars Surrounding the Polestar'

In the late High Qing Period, the government reconfigured the site as 'myriad stars surrounding the polestar' (Fig. 4.9). This vision, initially proposed by Wu Cunli in 1710s, was fully realised through subsequent officials' construction efforts, with the endorsement from the emperor's visits in the 1750s (Table 4.1). A 3D model is created by coordinating textual descriptions from official records to verify the pictorial accuracy of the 1753 *Gazetteer of Changzhou County* (Fig. 4.10) and the painting leaf from the album *Grand Ceremony of the Southern Inspections*.[34]

A comparison between the late High Qing site model in Figure 4.9 and the early High Qing site model in Figure 4.6 reveals a significant enlargement of the site, almost doubling the site's area. This expansion involved the incorporation of two courtyards from Gathering Clouds Temple at the west, the integration of farmland from the Grand Clouds Cloister at the north, and the acquisition of additional land at the site's eastern entrance.

The site as 'myriad stars surrounding a polar star' was first envisioned by Wu Cunli. The polar star refers to the imperial stele Wu established in 1747 (Fig. 4.9(5), Fig. 4.10(5)). On his sixth southern inspection tour in Suzhou, emperor Kangxi honoured Wu with a poem and a couplet. The poem reads,

> In the past twelve years, Suzhou has transformed into a prosperous city with outstanding literary talents. I often reminisce about how, years ago, it was impoverished and beset with difficulties. I thus exhort you bureaucrats to be diligent and take responsibility of the city's management.[35]

The couplet was 'When ointment rain is plentiful, the farmers rejoice. When the flowers of the county bloom brightly, the integrity of the officials is testified.'[36] Wu immediately engraved these imperial words on a stele, and built a pavilion to enshrine it in front of Surging Wave Pond.[37]

To elevate the stature of the imperial stele pavilion, Wu undertook significant renovations of the existing waterfront artefacts, deeming the earlier creations by Song Luo as modest and lacking the requisite splendour for an imperial presence.[38] A comparison between the 3D representation in Figure 4.9 and the model of Song Luo's prior constructions in Figure 4.6 illustrates Wu's constructional efforts along the waterfront. Wu viewed the original Place for Watching Fish (Fig. 4.7b(13)) as undersized for its purpose and thus constructed an expansive courtyard with five bays, enhancing its waterfront façade with lattice windows of intricate patterns. This new edifice was named 'Gazing from the Bank of River Hao (*Haoshang guan*)' (Fig. 4.9(2), Fig. 4.10(2)). Overcome Self Gazebo (Fig. 4.7(12)) was replaced by a new structure named 'Roving in Mirror (*Jingzhong you*, Fig.

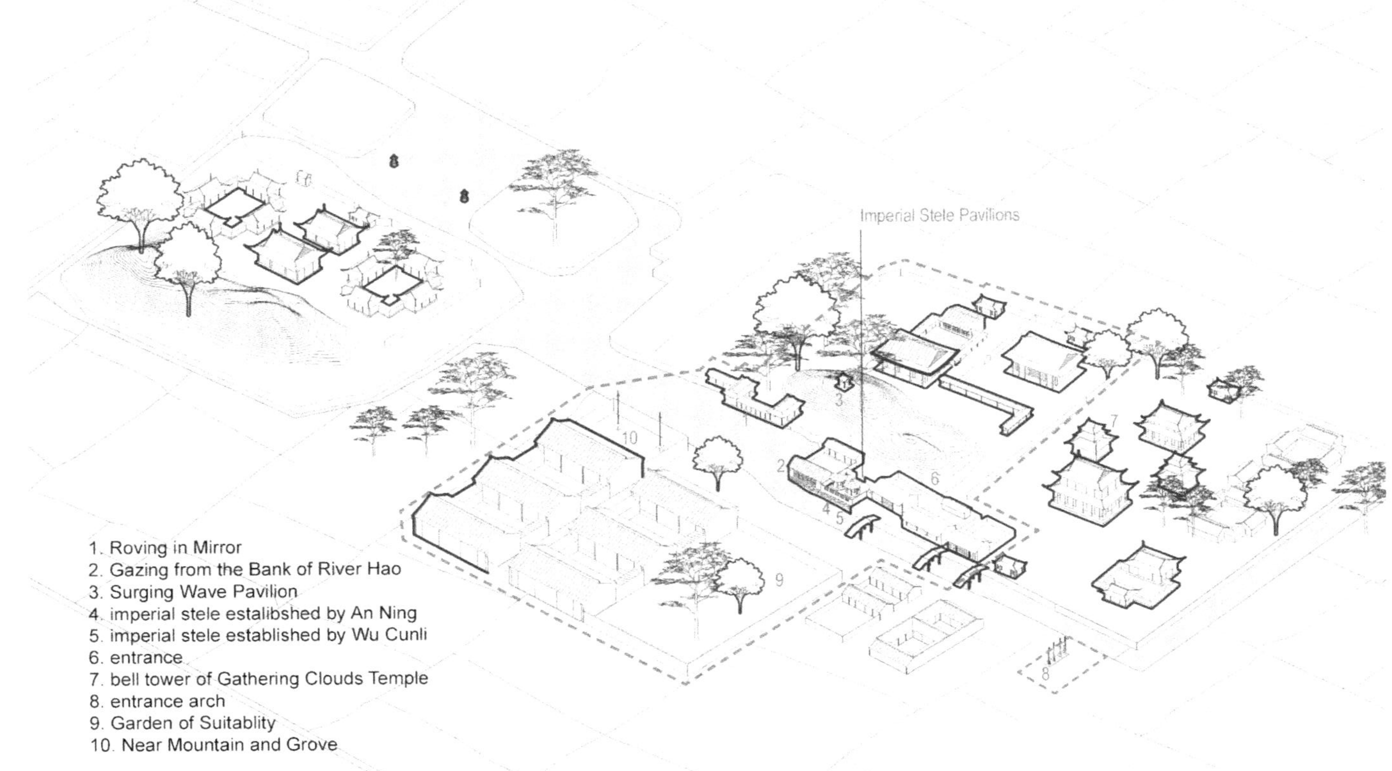

Figure 4.9 Shrines and academies surrounding the Surging Wave in the late High Qing, restored northwest isometric view. The dashed line indicates the area occupied by the government. Author's modelling and drawing.

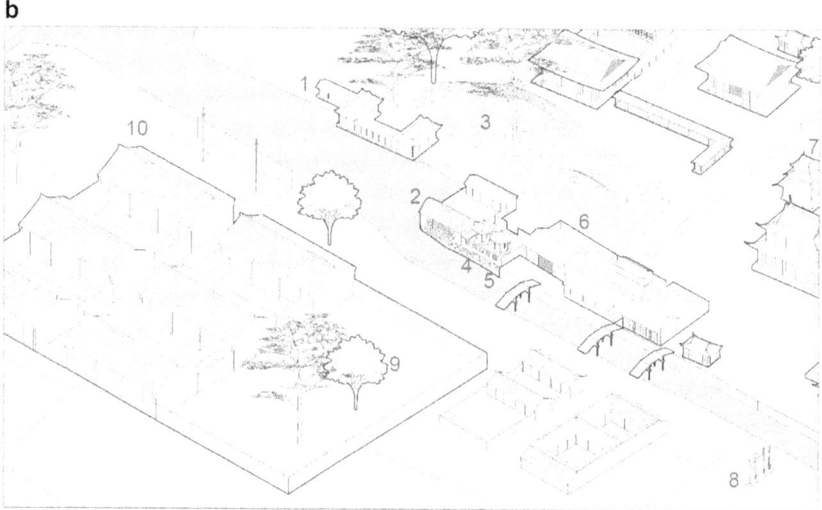

1. Roving in Mirror
2. Gazing from the Bank of River Hao
3. Surging Wave Pavilion
4. imperial stele established by An Ning
5. imperial stele established by Wu Cunli
6. entrance
7. bell tower of Gathering Clouds Temple
8. entrance arch
9. Garden of Suitability
10. Near Mountain and Grove

Figure 4.10 Translating artefacts in *Painting of Surging Wave Pavilion* into 3D model. (a) Painting of Surging Wave Pavilion in *Changzhou County Gazetteer*, 1753. Author's traced drawing following the gazetteer painting. (b) Author's modelled drawing simulating the scene.

4.9(1), Fig. 4.10(1)),' offering visitors an uninterrupted waterfront experience from the interior.

Wu envisioned the site as a constellation of artefacts encircling the imperial stele as the centre, as he articulated:

> Some have been expanded, some buildings were added, The configuration was altered, with all arrayed orderly to encircle. As if myriad stars surrounding the polestar, like hundreds of rivers converging towards the origin of the sea. Majestic and splendid, the place acquired the magnificent sight for ten thousand years.[39]

Two spatial reasons could be deduced for why Wu Cunli chose to position the imperial stele to the site of Surging Wave Pavilion. Firstly, renovated by Hu Zuanzong and Song Luo, the site had evolved to embody Confucian ideals and literary excellence, making it a fitting venue for the emperor's didactic literary works. The second reason could be the site's public accessibility. Unlike the prefectural school, which was primary for students, teachers, and officials, the Surging Wave site was open to the public. Wu strategically placed the stele pavilion along the shore, as illustrated in Figure 4.10, rather than within the secluded garden, ensuring a high visibility of the imperial stele pavilion to passersby.[40] For those boating or walking along the canal, the stele pavilion was constantly in sight, a visible reminder of the imperial presence and its admonishment.

The renovation of the site as being imperial likely contributed to an increase in visits from provincial officials. To accommodate them, Yin Jishan (1689–1755) established a lodging complex to the north of the Surging Wave Pavilion around the 1730s. The lodge was named 'Near Mountain and Grove,' evoking the scenic beauty of the 'mountain and grove' featured by the southern part. This complex was built on the paddy fields previously owned by Grand Clouds Cloister.[41] It featured two main axes with grand buildings, including shrines, meeting halls, guest rooms, book storage facilities, and a garden. Its entrance courtyard facing the river was adorned with two flags, enhancing visibility and prominence. It could be deduced that the second floor of its book storage tower commanded the view of the southern shore. The added lodging complex fulfilled Wu Cunli's vision of the imperial stele as the centre of the site.

Following Wu Cunli's installation of the stele of Kangxi emperor's poem, An Ning, introduced another imperial stele to the site in 1747 (Fig. 4.9(4)). The stele bore the inscription of Qianlong emperor's seven poems *On the flood disasters happened in Jiangnan area*.[42] The floods of 1747 devastated cities along the lower Yangtze river and caused significant loss of life. In response, the emperor wrote poems expressing his profound empathy. As the responsible official, An Ning inscribed Qianlong's compassionate verses on a stele, installing it adjacent to Kangxi emperor's stele (Fig. 4.9(4)). In the 1753 gazetteer painting, there were two stele pavilions aligned along the waterfront (Fig. 4.10a(4,5)).[43]

An Ning's addition of the imperial stele augmented the meaning Wu Cunli previously imparted, the imperial emphasis of the local officials' duty to tackle social challenges. Setting the stele in front of the Surging Wave highlighted the need for effective canal management, since the Late Ming canal integration of Surging Wave Pond was deemed as a successful hydraulic precedent.

After Wu Cunli, Yin Jishan, and An Ning's site renovations during the late High Qing period, the site of Surging Wave Pavilion encompassed an extensive area centring around the Surging Wave. The government established an arch to the west of the site, marking its main entrance. The archway bore the inscription 'Extraordinary Traces of Surging Wave (*Canglang shengji*).'[44] The astral concept of 'myriad stars surrounding the polestar' was fully realised, which was visually captured in the gazetteer *Painting of Surging Wave Pavilion* (Canglang ting tu) in 1753 (Fig. 4.10a). In the painting, the imperial stele was surrounded with other artefacts, each contributing to the overarching imperial vision.

Shrines and Academies Surrounding Surging Wave

During the Middle Qing, Suzhou was administered by a group of officials known for their statecraft-oriented approach. Prominent figures included Tao Shu, Liang Zhangju, and Zhu Jian. They were scholars dedicated to evidential studies addressing social issues. Before moving to Suzhou, in around the 1810s, they founded Xuannan Poetry Club (*Xuannan Shishe*) in Beijing, a club geared towards discoursing evidential scholarships and statecraft.[45]

These officials questioned the prevailing imperial notion that underscored the emperor's control over local affairs, advocating instead for the accountability of local elites in ensuring regional prosperity.[46] This rising local activism was manifested architecturally in the renovation of the Surging Wave Pavilion site. The primary hall on the western axis was transformed into Shrine of Five Hundred Worthies.[47] The visiting officials' lodge was renovated into Rectify Meanings Academy and Garden of Suitability (Fig. 4.11).

Shrine of Five Hundred Worthies, situated on the western axis of the site and shown in Figure 4.11, was created by adapting the corridor and assembly hall previously established by Song Luo. This shrine paid homage to a selected group of five hundred local elites, whose legacies spanned from esteemed ancients of the 6th century B.C. to the notable contemporaries. Each individual was commemorated with a stele inscribed with

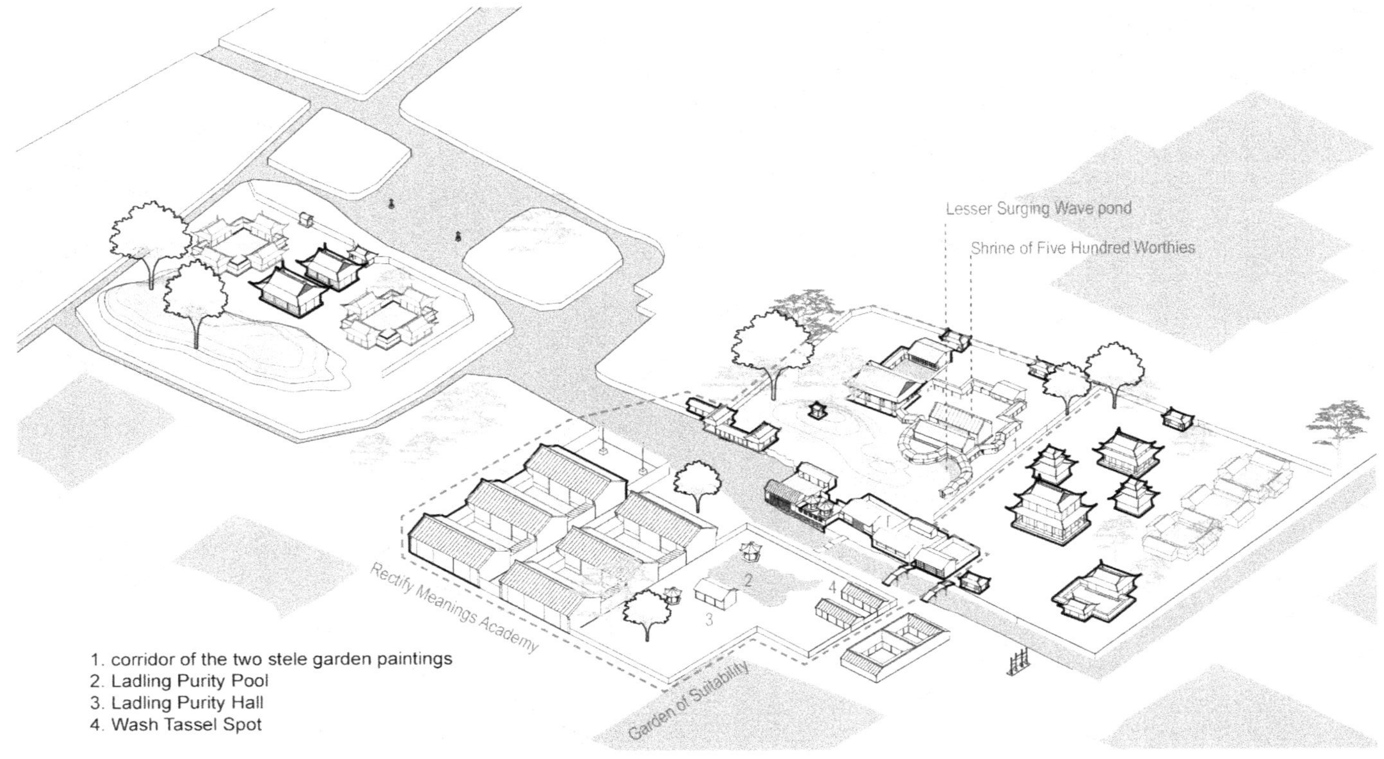

Figure 4.11 The site of Surging Wave Pavilion in the Middle Qing, restored northwest isometric view. The dashed line indicates the area occupied by the government. The Author's modelling and drawing.

their portrait in official regalia and an encomium praising their virtues, accomplishments, and social contributions.

The commemorative steles in the hall were organised chronologically in rows onto the rear wall. The initial steles bore the inscription 'Exalted Go Only the Wise and Good (*Jingxing weixian*),' setting a reverential tone. Below them, tables were set for daily offerings and the performance of seasonal ceremonies. Above the steles, a plaque bearing the inscription 'Teachers of Deeds (*Zuo zhi shi*),' accentuated the hall's role for moral education. Standing in front of the wall, a visitor would confront the gaze of these historical elites, being admonished to emulate their virtuous conduct.

Tao Shu and Liang Zhangju's group also represented themselves as contemporary local worthies through two garden paintings, which were inscribed onto steles and embedded in the wall of the corridor adjacent to Shrine of Five Hundred Worthies (Fig. 4.11(1)).[48] The first, *Painting of Five Old Men* (Wulao tu) was commissioned following the shrine's completion to illustrate their gathering at Surging Wave Pavilion.[49] In the painting, Han Feng (1783–1860) and Pan Yijun (1740–1830) are depicted as playing chess within the pavilion, where Shi Yunyu (1755–1837) fishes to their left, and Wu Yun (1811–1883) reads, seated on a stone to the right. The figure approaching them was Tao Shu. Comparing the depiction in the painting to the site model, it becomes clear that the pavilion, originally atop a mound, was depicted in the painting as along the shore of the Surging Wave. This spatial manipulation creates an impression of the figures as closely in contact with the pure water of the Surging Wave.

The second painting, *Painting of Seven Friends* (Qiyou tu), sets the scene at the Lesser Surging Wave pond (*Xiao Canglang*), which was newly dredged to the south of Shrine of Five Hundred Worthies.[50] This painting captures seven gentlemen deeply involved in intellectual and leisurely pursuits: playing chess, strumming the zither, composing poetry, and reading. Together with portrait steles, the stele paintings advocated the value that personages acknowledged by the locals, rather than by the emperor, truly represented the essence and purity of the Surging Wave.

Tao Shu's group promoted Rectify Meanings Academy to the north of Surging Wave. The academy was initially established by Tiyeboo in 1805 through repurposing the then-vacant official lodge, Near Mountain and Grove, as shown in Figure 4.11.[51] As Jiaqing emperor did not frequent southern inspections to Suzhou as his father Qianlong did, the visits to Suzhou by officials decreased, leaving the lodge unoccupied. It could be inferred that the transformation of this complex into an academic institution was likely to be seamless, with the preservation of existing shrines and the conversion of former meeting halls into lecturing spaces. The pre-existing book storage naturally complemented the educational function.

Tiyeboo advocated the aim of Rectify Meanings Academy as promoting evidential studies. This focus was explicated by its name, 'Rectify Meanings,' meaning to verify traditional knowledge with evidential studies. Tiyeboo also set forth three principal objectives for the academy in his record. The first was to advance 'solid learning (*shi xue*),' or evidential study, which challenged the abstract philosophy of neo-Confucianism in the Song dynasty.[52] The second goal was to produce more degree holders, and the third was to cultivate morality.[53]

With the backing of Liang Zhangju and Tao Shu's group in the 1820s, the academy gained its momentum in development. Liang's group donated ten thousand taels for students' subsidies and established awards for the excellent.[54] To enhance the academy's standing, Tao Shu invited the nationally renowned scholar Zhu Jian from Zhongshan Academy (*Zhongshan shuyuan*) in Nanjing as the director.[55]

To accommodate Zhu Jian, Liang renovated Garden of Suitability, situated to the west of the former official lodge (Fig. 4.11). The naming of garden's scenes was inspired by the Surging Wave's connotations of purging the impurities. The garden features 'Ladling Purity Pool (*yiqing chi*, Fig. 4.11(2), Fig. 4.1a(3))' at its centre, with a hall to its north. Another principal hall, oriented towards the waters of Surging Wave, was titled 'Wash Tassel Spot (*zhuoying chu*, Fig. 4.11(4), Fig. 4.1a(1)).'[56] Zhu Jian named the garden 'Garden of Suitability,' considering the garden's unpretentious style and the sceneries resonated with his personality.

Under the guidance of Zhu Jian and bolstered by Liang Zhangju's group, the academy rose to a provincial centre of learning. Its teachers and students published extensively, contributing significantly to the evidential study of *The Four Books*.[57] In 1873, Tongzhi emperor granted the institution a name plaque, honouring its esteemed status.[58]

The academy fostered an environment encouraging students to criticise contemporary political and social issues. A number of reformist scholars and officials attended its academic activities. Notably, in 1834, Lin Zexu (1785–1850), the renowned statesman who later waged a campaign against the opium trade, delivered lectures to students and evaluated their writings.[59] In his preface to published outstanding essays, Xu extolled the academy's members' rigorous study of classical texts.[60]

The sceneries of Surging Wave Pavilion nurtured students' literature talent. Teachers and students convened in the southern scenic site for poetic compositions, drawing inspiration from its natural beauty. They also maintained the seasonal rituals at Shrine of Five Hundred Worthies. This shrine encouraged students to aspire beyond academic and literary success, emphasising moral development. This ethos is captured in a couplet mounted on the columns of the shrine: 'The three regions of Wu maintained the custom of worshipping by setting up ritual tables and vessels.

The good local custom should not be sustained only by encouraging talents in imperial examinations and literary composition.'[61]

The Last Confucian Garden

During the 1820s to 1860s, the Qing empire was besieged by foreign invasions and internal rebellions. The Opium War in the 1840s and the Taiping Rebellion from 1850 to 1864 deeply shuttered the Confucian social order and values. Suzhou, deeply affected by the Taiping wars, saw its population halved and many of its structures ruined. The Surging Wave site and Rectify Meanings Academy were severely destroyed. After the war, Zhang Shusheng, a leading official in Tongzhi Restoration movement, undertook a comprehensive reconstruction of the site to restore its former glory as a paragon of Confucian gardens.

Zhang Shusheng's commitment to reinstate the Confucian ethics was manifested in the construction of Illuminate the Way Hall and the site's ornate waterfront, a vision captured in monk Jihang's (fl. late 19th c.) painting (Fig. 4.12a). Constructed along the eastern axis, Illuminate the Way Hall still stands today (Fig. 4.1a(9)). It was lauded by a Qing visitor as 'Lofty, spacious, and grandly broad, which truly claims its space as the main building of the whole garden.'[62] Couplets, inscribed on vertical plaques and hung on columns, imbued the hall with didactic messages and literature allusions. The name 'Illumining the Way' was an excerpt from Su Shuqing's *Record of Surging Wave Pavilion*, 'My body at ease, my mind thus becomes untroubled; nothing perverse seen or hear, the way thus becomes luminous.'[63]

During the Northern Song dynasty, the thing inflicted Su Shunqin was the political injustice. This serene and untamed garden provided him an escape, a place of retreat and solace. Differed from Su Shunqin, the adverse things Zhang Shusheng confronted at the Late Qing were challenges to the Confucian society: inner rebellions, foreign invasions, and the dissolution of Confucian values and literati culture. For Zhang, the restoration of the garden was a testimony to the enduring Confucian values, affirming their relevance amidst the upheavals and reformative pressures of the time.

Zhang Shusheng crafted a continuous and ornate waterfront façade as a visual testament to the grandeur of Confucian heritage. The façade was designed with exquisite details. The embankment was piled with yellow rocks, and the mound was adorned with Lake Tai rocks, presenting a visual contrast (Fig. 4.12a(2,3)). This picturesque waterfront façade rendered the site appealing to the public.[64]

Zhang admitted that the site's reconstruction was not an easy task. The government meticulously oversaw it to ensure appropriate spending.[65] From Zhang's conservative perspective, the site not only manifested Confucian ideology and literati culture, but also testified their group's

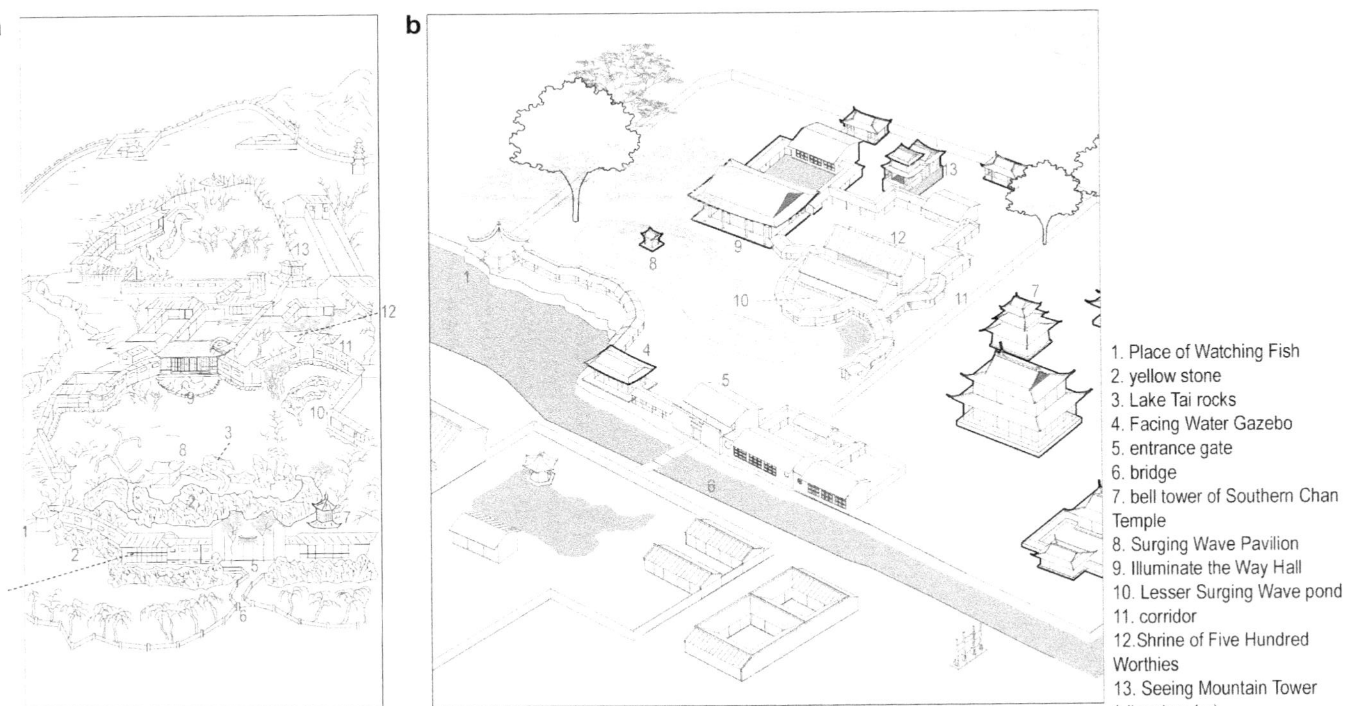

Figure 4.12 Translating Jihang's painting of the Surging Wave Pavilion into 3D model. (a) Author's traced drawing following Jihang. (b) Author's modelled drawing simulating the scene.

ability to restore the city to its previous social order, peace, and prosperity.[66] The Suzhou gentry resonated with Zhang's perspective, as one visitor wrote: 'upon our arrival, we observe crowds of visitors there as if it was a marketplace. It is truly a scene betokening an aged of peace and plenty.'[67]

Zhang Shusheng also restored Rectify Meanings Academy demolished by Taiping soldiers. He recruited former graduates of Rectify Meanings Academy, such as Feng Guifen (1809–1874) and Gu Wenbin (1811–1889), to take up the teaching roles.[68] In 1888, another academy, Learn the Classics Hall (Fig. 4.13(2)), was adapted from Garden of Suitability.[69] Although the academy offered courses in the western learning such as mathematics, its founder Zhu Kebao (fl. 1900s) integrated traditional pedagogical practices as well. The concern of preserving Confucian tradition persisted. In the 1900s, En Shou (d. 1911) adapted Illuminate the Way Hall to enshrine the neo-Confucianism scholars Zhu Xi (1130–1200), Chen Hao (1032–1085), and Chen Yi (1033–1107) in the Song dynasty.[70] This teaching method emphasised the indoctrination of principle (*li*), guiding students to behave according to their respective social role.

Summary

The transformation of Surging Wave from a Buddhist site into a Confucian one showcases how the government progressively encroached upon a temple. It began with Hu Zuanzong's violent intrusion into the Buddha hall. This encroachment was legitimised in the early High Qing, as the government purchased the monastic land and established the official estate with a clear boundary. In the middle High Qing dynasty, the government appropriated more Buddhist lands, which was further legitimised by emperor's visit. Since then the government controlled the central lands surrounding the Surging Wave, while Buddhists were relegated to the peripheral, distant from the Surging Wave. The gradual shrink of Southern Chan Temple's monastic land from the High Ming to the Late Qing indicates the vulnerability monastic properties to be transferred since Buddhists were marginalised from political arena in the Early Ming. The fate of other Buddhist estate likely followed a similar trajectory, contributing to an overall decline of the temple-scape.

The surging Wave site, which was inhabited by Buddhas and Buddhists, transitioned to serve Confucian idols and their disciples. The spatial privilege of appreciating the Surging Wave shifted from Buddhists to Confucians. The meaning of Surging Wave evolved from being a Buddhist 'mirror of wisdom' to symbolising official responsibility, and eventually, to representing the Confucian social order. The ideological shifts were manifested in the site's reconstructions and illustrated in model drawings.

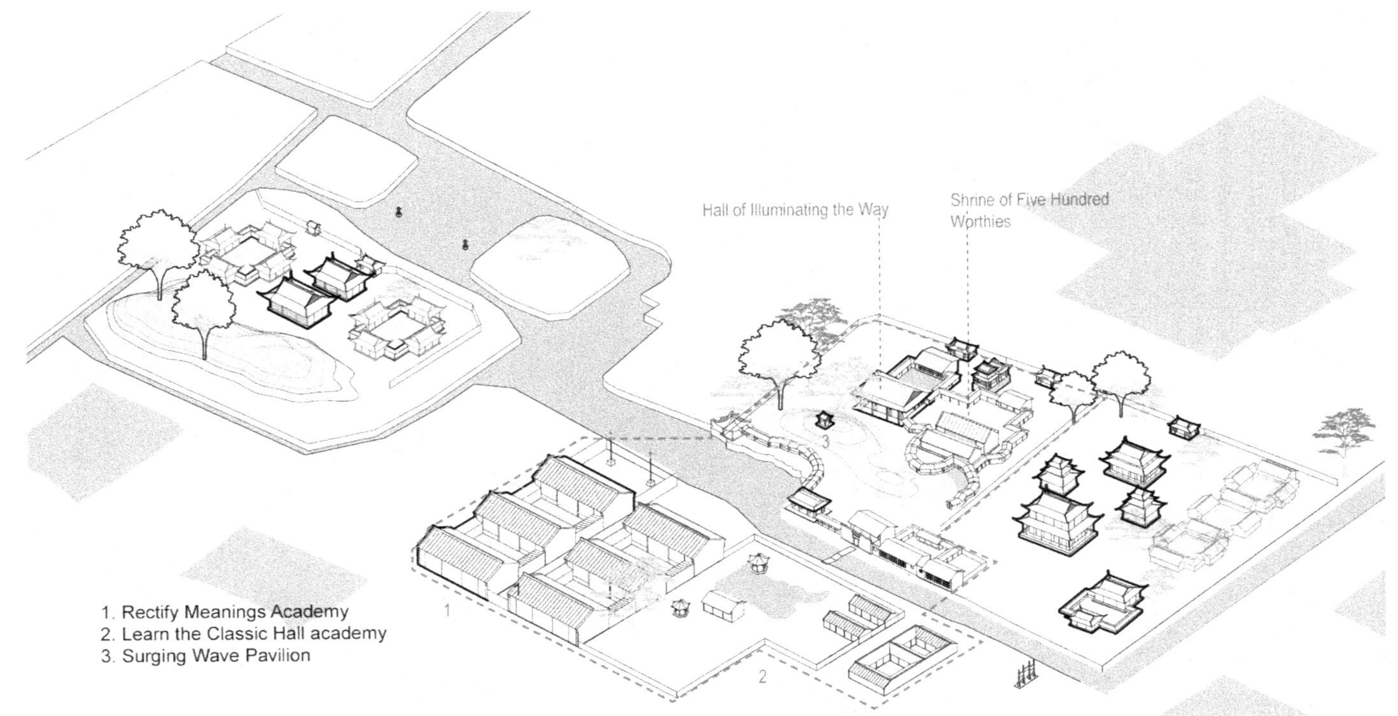

Figure 4.13 Shrines and academies surround the Surging Wave in the Late Qing, restored northwest isometric view. The dashed line indicates the area occupied by the government. The Author's modelling and drawing.

Although Hu Zuanzong, Zhang Guowei, Song Luo, Wu Cunli, Liang Zhangju's group, and Zhang Shusheng each had their unique perspectives of the site, they all leveraged two spatial strategies. The first strategy was to maintain the axial order of memorial halls. The Confucian shrines orienting towards Surging Wave Pond followed the spatial axes previously established by Witty Retreat Cloister. Tracing spatial changes in the six isometric model drawings, it is evident that Illuminate the Way Hall from the late Qing was built on the eastern Buddha hall of Witty Retreat Cloister and Shrine of Five Hundred Worthies was built on the western residential hall of the cloister. Both Rectify Meanings Academy and Learn the Classics Hall academy were constructed on the fields of Grand Clouds Cloister.

The second spatial strategy was choreographing the waterfront. Since the private Surging Wave Pond was integrated into the urban canal in the Late Ming, the pond's waterfront became publicly visible. In the early High Qing, Song Luo developed the site's waterfront façade to invite citizens (Fig. 4.6). In the late High Qing, the government deliberately designed a continuous waterfront façade with bridges across Surging Wave Pond to promote imperial messages to passersby (Figure 4.9). By the Late Qing, the site's intricate waterfront façade became a symbol of the past glory of a Confucian society (Figure 4.13).

The transformation of the Surging Wave site during Ming and Qing dynasties showcased how temple-scape gradually declined with Confucian landscape established. In the 1950s, the last land piece of Grand Clouds Cloister, located to the east of Surging Wave, was converted into a hospital. Meanwhile, the remaining lands of Gathering Clouds Temple were converted into residences. From then onwards, all monastic traces were eradicated around Surging Wave, except two sutra pillars (Fig. 4.1a(10,11)).

Confucianism's Urban Control: Converting Temples into Academies and Shrines

The government's intrusion into Southern Chan Temple in 1523 is indicative of the anti-Buddhism campaign happened in Suzhou during the 1520s to 1540s. This anti-Buddhism movement was led by Hu Zuanzong, who transformed four temples into Confucian institutions within three years (Fig. 0.2). In 1523, Forever Pacification Temple was converted into Golden Country Academy, Dragon Revival Temple (*Longxing si*) into Hejing Academy, and Witty Retreat Cloister into Shrine of the King of Qi. Two years later, Hu replaced Bright Virtue Temple with Learn the Way Academy.[71]

What social conditions prompted Hu Zuanzong's reformation of Suzhou's Buddhist tradition? This section investigates how Hu Zuanzong re-adapted monastic buildings into academies. The ideology Hu Zuanzong advocated is revealed through analysing the stele texts he inscribed. By converting temples adjacent to major canals, the government effectively exerted control over key urban nodes, facilitating a transformation of citizen's believes.

Prior to the anti-Buddhism movement, Confucian institutions in Suzhou were limited, consisting of one prefectural school and a few academies (Fig. 0.1).[72] Education for distinguished gentry degree holders, such as Wu Kuan and Wang Ao (1450–1524), primarily took place within family circles and via personal connections. For example, Wen Zhengming received his early education from his parents. Later he was tutored in literature, calligraphy, and painting by his father's friends, including figures like Wu Kuan, Li Yingzhen (1431–1394), and Shen Zhou.[73] Temples and gentry gardens hosted gatherings that nurtured young literary talents.[74] Prefectural schools thus served as preparatory grounds for degree entrance examinations. In contrast, Confucian institutions in northern China were more prevalent and exerted a substantial role in education.[75]

Contrast to the prosperity of Confucianism in the north, the mainstream of Suzhou's culture was Buddhism, literature, and art.[76] Buddhism was popular among Suzhou citizens, after its one-century accumulation in the Yuan dynasty.[77] Suzhou's literati culture in the Ming period emphasised the free expression of mind and feelings, diverging from the moral persuasion and official pursuits. This ethos is captured in Tang Yin's (1470–1524) poem *The Song of Peach Blossom Cloister* (Taohua An ge), which declares, 'I wish to grow old and die among flowers and wine. I do not wish to bow down to official carriages.'[78] The elite gentry in Suzhou identified their literati lineage as originated from the Yuan masters, such as Gao Qi, Ni Zan, and Zhao Mengfu. Emulating these masters' close affinity with Buddhism, the Ming elite gentry in Suzhou frequently gathered in temples, absorbing Buddhist ideas and creating art works. They identified Buddhism as a part of the local tradition.

The divergence from Confucianism in Suzhou's society, deeply conditioned by the region's rapid economic growth and commercialisation since the Middle Ming, was further exacerbated by diminishing opportunities to obtain academic degrees.[79] The gentry class engaged in economic ventures apart from seeking official success. Shen Zhou's poem *Odes on Money* (yong qian) reflects the growing aspiration of accumulating wealth.[80] As land remained a lucrative investment, many gentrymen actively acquired lands, rising to major landowners. This extensive land ownership increased the gentry's independence, leading to their growing detachment from the pursuit of government positions, and further, the traditional Confucian social order.

Gentry's diminishing interest in Confucian social order and their expanding land control were likely perceived by local officials as a threat to governance.[81] In an effort to steer society back towards Confucian principles, Hu Zuanzong attacked temples and converted them into academies and shrines. Within just three years, tablets of Confucian figures took place of Buddha statues at former monastic sites. Rituals worshipping these Confucian figures were established to promote their respect and reverence.

Forever Pacification Temple, a renowned Tiantai Buddhism centre during the Yuan dynasty, featured a typical courtyard layout as depicted in Figure 4.14(a). The temple's land was juxtaposed by canals to its north and south. Monastic buildings and garden elements were aligned in axis. The mountain gate, facing the major canal, symbolised the entrance to the Buddhist realm. Two lines of juniper trees flanked the passageway guiding visitors to the Buddha hall, adding solemnity to the worship. A delicately carved golden Buddha statue was enshrined in this hall.[82] Behind the Buddha hall stood the Dharma hall, named 'Sea Seal Hall (*Haiyin tang*),' where the Tiantai master Miaosheng lectured. Both the Buddha hall and the Dharma hall were flanked by corridors of monks' residence, a layout pattern inherited since the Song dynasty. Miaosheng's abbot's quarter 'Idling Chamber' was in the rear facing a pond, where the beautiful scenery inspired Miaosheng to write lots of poems. The temple was a popular destination for Yuan officials, as discussed in Chapter 2.

Yang Xunji's (1458–1546) poem in the Early Ming encapsulated the spiritual essence of the temple. The poet contemplated Buddhist philosophy as he was wandering in the temple's serene landscape and observing sensory details, as he wrote:

In the ancient temple within Wu city, it is said to have begun in the Tang dynasty. Tall halls reach up high where the eagle grinds its summit. Doors open to rows of Junipers.
The linen ground has witnessed a millennium. Lamps are lit in several divided rooms. Receiving the empty sky and clouds, the lake is tranquil. By the pavilion, wild plums grow.
Where dust stirs, there is no Dharma left. After chanting the sutra, the fragrance lingers. A brief visit, not a long stay, in this spiritual realm, I wander and drift alone. [83]

In the 1460s, the temple was devastated in fire and unable to recover, a situation described in Wu Kuan's poem: 'Idling Chamber was closed and empty beneath the spring tree. The pond and courtyard are filled with ashes after the catastrophe.'[84] The state of dilapidation left the temple particularly susceptible to conversion.[85]

Hu Zuanzong's conversion of the temple into Golden Country Academy exemplified a typical pattern of repurposing temples into academies (Fig. 4.14a,b). The Buddha hall was refurbished into a shrine dedicated to

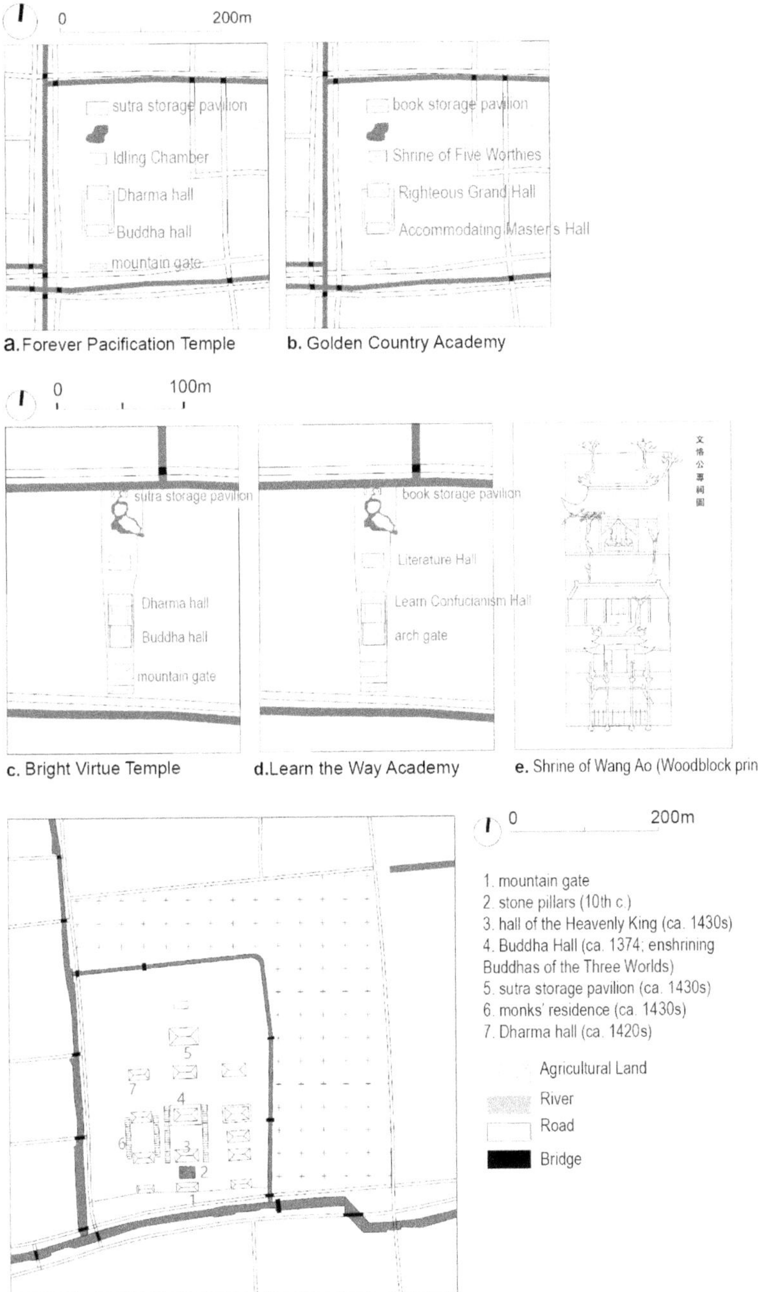

Figure 4.14 Temples converted into Confucian academies. Source: Author. The woodblock print drawing is traced following *Wangshi jiapu*.

Ziyu (b. 512 B.C.), whose posthumous title was 'Golden Country'.[86] The Buddha statue was replaced with a portrait of Ziyu, alongside portraits of other Confucian figures, Yin Tun (1071–1142) and Wei Liaoweng (1178–1237).[87] This newly established shrine was named 'Accommodating Master's Hall (*Yugong tang*)', with its prominent title displayed on a board at the entrance.[88] The former Dharma hall was converted into Righteous Grand Hall (*Zhengda tang*), a lecture space where Hu delivered speeches to students. The previous monks' residence flanking the two halls were converted into reading rooms and residences for students.[89]

Hu Zuanzong embedded a record tele into the wall to commemorate the founding of Golden Country Academy.[90] In the stele inscription Hu highlighted Ziyu's pivotal role in bringing Confucianism to Suzhou when the region was uncivilised during the 6th century B.C. He thus identified Ziyu's teachings as fundamental to Suzhou's Confucian heritage. He believed that enshrining Ziyu and creating an academy dedicated to this early Confucian patriarch would transform the society through education. As stated in the stele:

> Scholars need instructions, learning needs the root, governance needs a framework, and rituals will be rewarded. Instruct the students, they will follow. Start learning from the fundamentals, it will be deep. Govern with a framework, it will be uplifted. When rituals got reciprocated, they will be esteemed. To speak of governance and transformation without these is likely to be ineffective.[91]

In 1523, another temple subjected to Hu Zuanzong's conversion was Dragon Revival Temple. The temple likely featured a courtyard layout akin to Forever Pacification Temple, as suggested by *Pingjiang Map*. Founded in the 6th century by imperial decree of the Tang dynasty, it remained to attract literati in succeeding dynasties. For example, the esteemed poet Qi Wuqian (b. 692) wrote the poem *Lodge in Dragon Revival Temple*.[92] The temple proliferated in the Yuan and received poems from literati such as Zhao Mengfu, Ni Zan, and Yu Ji. They contemplated Buddhism in the monastic setting, reminiscent of Tang poetry. However, the temple seems to have fallen into disrepair during the Ming, as there are no records of renovations or literary works from this era. This state of dilapidation and lack of patronage rendered the temple particularly vulnerable to conversion.[93]

Hu Zuanzong repurposed Dragon Revival Temple into Hejing Academy, creating a space for students from suburban counties to study. The name 'Hejing' was from the posthumous title of Yin Tun, a distinguished Northern Song scholar in neo-Confucianism. As a leading disciple of the Chen Brothers, Yin Tun migrated to Suzhou and taught there in his later years. The conversion of Dragon Revival Temple likely mirrored the pattern seen in the conversion of Forever Pacification Temple. It could be deduced that the Buddha hall was converted into the shrine of Yin Tun, its Dharma hall into the lecture hall, and corridors to students' residences and study rooms. Students were instructed to learn classical texts and perform rituals in honour of Yin Tun.[94]

The stele inscription by Yuan Zhi (1502–1547) vividly depicted Hu's instructions inside the academy, presenting Hu as a dedicated and committed teacher:

> When had time, the master (Hu Zuanzong) would visit the academy. He discourses on the classes and resolves doubts. He closely supervises students' learning. He set up ritual vessels full of food and wines for worshipping in the hall and hangs the whip up at his teaching seat for punishment. He appointed educational intendants in the academy and established rules to make people from all directions to follow. If the later generations glorify this, they will say: the teaching of Suzhou and Huzhou has been revived. This is due to mater Hu's ambition.[95]

Yuan Zhi extolled Hu Zuanzong's efforts to rejuvenate Confucian teachings in Suzhou. He likened Hu Zuanzong to the innovative educator Hu Yuan (993–1059) of the Northern Song dynasty, known for his impactful teachings at Suzhou's prefectural school.[96] Yuan Zhi also claimed Buddhism as a heterodox and corrupt belief that ought to be supplanted by the veneration of Confucian sages.

In 1524, Hu Zuanzong's conversion of Witty Retreat Cloister into Shrine of the King of Qi identified Han Shizhong within Suzhou's genealogy of Confucianism.[97] The Buddha statue was replaced with the portrait of the King of Qi.[98] Hu Zuanzong regarded Han Shizhong as representative of the loyalty and bravery of Confucian ideal.[99] In the stele inscription, Hu praised Han Shizhong's military defence of the Song empire. He emphasised that reverencing the King of Qi aligned with rituals in *Rites of Zhou* (Zhou Li). Such worship would accumulate blessings from the deceased king. Hu's veneration of a military hero could also be interpreted as a response to the ongoing threats to the empire, such as the Japanese pirates along the southeast coast and Mongol invasions from the north.[100]

Hu Zuanzong also claimed the conversion of temples as a way to transform Buddhists. To be more persuasive, Hu invited Huang Xingzeng, a famous local gentry to compose the stele record. Huang wrote that people praised Hu's conversion, and Buddhists were pleased to be transformed, which might not be the fact.[101]

Hu Zuanzong's fourth target of conversion was Bright Virtue Temple. This temple's ground included the area of the present-day Shrine of Wang Ao and Mountain Villa of Embracing Beauty (*Huanxiu shanzhuang*), as shown in Figure 4.14c. The temple was situated between canals to its north and south. It featured sequential courtyards. Along the axis aligned two gates, followed by a Buddha hall, a Dharma hall, a sutra storage pavilion, a garden, and the abbot's quarter. The Buddha hall and Dharma hall was flanked by corridors of monks' residence, forming an enclosed courtyard.[102] The garden likely corresponds to the area of the artificial mound that is now known as Mountain Villa of Embracing Beauty.

Despite its fame in the Yuan dynasty, the temple faced financial crisis during the Ming dynasty. Monks struggled to maintain the temple independently. Its reconstruction in the 1400s relied on income mainly generated from monks' ritual services, rather than support from wealthy donors.[103] The scant literature works about this temple during the period indicates a possible lack of patronage from the local gentry. When Wen Zhengming visited in 1502, he noted that the temple was deserted and dusty, with no monks in sight, only a dog guarding the fence.[104] This combination of limited support and neglect made the temple particularly susceptible to conversion.

In 1525, Hu Zuanzong converted the temple into Learn the Way Academy (Fig. 4.14d), dedicating it to the teaching of Ziyou (506–443 B.C.). Hu emphasised Ziyou as pivotal in transmitting the northern civilised culture, such as literature, ritual, and music to Suzhou. This cultural lineage was further advanced by local figures such as Lu Zhi (754–805) in the Tang and Fan Zhongyan in the Song. The type of literature to which Hu referred—Confucian writings on moral discourse and official duty—sharply diverged from the literary tradition of Suzhou elites, which celebrated the expression of free spirit and retreat from officialdom. The tradition of Suzhou's literature outlined by Hu Zuanzong thus contrasts with local elites' account.

In his stele text, Hu Zuanzong explicated the necessity of founding an academy devoted to venerating Ziyou.[105] He argued that in Confucian teachings, ritual and music held precedence over literature. Drawing an analogy, he likened Confucian philosophy to a vast universe—literature represented the visible stars, while virtue and music constituted the unseen essence. Given Ziyou's renown for virtue and musical proficiency, Hu Zuanzong asserted that honouring Ziyou through the practice of ritual music represented the truest adherence to Confucian principles.

Hu Zuanzong converted extant monastic buildings to the academy's use, as illustrated in Figure 4.14d.[106] With the Buddha statue removed and the elaborate Buddhist decorations erased, the Buddha hall was repurposed into the shrine of Ziyou, aptly named 'Learn Confucianism Hall (*Xuekong tang*).' The hall's prime location expressed Hu's notion that venerating Ziyou was an entrance step to practice Confucianism. Behind, the Dharma hall was converted into 'Literature Hall (*Wenxue tang*)' for teaching.

The sutra storage pavilion was converted into Tower of Strumming and Singing (*Xiange lou*), where students would strum instruments and chant the classics there.[107] The monks' residential corridors were converted into dormitories for students. Living communally in the new academy, students were expected to be transformed not only through intellectual learning but also through disciplining.

Hu Zuanzong established two gates of the academy on the waterfront, displaying Confucian presence to the public prominently. The first gate bore the title 'The Southeast Hasten Towards the State of Lu (*Dongnan qulu*),' meaning that Suzhou, a city at the southeast, would follow the Confucian traditions originating from the northern State of Lu.[108] The second gate was inscribed with the academy's title 'Learn the Way Academy,' referring to the path of Confucianism. These ornate gates were intended to impress passersby about the grandeur and supremacy of Confucianism, a visual experience depicted in the woodblock print of Shrine of Wang Ao, into which the academy was later converted (Fig. 4.14e). Anyone pausing to look inside the academy would see Learning Confucian Hall and even catch a glimpse of the Ziyou statue within.

Hu Zuanzong justified his conversion of temples in three ways. First, he criticised monks' usurpation of the land of Learn the Way Academy during the Yuan dynasty. He recounted how the original academy, established in the Song dynasty near Jinfan canal in the city's southwest corner, was seized twice by influential Buddhists, leading to its eventual demise.[109] By reclaiming monastic lands for the new academy, Hu framed his actions as rectifying a historical wrongdoing. Second, he condemned the excessive decoration of the temple for violating ritual norms, arguing their removal was necessary to adhere to proper ritual standards.[110] Lastly, Hu portrayed the conversion as a considerate and pragmatic decision. Instead of constructing a new building, repurposing an extant temple into an academy would reduce the cost. He backed this conversion with approvals received from higher officials, such as Mr. Chen and Mr. Zhu, the provincial censors, and Mr. Lu, a superintendent of education.[111]

Hu Zuanzong's transformation of the four temples into three academies and one shrine followed a consistent pattern. The Buddhist statues were replaced with Confucian figures, and Buddhist decorations and furnishings were substituted with Confucian ones. Beyond the replacement and erasure, the original timber structure and axial layout of the temples were preserved for new purposes, thereby reducing costs. The Buddhist monastic communities previously residing in these temples were supplanted by disciples of Confucius, and new Confucian rituals and educational programmes were implemented. Stele inscriptions served to officially validate and promote the rationale behind these conversions.

The research finds that by enshrining five Confucian figures on the converted monastic sites, Hu Zuanzong crafted a continuous and evolving history of Confucianism in Suzhou. This lineage began with the transmission of Confucian teachings to Suzhou by Confucius' disciples Ziyou and Ziyu, dating back to the 6th century. The legacy was then embodied by Han Shizhong, a defender of a Confucian society. Yin Hejing and Wei Liaoweng in the Song dynasties promoted Neo-Confucianism and reformed Confucian pedagogy, making Suzhou's Confucian education competitive with other regions.[112] This Confucian lineage was advocated architecturally by shrines aligning in front of the city's urban canals.

By converting temples into academies, Hu Zuanzong achieved urban control. All the academies were located within the city, facing urban canals to maximise public accessibility and visibility. As citizens boated along the urban canal, the prominent gates, façades, and monumental structures of these academies visually communicated Confucian values. The academies hosted annual rituals that encouraged citizen's participation and offered educational programmes aimed at nurturing Confucian talents. Through these efforts, Hu Zuanzong realised his objective of 'transforming through teaching (*jiao hua*)' and effectively implanted the 'root (*ben*)' of Confucianism in Suzhou, reshaping the city's cultural and philosophical landscape.

Transforming Buddhist Landscape: Converting Ten Thousand Longevity Chan Temple into the County School

'Outstanding spirit preferred this site (Ten Thousand longevity Chan Temple. Finally it became a school, this was not by accident.'

> Zhu Xizhou (1473–1557), Record of Reconstructing Changzhou Confucian School[113]

'Thus all scholars could benefit from wiping off Buddhism and returning back to Confucius. They took full responsibility to drive off the demonic and the obscene. Once the way of Buddhism is put off, the way of the sage will be prominent.'

> Xu Jie (1503–1583), Record of Revere the Classics Hall in Changzhou County School[114]

The above two excepts are from the stele records of Changzhou County school, established in the 1540s on the site of Ten Thousand Longevity Chan Temple. These excerpts captured the government's intolerance of the monastic possessions of favourable geomantic lands within the city.

Officials explicitly advocated the elimination of Buddhism and the acquisition of monastic lands. The conversion of temple in the 1540s far-outweighs that in the 1520s in terms of the scale of violence involved, the extent of land seized, and the number of collaborators participated. The government not only refurbished monastic buildings but also renovated the temple's landscape to align it with the Confucian ideology.

Ten Thousand Longevity Chan Temple occupied an extensive area advantageously located with major canals flanking its west and south sides, as depicted in the map of Figures 0.1(a4).[115] Monks dredged a private river connecting these two canals, dividing the land into zone for building complex and agricultural fields (Fig. 4.14f).[116] This river provided water for daily living, irrigation, and Buddhist rituals. The temple owned the collection of Tang monk Guanxiu's (832–912) Sixteen Arhats' portraits and featured two ancient tenth-century stone pillars at its entrance (Fig. 4.14f(2)).[117] In the Southern Song dynasty, the temple was recognised as one of the ten notable monasteries in the southern Yangtze river region.[118] Its prominence continued to the Yuan dynasty, as monk Tianyin Yuanzhi (d. 1635) observed, 'at the sound of the drum, people came in like clouds.'[119]

The temple complex was structured along two primary axes, defined by Buddha hall and Dharma hall respectively (Fig. 4.14f(4, 7)). The layout was formed by monks' century efforts. The Buddha hall was reconstructed by Xingzhong Ren (fl. 14th c.) in the 1370s (Fig. 4.14f(4)).[120] Its grandeur was depicted by Song Lian (1310–1381) as 'its four eaves resemble the pleasant birds expanding their wings, its painted wooden decorations are so splendid which dazzle people's mind and heart.'[121] Within this hall enshrined Buddha statues of the Three Worlds, which was installed by monk Yingzhong Huan (fl. late 14th c.).[122]

The western axis was distinguished by the Dharma hall (Fig. 4.14(7)), constructed by Meng'an Nan (fl. 1420s–1430s) in the 1420s. He also built corridors connecting the Buddha hall and hall of the Heavenly King (Fig. 4.14(3)), creating an enclosed courtyard.[123] Monk official Jueyi Jieli (fl. early 15th c.) went to the court in Beijing and received a collection of sutras, prompting the construction of a two-storey sutra storage pavilion (Fig. 4.14f(5)).[124] Wang Ao's poem captured the panoramic view commanded by the pavilion's upper floor. He saw the twin pagodas of Determined Wisdom Chan Temple (*Dinghui chan si*) shimmering in the sunlight and smoke rising from dwelling courtyards.[125] Wu Kuan noted that after four Buddhist generations' effort, this temple was fully restored and became a supreme temple in Suzhou.

This prominent temple with a long tradition was targeted by teachers and students in Changzhou County School in the 1540s. They coveted the temple's prime location and its substantial structures. Positioned on the elevated ground and surrounded by flowing water, the temple's setting was both aesthetically pleasing and spatially advantageous. Its buildings were robust and spacious. This was in stark contrast to the situation of their school, which was situated in a lower area with limited space.[126]

The teaching intendant Xiao Wen (fl. 16th c.) thus reported to provincial authorities and obtained their approval to convert the temple.[127] However, this initiative was halted due to the government's pressing need to combat pirates.[128] In the following year, teachers and students of Changzhou County School, backed by the prefects and magistrate, appealed to the new provincial censor Shu Ting to resume this project.[129] Shu Ting inspected the temple and expressed his indignation when seeing monks occupying such a large area of land with favourable geomancy (*fengshui*). He remarked, 'Those monks' residence was so luxurious! A school was for cultivating the literati. How could the space of school be worse than the temple? The school must be relocated to this temple's site. This project is unstoppable.'[130] He thus assigned a group of officials to transform the temple.[131]

The swift conversion of the temple to school was completed in just four months, from December 1541 to March 1542.[132] The efficiency was achieved by reusing monastic artefacts while removing Buddhist spatial motifs.[133] Referring to the gazetteer map (Fig. 4.15a) and deducing from official records, the layout of Changzhou County School is reconstructed in Figure 4.15b. The Buddha hall was transformed into Primordial Teacher's Hall (*Xianshi dian*, Fig. 4.15(19)), where the once awe-inspiring Buddhas of the Three Worlds were replaced by the portrait of Confucius.[134] The converted Primordial Teacher's Hall remains standing today, which allows us to imagine its grandeur as a Confucian shrine in the Ming and as a Buddha hall prior to the conversion.[135]

The Dharma hall was repurposed into the lecture hall, renamed as 'Illuminate Ethics Hall (*Minglun tang*, Fig. 4.15(9)),' while monks' corridors were converted into students' dormitories. The hall of the Heavenly King became Gate of Confucius Temple (*Wenmiao men*, Fig. 4.15(20)), and the sutra storage pavilion was turned into Revere the Classics Pavilion (*Zunjing ge*, Fig. 4.15(17)).[136] In addition to these monumental buildings, the school's offices, warehouses, granary, and kitchens were also largely adapted from monastic structures.[137] The rapid and comprehensive conversion demonstrates the versatility of Chinese architecture in accommodating contradictory cultures.

Apart from fully utilising monastic structures, the officials meticulously transformed the temple's landscape to demonstrate Confucian ideals. On the waterfront, the newly erected Lattice Star Gate (*Lingxing Men*, Fig. 4.15(21)) and Confucian Gate (*Ruxue Men*, Fig. 4.15(7)) were elaborately designed to advocate the grandeur and authority of the Confucian school. The entrance arch, situated south of the public canal, bore the inscription

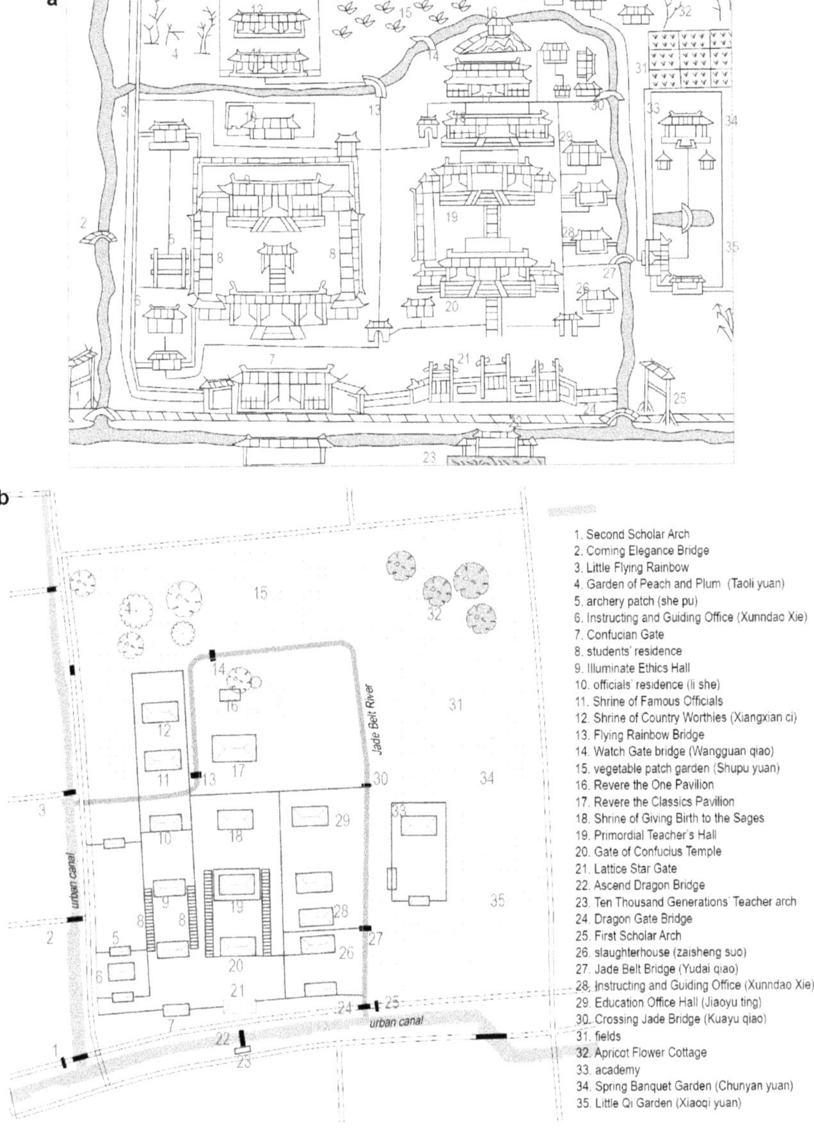

Figure 4.15 Translating gazetteer map into plan. (a) Map of Changzhou County School, the late 17th c., a woodblock print. Author's traced drawing following Zhang Defu. (b) Author's drawing.

'Ten Thousand Generations' Teacher (*Wandai Zongshi*, Fig. 4.15(23)),' referring to Confucius as the ultimate teacher. The Pan pond (*pan chi*) in front of the gate further enhanced the ritual grandeur of the entrance. The bridge in front of Lattice Star Gate was named 'Ascend Dragon Bridge (*Shenglong qiao*, Fig. 4.15(22))', and the one west of Confucian Gate was called 'Dragon Gate Bridge (*Longmen qiao*, Fig. 4.15(24))'. Additionally, entrance arches bearing the names 'First Scholar (*Zhuangyuan*)' and 'Second Scholar (*Jieyuan*)' were erected (Fig. 4.15 (25, 1)), symbolising the aspiration that the water from the canal would bring forth more degree achievers.

The river that monks had originally dredged was repurposed as a public canal, now named 'Jade Belt River (*Yudai He*).' Spanning this river and the public canal, a series of bridges were named with Confucian connotations. For instance, the name, 'Coming Elegance bridge (*Laixiu qiao*, Fig. 4.15(2)),' carried the notion that deep origins lead to clear waters, implying that a profound Confucian tradition clarifies further learning. This allusion was revealed in Wang Ruyu's (d. 1415) poem *Coming Elegance Bridge*, which describes the limpid water flowing into the Confucian school.[138]

The temple's agricultural fields were transformed into gardens infused with Confucian symbolism. One such garden, named 'Apricot Flower Cottage (*Xinghua cun*, Fig. 4.15(32)),' was imbued with the spirit of teaching and learning, as Confucius was said to have taught his students under an apricot tree.[139] The most laborious landscaping was the construction of the mound of Revere the One Pavilion (*Jingyi ting*, Fig. 4.15(16)). A pavilion atop a mound, often referred to as 'Way Mountain Pavilion (*Daoshan ting*),' was an iconic Confucian scene in schools.[140] The landscape of the school was both didactic and pleasing, which was glorified in the record: '[scenic sites] for touring and visit were all equipped.'[141]

Upon its rapid completion, Changzhou County School was honoured with stele inscriptions from Zhu Xizhou, a provincial officer from Nanjing, and Xu Jie, Grand Secretary from the central court. Zhu's inscription meticulously inventoried all individuals involved in the project, implying acknowledgement and appreciation of their efforts. He proclaimed that the temple's favourable landscape was ideally suited and destined for use as a Confucian school, a claim that is quoted at the beginning.[142] Zhu further encouraged students to aspire for high achievements, to fulfil the purpose of the new environment: 'The grandiose of the newly constructed school is unsurpassed. Is this just for appreciating beautiful architecture and sceneries? Only when the residing students improved their virtue and study did the magnificence of the school achieved its goal.'[143]

The second stele record from Grand Secretary Xu Jie was dedicated to Revere the Classics Pavilion. Xu tactically addressed the temple by its colloquial name, 'Blessed Peace Temple (*Funing si*),' avoiding its official name 'Ten Thousand Longevity Chan Temple.'.[144] The name choice omitted the temple's imperial recognition in the Song dynasty. Xu Jie praised those involved in the conversion for their adherence to Confucianism and commitment to abandoning Buddhism. He described Buddhism as heretical, highlighting its harmful influences and asserting that it cannot coexist with Confucianism. Bringing in historical anti-Buddhist sentiments, he advocated for the elimination of Buddhism:

> The teachings of heretical sects have sometimes led to confusion and delusion. Thus, dispelling the evil and banishing licentiousness are vital for the success of a generation. The teaching of revering the classics holds a significant position, and gentlemen too cannot shirk their responsibility in this matter.
>
> Since the Three Dynasties, the teachings of these sects have been sufficient to muddle people's hearts, and many are those led astray. Among these, none are more confounding of truth and order than Buddhism. In the past, Han Yu wanted to fire Buddhist texts and demolish their dwellings, and Ouyang Xiu had made the principles of the roots. The above testifies the intolerance between the ancient sage and the Buddha, which is intrinsic of the natural order.[145]

These anti-Buddhism actions, while lauded in the official stele records, were viewed as villainous by some local gentry. Qian Xiyan provided a vivid account recording the violent occupation of the temple:

Censor Shu Destroyed the Temple

In the Jiajing era, there were two and three literati. They were from bossy families. One day they intruded into the temple. They blamed the monks for not accommodating well. At that time, magistrate Shu Ting, who was from Fujian, governed Suzhou. He only believed in Confucianism and was opposed to Buddhism. He adopted suggestions from these literati and other officials and ordered the temple to be converted into the new Changzhou County School.

They got ropes from the official warehouse. No one knows what they would do with the millions of ropes. At midnight, labourers and soldiers marched into the temple. They dismantled tiles and destroyed beams, burned sutras and pulled down statues. Three hundred monks were dispelled from the temple. They cried and whined around the transporting carts. The Three Worlds Tathāgata Buddha (*sanshi rulai*) could not be torn down because it was sculpted from a single tree.[146]

The details in Qian Xiyan's account conveyed a strong sense of condemnation of these Confucian activists, portraying them as aggressive, intolerant, and excessively zealous. The student participants are described as coming from overbearing families, and the official Shu Ting was characterised as a fervent Confucianist with a strong bias against Buddhism. The swift conversion under the cover of darkness suggests a tactic to avoid public scrutiny or resistance, taking advantage of night hours to execute a controversial plan. The demolition of the temple was intensive and thorough. Soldiers and labourers, equipped with thousands of ropes, were involved in dismantling tiles, destroying beams, burning sutras, and toppling down statues. Carts were used to transport the robes and remove the debris and remnants of the destruction.

Another gentry, Xu Shupi (1596–1683), offered a similar observation as Qian Xiyan's. He referred to the eight leading students in the demolition as 'eight eccentrics' and vividly portrayed their arrogance and destruction of Buddha statues:

> However, only the smaller Buddha statues were removed. Despite being pulled by dozens of people, the large statue in the main hall remained steadfast. One of the Eight Eccentrics, whose name I forget, climbed atop and swung an axe, first chopping off the head and then smashing the body.[147]

Both Qian and Xu regarded these Confucian activists as brutal and over zealous, whose suffering from karmic retribution was inevitable. Their narratives suggest that this violent conversion was contentious and not been widely accepted by the society. Despite this, there was no record of anyone daring to stand out to intervene and stop this catastrophe. The Buddhists, oppressed and powerless, had no means to resist the conversion. As Qian Xiyan described, monks were forcibly exiled from their temple, weeping beside carts transporting ropes and relics.[148] One monk, Zhenguo (fl. 16th c.), refused to leave and built a hut at the eastern corner of the new school.[149] Contrasting with Hu Zuanzong's approach in the 1520s, which sought to transform Buddhists, the anti-Buddhist movement of the 1540s aimed for a complete eradication of Buddhism.

The degree of violence accelerated from Hu Zuanzong's conversions in the 1520s to Shu Ting's aggressive assault on a temple in the 1540s. As the number of anti-Buddhists grew, so too did their boldness, evident in their targeting of a much grander and substantial temple. This trend indicates that anti-Buddhist sentiment persisted and grew in the city after Hu Zuanzong's 1520's reformation. What required careful legitimisation in the 1520s turned into an overt and violent campaign in the 1540s. The government openly declared Buddhists undeserved to possess prime land, calling for eliminating Buddhism. The school established in the 1540s was much larger and splendid than its 1520s predecessors. It is also more enduring. While academies built by Hu Zuanzong were demolished in the late 17th century under the edict of destroying academies, Changzhou County

School survived and thrived through subsequent centuries. Primordial Teacher's Hall of the school remains standing today, reminding us of its contentious founding history.

Conclusion

The conversion of five temples illustrates the government's transformation of the temple-scape into a Confucian one in three steps. The first involves acquiring Buddhist lands through coercion, as seen with Hu Zuanzong in the 1520s, through violent invasion during the peak of anti-Buddhism in the 1540s, and via purchase in a seemingly legal manner in the Qing dynasty. Second, Buddhist symbols were replaced with Confucian ones: Budda statues were removed to install Confucian images, Buddha halls were converted into Confucian shrines, and Dharma halls into lecture halls. The final act was to establish steles to commemorate the conversions and articulate Confucian values.

Two spatial strategies were notable in the government's transformation of temple-scape. The first strategy established axial monumental artefacts facing the urban canal, whereby visitors would be imbued with Confucian values as entering into layers of halls inscribed with Confucian metaphors. This axial order, however, was inherited from the layout of previous temples. The second strategy was to choreograph the waterfront façade by crafting scenes along the shore. The government efficiently utilised the potential of former temples' waterfronts, both axial and sequential, to convey Confucian messages to the public. In both cases, citizens absorbed didactic meanings through movement.

Notes

1 See the portrayal of urban Buddhist scenes and sounds in Gao, 'Preface of the Twelve Odes of Shizi Lin.'
2 'I was so taken with it that I paced back and forth around it. Then I procured it for forty thousand cash.' in Su Shunqin, '滄浪亭記 Record of Surging Wave Pavilion,' in *CLXZ*, shang juan, 13–14. Translated by Xu Yinong in Hardie and Campbell, eds., *The Dumbarton Oaks Anthology of Chinese Garden Literature*, 245–247. The subsequent translations by Xu Yinong are from the same book. Researches on the site's evolution include Xu, 'Interplay of Image and Fact'; Ye, 'The Construction of the Confucian Garden.'
3 See Su, 'Record of Surging Wave Pavilion,' translated by Xu Yinong. Wang Jin proposes that Su Shunqin's pavilion was located at the northern foot of the mound, next to water where the shore bends. See Wang, 'How Could This Turbidity Clarify the Tassel.'
4 '形骸既適則神不煩，觀聽無邪則道以明.' In Su, 'Record of Surging Wave Pavilion,' trans., Xu Yinong.
5 Wei, *History of Classical Gardens of Suzhou*, 135.
6 Gathering Clouds Temple was constructed in the Tang dynasty by monk Yuanshui 元邃. Witty Retreat Cloister was constructed by monk Zongjing

during 1314 to 1320 in the Yuan dynasty. Great Clouds Temple was constructed by monk Shanqing during 1341 to 1370. In *GSZ*, vol. 29, 19–24.

7 In Minghe 明河, 'Shiying Zhuan 示應傳,' CEBTA, X77n1524_025, 0530a03. Also recorded by Wu Kuan, 'Records of the Reconstruction of the Great Buddha Hall of Gathering Clouds Temple of Southern Chan Abbey 南禪集雲寺重建大雄殿記 (Nanchan Jiyun Si Chongjian Dadian Ji),' in *CLXZ*, shang juan, 18–19.

8 '吳中諸蘭若多出前代褒崇. 至於親承聖渥, 惟此得之. Most temples in Suzhou were highly esteemed by previous dynasties. However, only this temple received imperial favour during the Ming dynasty.' in Yang Xunji, 'Records of Reconstructing Grand Clouds Cloister 大雲菴重建殿宇記 (Dayun An chongjian dianyu ji),' in *WDWCXJ*, vol. 30, 40–42.

9 Wu, 'Records of the Reconstruction of the Great Buddha Hall of Gathering Clouds Temple of Southern Chan Abbey.'

10 Pictorial evidence of the Bodhisattva Hall of Southern Chan Temple is recorded in Grand Ceremony of the Southern Inspections. The photo of Gathering Clouds Temple's bell and drum towers and its Buddha hall is in https://dfzb.suzhou.gov.cn/dfzb/gcty/200812/a129a6900cd348da98d6a5f0e134b38c.shtml, assessed in 2024. 05.

11 See Wu, 'Records of Reconstructing Grand Clouds Cloister.'

12 See Zhou Chen, *Painting of Surging Wave Pavilion* (Canglang ting tu), colour ink on paper, handscroll, 35.8 × 300cm, Beijing: National Museum of China.

13 See the fields in *Map of Gusu City*.

14 '其水從葑溪而西, 過長洲縣治, 由支港稍南折而東. 複南衍至庵左. 流入環後如帶, 匯前為池.' 'Preface of Poems on Visiting Thatched Cloister 草菴記遊詩引 (Cao'an jiyou shi yin),' in the post-inscription of Shen Zhou's painting 'Painting of Thatched Cloister.' Also see *WDWCXJ*, vol. 30, 42–43.

15 '今日滄浪休問主, 百年興廢本同波. 煎茶又勺吳僧鉢, 濯足空傳孺子歌. 只在城南自清徹, 車塵馬足有誰過.' in Shen Zhou, '滄浪亭故址為僧所居 (Canglang Ting guzhi wei seng suoju).'

16 '大雲菴下水,蕅子舊沧浪. 净眼看塵界, 洗心開道場. 芳蓮污不染, 慧鏡湛無方. 潭雨涵空碧, 雲天自景光.' in Zhou Lun 周倫, 'Grand Clouds Cloister大雲菴 (Dayun An),' in *WDFC*, 625. Thanks for Xu Zhu's discussion with me on the poem's allusion to Buddhist philosophy.

17 '嘉靖二年, 知府胡纘宗因於其寺前殿撤佛像, 塑王冕旒七章位其中. 凡用仲秋一行, 特牲豕一. In the second year of Jiajing's reign (1524), the prefect Hu Zuanzong removed the Buddha statue from the Buddha hall in front of the temple and sculpted the statue of the King of Qi wearing crown with seven strings of tassels there. The annual worship was arranged every mid-autumn, using one sacrificed swine in particular.' in Yang, *Gazetteer of Wu County and Changzhou County*, 52.

18 See Huang Xingzeng, 'Han Qi Wang Miao beiji 韓蘄王廟碑記 (Stele Record of Shrine of the King of Qi),' in *CLXZ*, shang juan, 15–16.

19 See Hu Zuanzong, 'Record of the Newly Constructed Shrine of the King of Qi 新建韓蘄王廟記 (Xinjian Han Qiwang Miao Ji),' in *GSZ*, vol. 29, 20–22.

20 '遂俾良定輩啟扃居守, 悉蠲里役. 乃至屢給符帖焉. 可為釋子之風勸矣乎. Thus [Hu Zuanzong] ordered Liangding and other monks to open the temple's gate [to the public] and to guard the shrine. All the corvee labour for this shrine was exempted, and many official seals and documents were issued. This exemplifies as encouraging proper conduct among Buddhists.' in Huang, 'Stele Record of the King of Qi.'

21 '懼久而輪復, 乃磨堅石, 用紀始事. Fearing that in the future the shrine might be restored to its former state as a temple, I polished the hard stone to inscribe this event.' in ibid.

22 '寺住持良定輩樂而從焉 ⋯ 郡縣嘉之. The abbot of the temple, Liangding, and others were pleased to comply⋯The prefectural and county officials praised.' in Huang, 'Stele Record of the King of Qi.'

23 Gui Youguang, 'Record of Surging Wave Pavilion 滄浪亭記,' in *CLTZ*, 172–173.

24 Wang Jin noted that the renovation of public canal in the Late Ming connected Surging Wave pond with the pond of the prefectural school. He considered that this renovation occurred in the late reign of Jiajing. In Wang, 'How Could This Turbidity Clarify the Tassel.'

25 See Zhang, 'Pictorial Explanation of Water Hydraulics of the Entire Region of Suzhou Prefecture (Suzhou fu quanjing shuili tushuo)' in Zhang, *Complete Record of Hydraulic Management in the Region of Wu*, 318–319.

26 See Song Luo, 'Juanzhi Canglang Ting tian shu shike 捐置滄浪亭田數石' and 'Xuzhi Canglang Ting tian shu jidi geshu bei 續置滄浪亭田數基地各數碑 (The second stele of adding several fields and plots to the site of Surging Wave pavilion),' in *CLTZ*, 96–98.

27 '從廊門出, 有堂翼然. 祠子美木主其中. 而榜其門曰蘇公祠. 則仍舊屋而新之. Exiting from the corridor gate, there is a hall with extended eaves. Inside, memorial tablets of Zimei was enshrined. The plaque hanging at the door reads 'Shrine of Master Su'. The shrine was adapted from the old building.' in Song Luo, ed., 'Record of Reconstructing Surging Wave Pavilion 重修滄浪亭記 (Chong Xiu Canglang Ting Ji),' in *CLXZ*, juan xia, 1–2.

28 '亭之南, 石磴陂陀, 欄楯曲折. 翼以修廊, 顏曰步碕. To the south of the pavilion, stone steps wind up the slope, the railing twist and turn. Flanked by the long corridor, its plaque reads "Walking Along the Winding Bank".' See Song, 'Record of Reconstructing Surging Wave Pavilion.'

29 '蘄王英魂定來此, 會與長史相周旋.' in Song Luo, 'Surging Wave Pavilion in the Rhyme of Master Ouyang 滄浪亭用歐陽公韻 (Canglang Ting Yong Ouyang Gong Yun),' in *CLXZ*, juan xia, 2–3.

30 Huang Xiao and Liu Shanshan have identified several artefacts in Wang Hui's painting in correspondence with the current site artefacts. In Huang and Liu, 'Garden Paintings.'

31 See Song, 'Record of Reconstructing Surging Wave Pavilion.'

32 Ibid.

33 Xu, 'Interplay of Image and Fact.'

34 See the painting of Surging Wave Pavilion in Gao Jin, *Pictorial Anthology of the Famous Scenic Sites from the Grand Ceremony of the Southern Inspections*, 65.

35 See '康熙御賜吳存禮詩聯碑 Stele of Kangxi's Imperial Poem Couplet endowed to Wu Cunli' in *CLTZ*, 98.

36 See '膏雨足時農戶喜, 縣花明處長官清.' in ibid.

37 '遂飭工庀材, 建禦書碑亭於其中. Therefore I ordered labourers and prepared materials, built an imperial pavilion inside.' in ibid.

38 '而其旁屋宇亦令增修, 以助亭之壯麗.' in ibid.

39 '或擴之, 或增之, 移易其勢, 皆森列環向. 如眾星之拱極, 如百川之朝宗. 崇閎巍煥, 為萬載之雄觀.' in ibid.

40 The high public visibility of Wu Cunli's imperial stele was noted by Zhang Shusheng, 'Record of Reconstructing Surging Wave Pavilion 重修滄浪亭記 (Congxiu Canglang Ting Ji),' in *CLTXZ*, vol. 2, 4–5.

41 See Fu and Jueluo, *Gazetteer of Suzhou Prefecture*, vol. 14, 9.
42 'Imperial inscription Stele "On the Flood Disasters Happened in Jiangnan Area" (the 12th year of Qianlong's reign) 乾隆十二年御書江南潮災嘆碑' in *CLTZ*, 98–100.
43 The locations of the two steles were confirmed by Liang Zhangju: '亭左右刻有康熙年間御書 "膏雨足時農戶喜", 縣花明處長官清聯語, 及乾隆年間御書 "江南潮災嘆".' in Liang Zhangju, 'Record of the Event of Reconstructing Surging Wave Pavilion 重修滄浪亭紀事 (Chongxiu Canglang Ting Ji Shi)', in *Gazetteer of Surging Wave Pavilion*, 'Stele Records (beiji),' 1–2.
44 The current archway was constructed in Kangxi's reign, as recorded in *CLTZ*, 22.
45 Han, *After the Prosperous Age*, 203–205.
46 These scholar officials' local activism is discussed by Han Seunghyun, in ibid.
47 See photos of Shrine of Five Hundred Worthies in *CLTZ*, 68, 264–265.
48 '陶澍 "滄浪五老圖" 位置: 仰止亭南. 數量: 一塊. 尺寸: 1.5米 × 0.4米. 年代: 道光間所刻. 朱珔 "七友圖" 位置: 仰止亭南. 數量: 一塊. 尺寸:1.5米 × 0.45米. 年代: 道光中所刻.' In *CLTZ*, 104.
49 See the stele paintings in Ye, 'The Construction of the Confucian Garden,' 269.
50 See the painting in ibid., 270.
51 See Tiyeboo, 'Record of Rectify Meanings Academy 正誼書院記 (Zhengyi Shuyuan Ji)'; The academy was also mentioned in Zhu Jian, 'Record of Garden of Suitability 可園記 (Ke Yuan Ji),' in Cao and Li, eds., *Gazetteer of Wu County*, vol. 27, 4–5.
52 '茲當聖天子敦崇實學, 雅化作人巽命, 重申諄諄. At the time, the holy emperor earnestly encourages practical learning and cultivates the people with gentle command, repeatedly emphasising it (the practical learning) with genuine sincerity.' in Tiyeboo, 'Record of Rectify Meanings Academy.'
53 In the record, Tiyeboo first listed the degree holders produced by newly founded academies to highlight how academies could contribute to the increase in degree holders. In ibid.
54 '道光二年, 布政使廉敬率屬捐銀一萬兩生息充費. In the second year of Daoguang's reign (1822), Provincial Administration Commissioner Lian Jing led officials to donate ten thousand taels of silver to generate income for students' living expenses.' in Cao and Li, *Gazetteer of Wu County*, vol. 27, 2.
55 See Lu, *Comprehensive Collection of Academy Curriculum and Arts in the Qing Dynasty*, 270.
56 Zhu, 'Record of Garden of Suitability'.
57 See 'Selected Course Works of Rectify Meanings Academy正誼書院課選 (Zhengyi Shuyuan kexuan),' in Lu, *Comprehensive Collection of Academy Curriculum and Arts in the Qing Dynasty*, 269–303.
58 '十二年, 巡撫張樹聲重建舊地, 奏頒帝書正誼明道額. In the twelfth year of Tongzhi's reign (1873), Provincial Governor Zhang Shusheng reconstructed the academy and recommended it to Tongzhi emperor. The emperor inscribed "Rectify Meanings and Clarify the Way" on the name plaque bestowed to the academy.' in Cao and Li, *Gazetteer of Wu County*, vol. 27, 2–3.
59 Zhongshan daxue lishi xi, *Anthology of Lin Zexu's Diaries*, 139, 145.
60 See Lin Zexu's preface to 'Selected Course Works of Rectify Meanings Academy.'
61 '三吳崇俎豆, 維風勵俗, 豈徒在科第文章.' in Anonymous, 'Record of Touring Surging Wave Pavilion.' Translation made with reference to Xu Yinong.
62 '誠足為全園主屋.' in Anonymous, 'Record of Touring Surging Wave Pavilion,' translated by Xu Yinong.

63 '形骸既適則神不煩, 觀聽無邪則道以明.' in Su Shunqin, 'Canglang Ting ji,' translated by Xu Yinong.

64 See Anonymous, 'Record of Touring Surging Wave Pavilion.'

65 '咸樂還承平之舊, 亦遂忘其勞費之多. 蓋諸君之用心, 唯兢兢焉. 以作無益害有益是懼. 故考成而勸勤, 用嗇而度豐, 其慊于人心也如此. All happy to see that the site had regained something of what it used to be in the previous peaceful times. People seem to not realise the great amount of labour and cost spent on in Zhang Shusheng, 'Record of Reconstructing Surging Wave Pavilion.' translated by Xu Yinong, with author's modification.

66 '群有奮乎百世之心, 治道懋而風化興. 於以上迓天和, 阜成民裕, 百廢具舉, 將遠追百年以前之隆. 則是亭之成, 實為之兆. 其非奢望也夫, 其尤當共勉也夫. The hearts of a hundred generations are in accord, as right principles govern the state and good morals and manners flourish, responding to the cosmic harmony on high. With good harvests and a prosperous populace, and with all neglected tasks being now undertaken, this region will eventually regain the prosperity of a hundred years ago. In this sense, the successful reconstruction of the pavilion is an auspicious omen. May it be that this is an extravagant hope? To realise this hope, should we not make efforts all together?' in ibid., translated by Xu Yinong. Mary Wright noted that neo-Confucianism 'offered both solace and a means of action, especially after the Opium War.' See Wright, *The Last Stand of Chinese Conservatism*, 59.

67 '乙亥春, 群工毕事. 余与数友同游于斯. 至則见游人如市, 诚升平景象也. In the spring of the year yihai [1875], when all the craftsmen have completed their work, I roam around the place together with a few friends. Upon our arrival, we observe crowds of visitors there as if it is a marketplace. It is truly a scene betokening an age of peace and plenty.' in Anonymous, 'Record of Touring Surging Wave Pavilion,' translated by Xu Yinong.

68 Zhang Shusheng. 'Gai Jian Zhengyi Shuyuan Ji 改建正誼書院記 (Record of Reconstructing Rectify Meanings Academy)'. In Cao and Li, *Gazetteer of Wu Prefecture*, vol. 27, 3.

69 Zhu Kebao 諸可寶 (1845–1903), 'Record of Learn the Classics Hall 學古堂記 (Xuegu Tang ji),' in Cao and Li, *Gazetteer of Wu County*, vol. 27, 4.

70 En Shou, 'Shrine of Two Cheng Brothers newly constructed near Surging Wave Pavilion 滄浪亭增建二程子祠 (Canglang Ting zeng jian Er Chengzi Ci),' in *CLTXZ*, vol. 2, 6–7.

71 '金鄉書院在西市坊内. 亦國朝郡守胡纘宗以長洲學諸生多家川澤來, 則就媚艄, 莫獲棲息. 乃示永定寺僧曰: 僧殘寺廢, 今改建書院. 庶寺廢而僧存. 遂因殿宇為門廡亭館. 窮鄉僻處始入城郭者, 若寧家焉.' in *GSZ*, vol. 24, 39.

72 '國朝嘉靖二年, 知府胡纘宗閱郡乘載書院者四. 今獨存文正. 鶴山爲祠, 餘悉結民廬.' in *GSZ*, vol. 24, 35–36. One of the four academies was in the suburb, and the other three academies are identified in Figure 0.2.

73 Clunas, *Elegant Debts*, 19–38.

74 Social gatherings in gentry estates introduced young learners to teachers and fostered literature and artistic exchange. Temples such as Bamboo Hall Temple also accommodated many gatherings of Suzhou elites, where the juniors showcased their talents to masters. These gatherings, serving as cultivating occasions for the emerging talents, can be glimpsed from Wang Chong 王寵 (1494–1533), 'Xue Yan Zhutang Si 雪宴竹堂寺 (Banquet in Snow in Bamboo Hall Temple),' in *Gazetteer of Bamboo Hall Temple*, vol. 43.

75 The numerous academies and schools could be found in northern cities such as Jiujiang 九江 in Shanxi 山西, where White Deer Grotto Academy 白鹿洞書院

was founded. See Walton, *Academies and Society in Southern Sung China*, 25–49.

76 Lin Cho-Ying points out that Suzhou literati had scarce achievements in Neo-Confucianism in contrast to their notable literature and artistic achievements. Li analyses that Suzhou literati valued the practice of literature as equal to or even superior to the practice of Neo-Confucianism. See Li, 'Locality and Trans-locality.'

77 Until the Mid-Ming, many temples still existed, as recorded by Wang Ao in the preface of chapters 29–30 in *GSZ*: '吳中多佛老之區, 雖更洪武歸併, 而其廬故在也. 既不可廢. 則列其叢林而諸歸併者各附見焉. In Suzhou there used to be many Buddhist Temples. Though the amalgamation rule of Hongwu's reign brought many changes, these temples sites still existed and should not be abandoned. Thus, I list these abbeys and monasteries, along with their incorporated temples and cloisters, as follows.'

78 See Chang, 'Literature of the Early Ming to Mid-Ming (1375–1572),' 37–38.

79 See Marmé, *Suzhou: Where the Goods of All the Provinces Converge*, 179–181.

80 See Chang, 'Literature of the Early Ming to Mid-Ming (1375–1572).'

81 Hu Zuanzong expressed concern about the local power of wealthy families in Suzhou, which rivals the government: '蘇郡也, 自今觀之其大也. 若省東南財賦, 衣食京師, 蘇州之出, 百倍他所. 豪右富強, 與官為家. 煢獨單微, 靡所依控. 故舊守蘇者, 多以賄敗. Suzhou is now a prosperous prefecture. In terms of taxation and the transmission of food and cloth to the court in Beijing, Suzhou contributed a hundred times more than other regions. The rich and powerful treat the government as their own family. The government is isolated and powerless, lacking dependency and control. Consequently, many previous officials in Suzhou have become corrupt.' See Hu, 'Preface to the Departure of Suzhou Prefect Mr. Lin 贈蘇州太守林先生序 (Zeng Suzhou Taishou Lin xiansheng xu),' in *Small Collection of the Bird and Mouse Hermit*, vol. 11.

82 This detail of golden plating was mentioned by Yao Guangxiao as: '命善工團造釋迦佛像, 布以膠漆, 飾以金粉. 輕堅美妙, 非木刻泥塑者可並.' Yao Guangxiao, 'Record of the Reconstruction of Sea Seal Hall in Forever Pacification Universal Benevolence Tiantai Teaching Temple 永定普慈天台講寺重建海印堂記 (Yongding Puci Tiantai Jiangsi Chongjian Haiyin Tang Ji),' in *WDFC*, vol. 10, 73–74.

83 Yang Xunji, 'Villa of Forever Pacification 永定精舍 (Yongding Jingshe),' in *WDWCXJ*, vol. 30, 37.

84 See Wu Kuan, *Anthology of Family Treasures*, vol. 24, 16–17.

85 A corner of Forever Pacification Temple was converted into the Office of the Judicial Commissioner for Administering Punishments by prefect Lin Er 林鶚 (1423–1476) in 1461. See Li and Feng, *Gazetteer of Suzhou Prefecture*, vol. 22, 57.

86 '金鄉固子羽之封也.' in Hu Zuanzong, 'Record of Accommodating Master's Hall in Golden Canton Academy 金鄉書院寓公堂記 (Jinxiang Shuyuan Yugong Tang Ji),' in *WDWCXJ*, vol. 13, 30–32.

87 '復得吳賢之寓.吳者曰尹和靖焞.魏鶴山了翁遂用祔之塞.門曰學得精華堂,曰學孔堂.塑吳公言子像于其中, 夾以兩廊. 各有三十餘號舍其後.' in ibid.

88 '中為寓公堂, 奉子羽像, 而嚴事焉.' in ibid.

89 '後為正大堂, 旁列書舍. 以郡之良子弟肆其中. 而以孔氏之道切劘焉.' in ibid.

90 '事既竣, 因文而記諸壁間. 俾後來者勿有斁.' in ibid.

91 '夫士必有倡也, 夫學必有本也. 夫政必有綱也, 夫禮必有報也. 士倡則從, 學本則邃, 政綱則舉, 禮報則稱. 舍是而言治, 未見其為善治也. 舍是而言化, 未見其為善化也.' in ibid.

92 See Qi Wuqian, 'Lodge in Dragon Revival Temple 宿龍興寺(Su Longxing si),' in *GSZ*, vol. 29, 28–29. The two works anthologising local records about temples, *GSZ* and *WDWCXJ*, only recorded the Tang and Song literature works of this temple.

93 In the Early Ming, part of the temple's land was encroached upon by the Wu County Storage House to its east. See '府倉在飲馬橋西, 唐龍興寺故基.' in Lu, *Gazetteer of Suzhou in the Reign of Hongwu*, vol. 8, 119.

94 '乃選縣膠弟子員弦誦其中, 修皇王之業, 趨孔孟之軌. 以無忘和靖先生之意. He selected students from suburb counties to strum and chant [the classics] in the academy, to learn the teaching of the emperors, to follow the path of Confucius and Mencius, and to not forget the intentions of Master Hejing.' in Yuan Zhi, 'The Record of Moving the Academy of Yin Hejing 尹和靖遷書院記 (Yin Hejing Qian Shuyuan Ji),' *WDWCXJ*, vol. 24, 24–26.

95 '公暇則游焉. 橫经析疑, 躬督肄习. 设俎豆龥竿于堂上, 悬夏楚朴棰于堂下. 董以学官, 申以科条, 将使四方则之. 后世传之曰苏湖之教, 其再见乎, 公之志也.' In ibid.

96 The biography of Hu Yuan as the founder of 'Teaching in Suzhou and Huzhou 蘇湖之教' is discussed in Hon, *The Yijing and Chinese Politics*, 50–51.

97 Milburn, 'From Hero to Ancestor, God, and Ghost'; Yue, 'Serpent and the Hero.'

98 See Yang, *Gazetteer of Wu County and Changzhou County*, 52.

99 Hu, 'Record of the Newly Constructed Shrine of the King of Qi.'

100 Isaac Yue pointed out that in the Ming dynasty, the need to portray Han Shizhong as a hero defending the nation aligned with the urgency to fight against foreign invasion and protect the country. In Isaac Yue, 'Serpent and the Hero.'

101 Huang, 'Stele Record of the King of Qi.'

102 Bright Virtue Temple in the Mid-Ming dynasty owned an area that includes the present-day Mountain Villa of Embracing Beauty 環秀山莊, Embroider Museum 刺繡博物館, and Shrine of Wang Ao 王鏊祠. See *Comprehensive Catalogue of Cultural Relics in Jiangsu*, 238–239.

103 Yao Guangxiao, 'Record of the Revival of Bright Virtue Teaching Temple in Suzhou 蘇州府景德教寺重新記 (Suzhou Fu Jingde Jiao Si Chongxing Ji),' *Anthology of Yao Guangxiao*, 293–294.

104 Wen Zhengming, 'Bright Virtue Temple 景德寺 (Jingde si),' *Anthology of Wen Zhengming*, 392. Zhou Daozhen dated the poem as written in 1502.

105 Hu Zuanzong, 'Record of Learning Confucianism Hall in Learn the Way Academy 學道書院學孔堂記 (Xuedao Shuyuan Xuekong Tang ji),' in Niu and Wang, eds., *Gazetteer of Wu County in the Reign of Chongzhen*, vol. 14, 5–6.

106 Hu Zuanzong, 'Record of Reconstructing Learn the Way Academy 重建學道書院記 (Chongjian Xuedao Shuyuan ji),' in ibid., vol. 14, 3–5.

107 '拔士之尤俊者絃誦其中. 洋洋乎渢渢乎, 若登闕里而遊洙泗也. He selected the most outstanding scholars, whose recitations filled the air, creating an atmosphere as if one were ascending the imperial court and wandering through the Zhu and Si rivers.' In Xu Jin 徐縉 (1482–1548), 'Record of Learn the Way Academy 學道書院記,' in Li and Feng, eds., *Gazetteer of Suzhou Prefecture*, vol. 26.

108 In *GSZ*, vol. 24, 35–36.

109 See Hu, 'Record of Reconstructing Learn the Way Academy.'

110 Ibid.

111 Ibid.
112 See Wang Chong's outlining of the Confucian lineage as established by Hu Zuanzong in Suzhou in 'Song Tianshui Hu Gong Xu 送天水胡公序 (Preface to the Departure of Master Tianshui Hu),' in *WDWCXJ*, vol. 47, 29–30. For Hu Yuan's biography, See Tze-ki Hong, *The Yijing and Chinese Politics*.
113 '靈秀所鐘, 殆若預為學宮設者, 豈為偶然.' in Zhu Xizhou, 'Record of Reconstructing Changzhou Confucian School 長洲縣重建儒學記 (Changzhou Xian Chongjian Ruxue Ji)', in Li and Gu, *Gazetteer of Changzhou County*, vol. 32, 13–15.
114 '而凡學者因益得去佛而歸儒公, 眞能以闢邪放淫為己責矣. 夫佛氏之道息, 則聖人之道著.' in Xu Jie, 'Record of Revere the Classics Pavilion in Changzhou County School 長洲縣學尊經閣記 (Changzhou Xianxue Zunjing Ge Ji),' in *Collection of the Shijing Hall*, vol. 14, 44–46.
115 The boundary of this temple is informed by *Map of Gusu City*.
116 Ten Thousand Longevity Temple derived its major income from agricultural production, as noted by Tianyin Yuanzhi, 'Record of the Firewood Pavilion of the Bathhouse in the Ten Thousand Longevity Temple, Pingjiang Prefecture 平江府萬壽寺浴院柴莊記 (Pingjiang Fu Wanshousi yuyuan chaizhuang ji).' See Li, ed., *Complete Anthology of Yuan Literature*, 42–43.
117 '寺有唐僧貫休所畫十六羅漢像, 頗著靈異. 吳越時邵思寶等共建尊勝二石幢, 今猶存. In the temple there was the portrait of the sixteen Arhats painted by monk Guanxiu in the Tang dynasty. The portrait of sixteen Arhats were efficacious. The two honorific stone pillars established by Shao Sibao in the time of Wu-yue still existed." Song Lian, 'Stele Record of the Reconstruction of Buddha Hall in Ten Thousand Chan Temple in Suzhou 蘇州萬壽禪寺重構佛殿碑 (Suzhou Wanshou Chansi Chonggou fodian bei),' in *Complete Works of Song Lian*, 1611–1613.
118 '南渡以來, 江南有十剎之號, 而萬壽居其一. 則其盛非特為吳中禪寺稱首而已. After the Song dynasty moved to the south, in the region of Yangtze River there were ten prime monasteries. Ten Thousand Longevity Temple was one of them. The temple was not just a supreme temple in the region of Wu but in the empire.' in Wu Kuan, 'Records on the Reconstruction of Ten Thousand Longevity Chan Temple 萬壽禪寺重修記 (Wanshou Chansi Chongxiu Ji)', in *WDWCXJ*, vol. 30, 4–6.
119 Yuanzhi, 'Record of the Firewood Pavilion of the Bathhouse in the Ten Thousand Longevity Temple, Pingjiang Prefecture.'
120 '重搆大雄殿五楹, 間鑿石于山. 市材于江, 陶瓦于郊. 工者奏技, 壯者獻力. 鞠明究晦, 不會而集. Chan master Xingzhong Ren reconstructed Buddha hall of five bays. Stones were quarried from the mountain, materials were purchased near Yangtze River, tiles were obtained in the outskirts. Craftsmen contributed their skills, and the strong contributed their effort. They gathered voluntarily and worked from dawn to dusk.' in Song, 'Stele Record of the Reconstruction of Buddha Hall in Ten Thousand Chan Temple in Suzhou.'
121 '四阿有嚴, 若翬斯飛丹. 臒絢爛炫, 人心目傸.' in ibid. Song Lian employed the metaphor of describing an ideal palace from *Book of Odes*. See Chen, *Chinese Environmental Aesthetics*, 86.
122 See Song, 'Stele Record of the Reconstruction of Buddha Hall in Ten Thousand Chan Temple in Suzhou.' The Buddha of the Three Worlds in Western Garden Temple in Suzhou could be a visual reference.
123 Wu, 'Records on the Reconstruction of Wanshou Chan Temple.'
124 '已而其徒僧録司覺義戒瑺請藏經自京師, 至建閣五間. 寺之既備. Bai'an Jin's disciple Jueyi Jieli requested sutra collections from the court. A pavilion of five bays

was constructed as sutra storage pavilion. All the artefacts were thus equipped in this temple.' In ibid.

125 '吳城第一好山川, 半醉來登意豁然. 天際雲開雙塔影, 城頭日出萬家煙. The best of Wu city is its mountains and rivers. When ascending it in half drunk, I open my mind to the view. The clouds passing away and the twin pagodas emerges in front of my eyes. The sun is rising, and smokes are emerging above ten thousand houses.' See Wang Ao, 'Ascending Buddha Pavilion of Ten Thousand Longevity Temple 登萬壽寺佛閣 (Deng Wanshousi foge),' in *WDWCXJ*, vol. 30, 6.

126 '顾其地犹为卑隘. 顷教谕萧君文佐谋于诸生为迁改之计. 咸谓城东有僧寺, 高爽宏壮, 建学惟称. Considering that their current location was still low and cramped, recently the educational instructor Xiao Wen devised a plan with the students for relocation. They unanimously agreed that there was a monastery to the east of the city, which was lofty, grand, and splendid, making it an ideal site for establishing a school.' in Zhu, 'Record of Reconstructing Changzhou Confucian School.'

127 Ibid.

128 Ibid.

129 Ibid.

130 '公乃躬詣僧寺而遍閱焉, 嘆曰, 彼僧徒之居, 若是其侈也. 學校為養士之地, 顧弗若耶? 是固所當遷改者, 不可已也.' In ibid.

131 '遂委通守邊君偲, 吳縣令張君道, 與吳君專董其事. 而貳守王君文儒, 通守包君梧, 牛君佐, 節推陳君一德亦共為之規畫. Thus Shu Ting appointed Bian Sai, Zhang Dao and Wu Shiliang to take charge of this event, Wang Wenru, Bao Wu, Niu Zuo and Chen Yide were also assigned for assistance.' In ibid.

132 '是役也, 經始於嘉靖辛丑十二月, 以明年三月落成. This task started in Lunar December of 1541 (The Xinchou year of Jiajing's reign) and finished in 1542's Lunar March.' in ibid.

133 '於是即其宮室之舊, 或飾而為新, 或撤而改造. 間以舊學之可用者合而成之. Some of the temple's halls and buildings were refurbished anew, and some were disassembled and renovated. Meanwhile available materials from the old Changzhou County School were also implemented.' in ibid.

134 '其制: 廟學皆南向, 其左由棂星門而入, 重之以戟門. 而中為先師殿. 殿之旁兩廡列焉. The new school followed this pattern: all the shrines and buildings were facing the south. The left axis begins with Lattice Star Gate and followed by Ji gate. In the middle there was Primordial Teacher's hall. This hall was flanked by two corridors.' in ibid.

135 See the remained hall in https://zh.wikipedia.org/wiki/%E9%95%BF%E6%B4%B2%E5%8E%BF%E5%AD%A6%E5%A4%A7%E6%88%90%E6%AE%BF.

136 'The right axis starts with Confucian Gate and followed by Ritual Gate (*Li men*). In the middle was Illuminate Ethics Hall. This hall was flanked by two chambers. One was called 'Respect Virtue (*Jing de*),' the another was called 'Cultivate Learning (*Xiu ye*).' Students' residences were aligned to the north and south of these two chambers. To the north of the shrine was Shrine of Giving Birth to the Sages (*Qisheng ci*), and the far north was Revere the Classics Pavilion.' In ibid.

137 '以至廨宇倉廩之類, 各量其地之所宜而建置焉. Buildings such as offices and granaries were constructed according to the land's suitability.' In ibid.

138 See Wang Ruyu, 'Suxue ba yong 蘇學八詠 (Eight Odes on Suzhou School),' in *WDWCXJ*, vol. 4, 14–15.

139 Ibid.

140 '堂之北为名宦,乡贤二祠.又北则累石为山.山之上为敬一亭,尊御制也.若夫棂星门之前,则树以绰楔.儒学门之外,则凿为泮池.' in Zhu Xizhou, 'Record of Reconstructing Changzhou Confucian School.'

141 Ibid.

142 '乃若茲地之形勝, 則山拱於西, 水匯於東. 前有橋曰升龍, 內有河曰玉帶. 靈秀所鐘, 殆若預為學宮設者, 豈偶然哉!' In ibid.

143 '若茲學之改建, 其宏麗壯偉無復加矣, 然豈徒為觀美而已乎? 將使士之居其所者, 德於是而進焉, 業於是而修焉, 斯不為虛設云爾.' In ibid.

144 '嘉靖壬寅春, 巡按直隸監察御史侯官舒公汀改長洲之福寧寺以為學. 遂葺其藏經之閣, 合六經群籍, 貯於其上, 而扁之曰尊經. In the Renyin year of Jiajing's reign (1542), Provincial Censor Master Shu Ting converted Fu'ning Temple into school. Then they refurbished its sutra storage pavilion. They collected books of six classics and stored in the pavilion's upper floor. The pavilion was with a title board named "Revere the Classics".' in Xu, *Collection of the Shijing Hall.*

145 '而異端之說, 或得以為之迷溺, 則夫闢邪放淫, 以翼成一代. 尊經之教在有位.' in Xu, *Collection of the Shijing Hall.*

146 '舒御史毀寺報. 蘇州城東積古有萬壽寺, 先朝所建. 每歲長至履端聖誕之辰, 守令衛尉而下先一日例用習儀, 必于斯寺.嘉靖間一日有二三廣文, 偕惡少子弟. 入寺中. 僧徒迎候稍遲, 唧之. 適有閩人舒汀為御史按臨蘇城. 此公專崇理學, 不事梵王. 遂聽澤宮之議.' In Qian Xiyan, 'Imperial Censor Wang Destroyed the Temple 舒御史毀寺 (Shu Yushi Hui Si),' in *WDFC*, vol. 28, 116–117.

147 'The Ten Thousand Longevity Temple was established during the Yixi period of the Jin dynasty. During the Jiajing era, there were eight scholars in Suzhou, known as the "Eight Eccentrics", who prided themselves on opposing heretical teachings. At that time, the state temples only revered Taoism and did not believe in Buddhism. Upon hearing this, the Eight Eccentrics pointed and swung an axe, first chopping off the head and then smashing the body. Not long after, he fell ill with typhoid fever and became delirious, eventually severing his own head with a sharp blade. He held it and stood upright for an entire day without falling. This event was witnessed by countless people and was reported by Liu Jianyu, a trustworthy person, and not a slanderer of Buddhism. In the summer of the Renchen year, prompted by the temple monks' request, I wrote this account to record the reconstruction of the Ten Thousand Longevity Temple.' Xu Shupi, 'Wanshou Si 萬壽寺 (Ten Thousand Longevity Temple),' in *Small Record of Observations.*

148 '僧徒三百眾並逐于外, 繞車號泣. Three hundred of monks were dispelled from the temple, they cried and whined around the transporting cart.' in Qian, 'Imperial Censor Wang Destroyed the Temple.'

149 Xu, 'Ten Thousand Longevity Temple.'

5 From Buddhist to Gentry Society

Buddhists were actively engaged in land investment, and they established monastic estates with hydraulic topography in the Yuan dynasty. Monks constructed extensive temples, supported by generous patrons, fostering a thriving Buddhist culture in Suzhou. In contrast, in the Ming dynasty, particularly amid the anti-Buddhism movement between the 1520s and 1540s, the gentry and government began to seize monastic lands, using tactics such as purchasing, coercion, and even violent invasion. On the appropriated estates, the gentry established an authority of literati garden culture and the government promoted Confucian values to the public.

Why were Buddhists dispossessed of their lands, and how did the gentry and government become predominant landholders in the Ming? This chapter contrasts the monastic policies issued between the Yuan and Ming dynasties, focusing on their local manipulation and the resulting political dynamics among Buddhists, the gentry, and the local government. It investigates how Yuan policies fostered investment in temple construction, whereas policy shifts in the Early Ming contributed to the marginalisation of Buddhists in land ownership in subsequent periods. Following the transformation of the monastic institutional structure in the Early Ming, local gentry and government officials leveraged the weakened political power of Buddhists and actively annexed Buddhist lands, especially those with valuable hydraulic features. The contention over hydraulic estates resulted in the gentry and local government replacing Buddhists at the pinnacle of the social hierarchy. A gentry society was formed within the landscape and estate transformation.

Monks as Landholding Class

The Yuan empire was a religious-oriented state which marked a significant departure from the Confucian statecraft and social order rooted in other dynasties of China.[1] Members of the imperial families were devout Buddhists, leading to an enthusiastic promotion of Buddhism throughout

DOI: 10.4324/9781003387169-6

the empire. Buddhism was integrated into the empire's political fabric. The Buddhist Bureau, being entrusted and empowered by the court, actively participated in politics. The Buddhist Bureau comprised of monk officials who operated independently without been overseen by the gentry official, unlike in the preceding Song and subsequent Ming dynasties. Buddhist officials often ranked comparably to, or even surpassed, their gentry counterparts. The appointments of Buddhist officials were often based on recommendations from the imperial family and court officials, enabling these Buddhist officials to bypass the traditional examinations required.

The empowered Buddhist Bureau diffused power to Buddhist communities by providing easy access to becoming a monk, enforcing fewer restrictions on monastic activities, and granting privileges to monks. Obtaining the monastic certificate was relatively straightforward, primarily involving registering and residing at a temple with general adherence to Buddhism.[2] Notably, there was no stringent requirement to pass sutra examinations. The issue of monastic certificates was not limited to a specific quota.

Buddhists enjoyed economic privileges such as exemptions from corvee labour and taxes, which facilitated wealth accumulation.[3] In contrast, Confucian families and commoners faced significant tax and labour burdens, posing substantial challenges to their economic stability. Xie Chongguang noted that during the Yuan dynasty, monks actively participated in various economic activities, including agriculture, commerce, transportation, and printing.[4] Given the profitability of agriculture, monks acquired massive lands for farming. Instead of working the fields themselves, many monks employed tenant farmers or even purchased slaves. The revenue from these ventures often became the personal property of the monks.

As monks in the Yuan dynasty evolved into skilful investors and prominent landholders, they also became active consumers. The Buddhist Bureau did not censor monks' income or regulate their adherence to asceticism–some Buddhist leaders were notorious for their lavish and corrupt lives.[5] Monks often consumed alcohol, disregarded vegetarian diets, and even married and raised families.[6] Their conspicuous consumption was most evident in the grand temples they resided in, the elaborate gardens they constructed, and the valuable artworks they collected.[7] Modern scholars characterised monks in the Yuan as secularised: although dressed in Buddhist robes, they engaged in savvy investment and consumerism, pursuing secular goals such as personal gain, wealth accumulation, and enjoying life.[8]

The combination of multiple privileges and lenient restrictions rendered the Buddhist clergy a highly desirable profession. Zhu Derun's poem, *The Lady Living in the Separate Residence* (*Waizhai fu*), juxtaposes the affluent lifestyle of a monk with the destitute circumstances of gentry families.

Free from corvee labour, the monk in the poem acquired land and built a separate residence from his temple. He purchased a beautiful woman, providing her with a comfort life in his off-temple abode. He even helped to refurbish the dilapidated house of his parents-in-law. In stark contrast, two gentry members, registered as Confucian households, were compelled to perform corvee labour and lived impoverished lives with limited income.[9] The poem concludes by highlighting the shifting attitudes towards the Buddhist profession: being a monk was no longer a backup option but a tempting occupation.[10]

The Yuan dynasty witnessed a notable increase both in the Buddhist population and the number of temples. Many pursued the monastic career not for spiritual reasons, but for the tangible benefits it offered, such as safeguarding and expanding family assets and improving social standing. The rise in the number of monks coincided with the surge in temple construction, as new temples provided more opportunities for monk registration. In 1291, the number of registered temples through the empire soared to 42,318, significantly outnumbering Confucian institutions.[11]

This national trend was observed in Suzhou, where the number of temples reached 372 (see Fig. 0.1 and 0.3), in stark contrast to the mere seven Confucian institutions, with four inside the city (Fig. 0.1). Many families transferred their lands to Buddhist ownership and converted themselves to Buddhism. Patronising temples became a strategy for registering properties under the name of monastic institutions, which allowed for tax reductions and enhanced financial security. The result, as noted by Wu Kuan, was that 'In the former [Yuan] dynasty, temples possessed many fertile fields, virtually without any restriction.'[12] Monastic estates became favoured investments and sources of consumption for both monks and patrons, providing not only financial benefits but also social and cultural capital.

These new temples and cloisters were established by transforming wetlands into well-designed hydraulic infrastructure. As observed by the gentry, Yuan dynasty temples often occupied lands with excellent geomancy. Monks created these hydraulic estates through intensive earthwork labour spanning over a century. The institutional stability and collective wealth enabled monks to undertake substantial long-term investments in earthworks and hydraulic infrastructure. This long-term and financially costly investment would be less viable for gentry family estates, due to uncertainties such as family disputes, economic fluctuations, and changes in social status.

Restructuring Monastic Institution and Lands

When Zhu Yuanzhang ascended to the throne, he faced with excessive number of temples and what he considered as fraudulent monks throughout the country. Having once taken up a monastic life for survival, Zhu was acutely aware of the economic and social influence wielded by monks, which led him to regard them as a potential threat to the throne. Orienting the state on Confucianism, Zhu implemented several acts to regulate, restructure, and marginalise Buddhist communities.

In the 1380s, the government carried out a comprehensive survey of monks, temples, and monastic lands. In the next decade, the state reorganised temple institutions into an 'abbey-cloister system' and imposed regulations to restrict interactions between Buddhists and other social groups. Zhu Yuanzhang was likely the emperor who issued the most multiple and meticulous edicts against Buddhism. Scholars regarded his reform as the most intense suppression of Buddhism prior to the 1950's Land Reform and Cultural Revolution (1966–1976) of the Maoist era.[13]

The Buddhist Bureaucratic administration was renovated in 1381. This new administration comprised of four levels at the central court and three local levels, extending through prefectures, subprefectures, and counties.[14] Unlike the autonomous institutions of the Yuan dynasty, the Buddhist Bureau in the Ming dynasty was under the supervision of gentry officials. Buddhist positions were assigned lower ranks and reduced salaries, and a quota system was introduced to limit their numbers.

Rather than distributing power, Buddhist offices in the Ming primarily served the purpose of monitoring Buddhists and their temples. They maintained detailed registries of monks, including names, the temple residency, ordination dates, and certification details.[15] These registries were instrumental in identifying and weeding out uncertified, fraudulent monks, who would be expelled from the Buddhist community unless they passed the required examinations.[16] Candidates for abbacy were also required to pass examinations and obtain endorsement from the Bureau of Ritual (*Li Bu*), staffed by gentry officials.[17] Additionally, temples were obligated to register information regarding their founders, patrons, and any former imperial recognitions.[18] The monastic registry provided the court with a comprehensive inventory of privately founded temples, aiding in the subsequent amalgamation of temples.

Buddhist offices in the Ming dynasty closely supervised monks' adherence to celibacy and other religious regulations. To prevent monks accused of crimes from seeking refuge within the temple, it was decreed that any Buddhist suspect should be judged by gentry officials rather than the Buddhist Bureau.[19] The stringent regulations and diminishing privileges made the monastic life less appealing, leading to a dramatic reduction in the number of monks.

In the subsequent year, 1382, all monastic lands were registered with the local government. Laws were enacted to prevent monks from selling monastic lands.[20] To ensure adherence to the new land regulation, a foundation cleric was appointed to each temple, tasked with overseeing monastic lands.[21] This legislation significantly curtailed monks' ability to possess or personally benefit from temple property. While these laws effectively prevented monks from selling monastic lands, they inadvertently paved the way for the encroachment on monastic lands from the gentry and local government.

The edict of amalgamating temples issued in 1391 significantly reduced the number of temples in Suzhou, bringing them down to a mere quarter of the previous count. This edict called for smaller temples to be merged into large monasteries, each requiring at least 30 monks.[22] As a result, the 119 intra-mural temples in Suzhou were incorporated into 22 monasteries (Fig. 0.2). In this new structure, one temple in the monastery would be nominated as an abbey with legitimation and privilege granted, while others were relegated to lower-status cloisters. The abbey was eligible for receiving government funding, refurbishment subsidies, and their lands were legally protected. In contrast, the land and properties of cloisters were susceptible to appropriation.[23] Regarding this situation, monks exerted strenuous effort to elevate their temples to abbey status.

Since the amalgamation order was brief yet ambiguous, analysing the post-amalgamation outcomes provides insight into the practical criteria for selecting abbeys. First, grand temples with rich historical and cultural heritage were chosen as abbeys. For instance, the four selected abbeys—Opening Origin Chan Temple (*Kaiyuan chan si*), Repay Kindness Teaching Temple, Northern Chan Teaching Temple, and Auspicious Light Chan Temple—were all established predating the Song dynasty. They were renowned for their prominent artefacts, valuable historical relics, and extensive representations in literature.[24] Merely possessing prominent buildings did not ensure selection as an abbey. This is exemplified by Illuminous Celebration Temple, which, despite its elaborate architecture sponsored by the Yuan imperium, was not chosen due to a lack of historical relics and literary records. Consequently, it was categorised as a cloister and incorporated into Northern Chan Temple.

The primary focus of the Buddhist Bureau in enacting the amalgamation edict was to ensure that each monastery housed at least 30 monks. Consequently, lots of local manipulations in selecting abbeys were observed, involving local officials, monks, and the gentry. To comply with the edict, some monks were willing to merge their temples; for example, to prevent their temples from being abandoned, monks from nine temples merged, resulting in a collective of twenty monks. They registered as a single monastery, nominating the Sacred Longevity Teaching Temple at Bantang (*Bantang shengshou jiao si*) as the principal abbey.[25] Another approach to gaining abbey status was to directly appeal to the imperium. For instance, monk Baotan Shiying, the national teacher, petitioned the emperor to elevate Gathering Clouds Temple to an abbey. With imperial sanction, Gathering Clouds Temple merged with two adjacent cloisters along the Surging Wave, establishing the Southern Chan Temple.[26]

Local gentry, especially those holding court positions, were instrumental in promoting temples to abbey status. Their stele records narrating a temple's history provided persuasive support for a temple's selection as an abbey. Consequently, many monks journeyed to Beijing to solicit such records from influential Suzhou officials such as Yao Guangxiao, Wu Kuan, and Wang Ao.[27] Upon returning, these records, accompanied by other relevant literature, were presented to the local Buddhist Bureau as part of the abbey status application. Court officials could also directly petition the emperor to designate a temple as an abbey. Recognising the effectiveness of gentry support, monks in the Ming actively sought opportunities to develop close relationship with the gentry. This indicates that Zhu Yuanzhang's edict in the early Ming, which prohibited monks from interacting with other social groups, became less stringently enacted in later periods.

The policies issued in 1391 and 1394 alienated monks from family ties, social connections, and economic pursuits. Severe penalties were stipulated to ensure monks' adherence to celibacy:

> If a monk were to be found to have a wife, people would be permitted to beat and humiliate him, and further demand fifty ingots. If he were not to have the money, it would be considered acceptable to beat him to death.[28]

Monks were prohibited from residing outside temples, and laypeople were not allowed to live within temples. The ordinance declared harsh consequences for violations, including decapitation and public display of the monk's head for those found residing in commoner's homes, and exile for anyone harbouring a monk for three thousand *li*.[29]

To address funeral needs, the court specified that only *yoga* monks were allowed to visit commoners' homes to conduct funeral rituals, with the payment for each service being prescribed by the government.[30] Other types of monks, such as the Chan and Teaching monks, were restricted to activities confined within the temple, such as meditation and studying sutras.[31] Timothy Brook pointed out that the underlying intention of this edict was to isolate monks from secular life, thereby preventing any potential conspiracy against state authority.[32] However, in practice, the rule intended to separate Buddhists from social interactions was not rigorously followed after the reign of Hongwu.

Early Ming regulations dispossessed monks' individual ownership of monastic lands and inhibited their economic activities. The imposition of heavy taxation on monastic lands and the burden of corvee labour on monks during the Middle to High Ming periods positioned monks in the financial crisis, diminishing their interest in land investment.

The Gentry and Local Government's Land Control

In contrast to the multiple restrictions imposed on Buddhists, the Ming state granted tax exemptions and corvee labour privileges to gentry families, incentivising their acquisition of lands for agricultural and industrial production. The local government competed with the gentry for land ownership and spatial control. Appropriating the court's edicts of 'destroying profane temples' in the High Ming period, the gentry and local government collaborated to convert temples into their own. They rose to the landed class, possessing a significant portion of the best lands. How the four groups—Buddhists, local officials, anti-Buddhist gentry, and elite gentry supporting Buddhism—responded to land policies shaped the power dynamics within the Ming society.

Since the Early Ming, monks were unable to maintain monastic buildings due to the heavy taxation and corvee labour imposed. Temples typically faced economic crisis and the shrinking of monastic community. This predicament was vividly captured in Wen Zhenmeng's (1574–1636) record of Benevolence Origin Temple (*Fuyuan si*):

Entering the Ming Dynasty, the temple monks, burdened with heavy labour, dispersed in all directions, leaving the temple's pearl grove and blue hall collapsed into lush grass. By the Jiajing period (1540s), a Bhikkhu named Dongran tied together a hut of three rooms with straw, burned incense there, and practiced asceticism for over thirty years. Starving weasels and squirrels lived in hunger, whistling amidst the white grass. [In this destitute situation] Dongran passed away while adhering to precepts.[33]

Wen's description highlights the shrinking Buddhist community, the deterioration of temple structures, and the resilience and devoutness of monk Dongran despite the impoverished condition. Timber structures were left in ruin due to lack of maintenance and renovation. This suggests that the shaky structures of the Buddha halls were no longer fit for use, thus Dongran had to construct a straw hut for living and worship.[34] This state of disrepair was not unique to Benevolence Origin Temple but was a widespread issue among Ming temples. Additionally, the temple did not receive patronages in the Ming, as people realised that monastic patronage yielded only spiritual merit rather than the tangible economic benefits during the Yuan dynasty.

The oppressive burden of heavy taxation and corvee labour discouraged monks from owning and investing in lands. As land ownership shifted from being an asset to a liability, monks even began refusing land donations. This shifting attitude was exemplified by Witty Wisdom Cloister (*Miaozhi an*). After Yao Guangxiao was nominated as the national Buddhist teacher for his role in assisting the Yongle Emperor to seize the throne, monk Xiuxue (fl. early 15th c.) from the cloister approached him seeking support. Xiuxue told Yao that the cloister was virtually abandoned, with him being the only monk remaining. Long-neglected, the timber beams of structures were rotten, decaying, and beyond repair. Xiuxue expressed a desire to renovate and renew the cloister but lacked the means to do so.

Yao immediately donated funds, purchased materials, and hired labourers. The old buildings were removed, and new ones were built, ever more magnificent than before. Upon completion, Yao intended to buy fields to provide a stable income for the cloister, but this generous proposal was declined by the monk community. They reasoned, 'owning fields would bring corvee labour, and the monks would be unsettled. It is better to not have them.'[35] As an alternative, Yao decided to offer monthly donation of 50 paper ingots for the temple's ritual services. This episode also revealed that Yao Guangxiao, as a Buddhist official at court, was unaware that land ownership had become more burdensome than beneficial to the temple. This also prompts the question of whether the heavy taxation and corvee labour imposed on temples were mandated from the central court or enacted by local officials.

During the High Ming Period, the extreme burden of corvee labour forced monks to sell valuable monastic assets in order to fulfil these demands, even trees could not escape. Wang Ao's *Lamenting Pines* recorded the loss of nature and tradition of Emerald Peak Chan Temple (*Cuifeng chansi*) under the pressure of harsh taxation in 1520:

Emerald Peak [Temple], located in Dongting, is an ancient temple. From the temple gate to the official road, it was flanked with pairs of towering pines, each large enough to encircle several times. Their crowns like umbrellas and their branches twisting like dragons. Stirred by the wind, their rustling could be heard for miles. [These trees were relics] from the Song and Yuan dynasties. I greatly cherished them and often sat beneath them whenever I visited. However, by the summer of this year, not a single one remained. Astonished, I summoned the monks and inquired about this. The monks said: 'The county officials demanded corvee labour urgently. We could barely survive, how could these pines remain? They were sold to pay for the corvee labour.'[36]

In contrast to the Buddhists' predicament, elite gentry enjoyed substantial privileges in taxation and corvee labour. In the Middle Ming, lands owned by gentry lineages with academic degrees were exempt from taxation.[37] Incentivised by this advantage, the gentry became interested in land investment and competed for land acquisition. Scholars have estimated that the gentry collectively own 50 to 90 percent of the country's land.[38] They coveted monastic lands with prime locations and hydraulic infrastructure. Being aware of the vulnerability of monastic lands, they were constantly looking for ways to acquire at minimal cost. The central court's edict in the 1450s and 1520s, which called for the destruction of profane shrines, provided such opportunities. Despite the vague definition of what constituted a 'profane' shrine, the local government and gentry actively responded, asserting temples as such in order to repurpose them.

Responding to the 1450s edict of destroying profane shrines, the local government in Suzhou began targeting temples. During this period, the government's actions against cloisters were restrained, considering the supportive attitudes of some gentry toward the temples. These gentry elites, including Wu Kuan and Wang Ao, were both patrons of temples and held prominent court positions at the time. Lower-ranking local officials had to consider their views, even seeking to please them. The case of Assemble Benevolence Cloister (*Jifu an*) illustrates this tension between local officials and the elite gentry:

> In the reign of Hongzhi (1488-1505), along with the issue of destroying profane shrines, the magistrate of Suzhou considered making this cloister as an estate of Wu Kuan. However, Wu refused, stating: 'this cloister used to be my neighbour for centuries. I cannot bear that it will be destroyed. How could I even occupy it as my own?'[39]

This record shows local government's designation of cloisters as 'profane shrines' and its intention to appropriate monastic land. Although Wu Kuan prevented the conversion of Assemble Benevolence Cloister, many other cloisters without the protection of elite gentry faced government acquisition.

The trend of destroying profane shrines escalated in the subsequent years. In the 1520s, temples were frequently categorised as 'profane shrines' and overtly converted. As detailed in Chapter 4, magistrate Hu Zuanzong transformed four temples into three Confucian academies and one shrine. The 1540s saw an intensification of anti-Buddhist sentiment, with actions becoming increasingly violent. Collaborating with officials, teachers, and students, Magistrate Su Ting attacked Ten Thousand Longevity Chan Temple, repurposing it into Changzhou County school. The anti-Buddhism campaign during the High Ming period drastically reduced the number of temples in Suzhou to a mere quarter of their former count.

Local gentry actively engaged in acquiring Buddhist lands through various means. Very few transactions, such as the Lu family's acquisition of Shizi Lin Temple, were formally recorded as purchases. Others involved violent takeovers, such as Wang Xianchen's conversion of Great Propagation Temple into the Unsuccessful Politician's Garden. Some conversions, such as the Xu Family's transformation of Return to the Origin Temple into the Western Garden, went unrecorded. Despite the scant historical records, a petition from the local gentry Huang Xingzeng provides clues about how these transactions were carried out. Huang explicitly stated that the prosperous estates of the gentry originated from monastic lands, though the gentry often lacked gratitude. He noted that lands acquired from temples were used to expand spaces for the gentry's recreational grounds, which very much alludes to the fact that gentry gardens were established on the monastic lands.

Research on the land ownership in Suzhou shows that the gentry occupied a majority of urban lands by incorporating other estates. The total landholdings of upper gentry family were significant.[40] As Dardess pointed out, while the upper gentry were exempt from land taxation during the High Ming period, the tax burden was shifted onto peasants, resulting in their bankruptcy.[41] Suzhou Buddhists faced a predicament similar to that of peasants, as noted by Huang Xingzeng.

The local gentry Huang Xingzeng criticised the legitimacy of the government's severe and prevalent attacks on temples through his petition titled 'Appeal to Magistrates Nie and Cai, Ministers Huang and Zhu to Stop Seizing Temples.'[42] He recorded that lots of destruction and seizure of temples occurred following the edict of destroying profane shrines. He questioned that 'how it could be tolerable that the bones buried under pagoda for years were excavated, and the temples' plaques being endowed by former emperors were destroyed.' The details he observed in destructing temples were astonishingly violent. He noted that many ancient temples were destroyed, with monks being expelled. He poignantly reported the severe suppression of monks, questioning why the government left not even one inch of land for Buddhists to rely on, driving them to such despair that some considered suicide by sea. He also pointed out that the prosperous estates of the gentry originated from monastic lands, yet the gentry showed no gratitude. Huang Xingzeng's observation closely aligned with my findings that monastic sites were often chosen for their well-established hydraulic infrastructure and privileged urban locations.

The practice of misappropriating the edicts of 'destroying profane shrines' to target temples was not confined to Suzhou but prevalent among local officials in other cities throughout the 1500s. Wang Jian's analysis indicates that local governments' conversion of temples was largely due to a shortage of government revenue. The confiscation and transformation of monastic lands into state-owned properties helped to alleviate financial crisis.[43] Sarah Schneewind's research suggests that local authorities intended to reorient the society from Buddhism towards Confucianism by converting temples into community schools.[44] Considering the strategic locations of these converted temples within Suzhou city—being popular communal spaces adjacent to major public canals—I argue that the local government exerted spatial control over the city through these conversions, thereby challenging the powerful gentry class.

Despite the widespread conversion of temples, a group of elite gentrymen stood by Buddhists, patronising and conserving temples. These patrons included elites such as Wu Kuan, Shen Zhou, Wang Ao, and other first-generation Suzhou literati of the Ming dynasty, followed by Wen Zhengming and his contemporaries in the second generation. They cultivated strong relationships with monks, providing financial support and creating literature and art works. While local officials and other gentry were attacking temples, these elite gentry funded the construction of a hall at Peaceful Governance Teaching Temple (*Zhiping jiao si*), which became their frequent gathering space.[45] Their sustained literature contribution also played a pivotal role in the revival of Clouds Spring Cloister (*Yunquan an*), elevating it from an unknown cloister to a renowned temple.[46]

Although unable to halt the anti-Buddhism, these elite gentry did their utmost to preserve Buddhist culture. Even for temples they were not familiar with, they would respond actively when the monks invited for a temple record or fund-raising appeal.[47] Their meticulous inventory of previous temples and cloisters in *Gazetteer of Gusu* has served as a valuable archive for the current research on Yuan dynasty temples and their reorganisation during the Ming. These gentry regarded Buddhism as an integral part of the local culture, a view they reasserted in *Gazetteer of Gusu*:

Suzhou held many areas of Buddhist and Taoist significance. Although they were amalgamated in the reign of Hongwu, their locations still existed, and they should not be abolished. Therefore, the abbeys and their affiliated cloisters are inventoried, and those that were amalgamated are also mentioned.[48]

Some gentry criticised the local government and other gentry's conversion of temples in stele records. Peng Nian wrote in his record for Auspicious Light Chan Temple (*Ruiguang chan si*):

The destruction and reconstruction of this temple testify to the prosperity and decline of Buddhism, further reflecting the affluence and corruption of the society. In recent years, exquisite and ancient temples were undermined by taxation and corvee labour and snatched away by the wealthy and empowered.[49]

Zhu Yunming denounced the government's conversion of temples into Confucian schools:

> Today, by enshrining hundreds of divines, constructing and renovating Confucian institutions, the government raises money and conscripts labour, forcing the people to submit to their authority. Do these officials not realize that the law declared the five religions as the cornerstones of our nation? Apparently, they do not know this and thus I advocate here. Even if they follow [this original intention], their compliance is only reluctant.[50]

Some gentry found themselves in complex relationships with the anti-Buddhists. They might be conflicted when witnessing their family members' conversion of temples. For instance, Huang Xingzeng appealed to officials to halt the conversion of temples, arguing that Buddhist temples and cloisters should not be classified as 'profane shrines.'[51] However, when Magistrate Hu Zuanzong converted Witty Wisdom Cloister, Huang authored a record to legitimate this conversion.[52] The contradictory writings of Huang Xingzeng reveal that, despite his general opposition to temple conversions, he might have acquiesced to a magistrate's request to legitimise the conversion.

A similar tension with anti-Buddhist officials was also witnessed in Wu Kuan's retirement. He composed a poem that appears to glorify magistrate Cao's demolition of profane shrines, but the title 'Beautiful Magistrate Cao destroyed improper shrines' hints at a mocking tone.[53] Wen Zhengming exhibited silent consent and perhaps a reluctant compromise in response to Wang Xianchen's conversion of the Great Propagation Temple. Although Wen must have known that Wang's garden resulted from the violent conversion of Great Propagation Temple, he chose not to openly oppose. Moreover, he concealed the site's Buddhist history and omitted any Buddhist traces in all his representations. Ironically, while Wang Ao endeavoured to preserve temples, his son constructed a shrine to honour him by converting part of Bright Virtue Temple.[54]

Facing the conversion of their temples, monks had no power to defend. Some magistrates converted the monks' religion, assigning them new duties such as maintaining the newly Confucian shrine that replaced their original temples. Qian Xiyan vividly captured Monks' vulnerability as officials marched into Ten Thousand Longevity Chan Temple. He wrote, 'the three hundred monks were expelled outside the temple. They surrounded the carts (which transported ten thousand ropes to destroy the temple), cried and sobbed.'[55] Similar was Wang Xianchen's invasion of Great Propagation Temple. Qian described the scene as 'in a moment, monastic buildings were demolished. Monks fled, steles were torn down, and the temple became desolate.'[56] Despite this onslaught, a few monks persisted in their Buddhist practices on site. They adapted by building huts at the corner or residing in the few remaining artefacts.[57]

Evolution of Social Structure

A new social structure took shape during the High Ming period. At the top of the pyramid, the local gentry and government emerged as the largest landholders, possessing prime lands with favourable geomantic and hydraulic features—much of which had been converted from monastic lands. Meanwhile, Buddhists and peasants, dispossessed from land ownership, were relegated to the bottom of the pyramid. They lost their lands, especially those with valuable hydro-topography. This social structure persisted in the subsequent Qing dynasty, during which the number of temples inside the city was reduced to around twenty. This number declined further during the 1950s reformation targeting Buddhists and their lands.[58]

The extensive transfer of estates in Suzhou was closely linked to shifts in garden practices, both in the ownership of hydraulic infrastructure and evolving perceptions of gardens. The intersection between garden practice and land ownership has been a recurring theme in history. For example, Henry VIII's conversion of temples into royal estates played a significant role in the development of gardens in 16th century's England.[59] Similarly, the gentry's widespread acquisition of lands during the Enclosure Movement provided spaces to develop the style of picturesque gardens,[60] while the dispossession of lands from Indian aboriginals' lands laid the groundwork for the creation of the wild landscape of the United States.[61] The rise of a social class was often facilitated by their large-scale acquisition of lands, but it is through the craft of the landscape that their social standings were presented and asserted. The land was transformed into landscape, and the gentry became cultural elites, the literati.

This final chapter delves into the political and economic dimension of landscape transformation, revealing how shifts in social structure interacted with the landscape and estate transformation. State's policies on Buddhism, taxation, and corvee labour reconfigured the economic activities of Buddhists, the gentry, and local government. A new social hierarchy emerged in the late Ming, with the gentry ascending to the top and Buddhists relegated to the bottom. This power dynamic in Suzhou took shape as the gentry and government contended with Buddhists over the possession of hydraulic estates and garden practice. The gentry and government erased Buddhist traces meanwhile largely inherited the hydro-topography established by Buddhists, grounding on which they established the literati and Confucian landscape.

Notes

1 Wang Jinping points out that the Mongols adopted a religious political system which deviated from the previous Song-Jin model, and the social order was transformed in the Yuan dynasty. Wang, *In the Wake of the Mongols*, 13.
2 Chen, 'Buddhism and Society in the Yuan Dynasty.'
3 Chen, 'The Development of Taxes and Corvee of Buddhist Temples in the Yuan Dynasty.'
4 Xie and Bai, *History of Policies of Buddhist Officials in China*, 209, 225.
5 '元仁宗時期有名囊加巴的僧錄, 侵吞大量四產, 被僧眾譏為 "獅子身蟲".' in ibid., 221.
6 Ma, 'Study on the Secularization of Monks in the Yuan Dynasty.'
7 The art works stored in temples of Suzhou could be glimpsed by reviewing *GSZ*, vol. 29. Ming literati in Suzhou often went to temples to appreciate the artworks of Yuan masters.
8 Wang, *In the Wake of the Mongols*, 164.
9 Ibid., 158–160.
10 Ibid., 164.
11 The number of registered monks and nuns soared to 213,100, and the unregistered were many. See Xie and Bai, *History of Policies of Buddhist Officials in China*, 209.
12 '蓋前代寺多腴田, 略無限制. 今則悉入於官.' In Wu Kuan, 'Records on the Reconstruction of Wanshou Chan Temple 萬壽禪寺重修記 (Wanshou Chansi Chongxiu Ji)', in *WDWXCJ*, vol. 30, 4–6.
13 Brook points out that Zhu Yuanzhang's amalgamation order issued in 1391 and the Buddhist reformation carried out in the 1950s were the two most severe suppressions of Buddhism in Chinese history. Brook, *The Chinese State in Ming Society*, 145.
14 This Buddhist Bureaucratic system was stipulated in the Buddhist law of 1381. see Ge, *Gazetteer of Temples in Jinling*, 55–56.
15 Ibid., 55–56.
16 See Ge, *Gazetteer of Temples in Jinling*, 57.
17 Ibid.
18 '供報各處有額寺觀, 須要明白開寫本寺本觀始於何朝何僧何道啟建, 或何善人施捨.' in ibid., 56–57.
19 Ibid., 57.
20 Ibid., 58.
21 Ibid., 62.
22 Ibid., 69.
23 A review of the records of temples in Suzhou reveals the different support for the abbey and the cloister from the state. The temples which nominated as abbeys got constant management, but cloisters did not.
24 Minxi, *Gazetteer of Repay Kindness Temple in Suzhou Prefecture.*
25 '國朝洪武辛未詔天下佛所有像設古蹟及僧眾者為叢林, 否則廢之. 寺僧南宗廼會緇流而告曰: 壽聖寺, 古名道場也. 可舍而弗居. 即聚指二百, 由是寺為叢林.' in Chen Ji (1370–1434), 'Record of Sacred Longevity Teaching Temple at Bantang 半塘壽聖寺記 (Bantang Shousheng Si ji),' in *WDFC*, vol. 10, 113–115. The other eight temples incorporated into Bantang Shousheng Teaching Temple were

Ultimate Benevolence Cloister 臻福菴 (Zhenfu an), Accumulated Benevolence Cloister 資福菴 (Zifu an), Pristine Brightness Cloister 原明菴 (Yuanming an), Kindness and Filial Piety Cloister 慈孝菴 (Cixiao an), Buddha Wisdom Cloister 佛慧菴 (Fohui an), Attained Achievement Cloister 得成菴 (Decheng an), Universal Light Cloister 普光菴 (Puguang an), Universal Benevolence Cloister 普福菴 (Pufu an). See *GSZ*, vol. 29, 43–44.

26 ‘吳有佛寺, 曰南禪集雲者, 國初賜額也. 寺之始建不可考. 自唐宋以來, 多名僧居之. 皇明又有若寶曇和尚者, 高皇帝知其名, 召赴闕下. 俾住蜀之峨嵋化行其地, 久之而還. 因奏先所居吳門集雲旁有妙隱大雲二寺, 乞合而一之. 為是上從之始賜今額, 寔洪武二十四年也.’ Wu Kuan, ‘Records of the Reconstruction of Great Buddha Hall of Gathering Clouds Temple of Southern Chan Abbey 南禪集雲寺重建大雄殿記 (Nanchan Jiyun Si chongjian dadian ji),’ in *CLXZ*, shang juan, 18–19. The two cloisters being incorporated into Gathering Clouds Temple are Witty Retreat Cloister and Grand Clouds Cloister, as discussed in Chapter 4.

27 See Yao Guangxiao’s multiple records for temples in Suzhou in *Anthology of Yao Guangxiao*.

28 Ge, *Gazetteer of Temples in Jinling*, 74.

29 See the 2nd law of Declaration of the Buddhist Register 申明佛教榜冊 (Shenming fojiao bangce) issued in 1391 in Ge, *Gazetteer of Temples in Jinling*, 66. This rule was reasserted in the first rule of the 1394’s law. See ibid., 73. See the 13th rule issued in 1394 in ibid., 75.

30 See the 8th rule of 1393’s Buddhist law in ibid., 74.

31 The first rule of 1391’s Buddhist rule classified Buddhists into three groups of Chan, Teaching, and Yoga. See ibid., 66.

32 Brook, *The Chinese State in Ming Society*, 146.

33 Wen Zhenmeng, ‘Record of Reconstructing Benevolence Origin Temple 重興福源寺記 (Chongxing Fuyuan Si ji),’ in *WDFC*, vol. 10, xia, 7–8.

34 Timothy Brook calculates that for the maintenance of timber buildings in temples, every five years required a refurbishment, for the replacement of some minor broken parts. Twenty years required a renovation, such as replacing the structural members. If a monastic building lacked maintenance for decades, it could not be recovered by renovation thus required a full-scale reconstruction. Brook, *Praying for Power*, 160–165.

35 ‘然欲置田贍僧眾. 議曰: 不可. 有田則有役, 僧反不安, 無之可也. 遂寢其事.’ In Yao Guangxiao, ‘Record of Witty Wisdom Cloister and Shrine of Yao Family in Xiang City 相城妙智庵姚氏祠堂記 (Xiangcheng Miaozhi an Yao Shi Citang Ji),’ in *Anthology of Yao Guangxiao*, 281–283.

36 Wang Ao, ‘Lamenting Pines 憫松 (Min song),’ in *WDFC*, vol. 20, 170.

37 See Dardess, *Confucianism and Autocracy*, 16–17.

38 Fan and Xia, *Social and Economical History of Suzhou Region, Ming and Qing Dynasties*, 187–193.

39 ‘蘇城有集福菴. 南鄰吳尚書飽菴, 東鄰施知州膚菴. 弘治中, 詔毀淫祠. 有司欲以爲飽菴後圃. 公曰: “僧菴吾世鄰也, 誠不忍其毀, 又安忍有之乎!” 有司以歸膚菴, 曰: “何不歸吳公, 而屬我?” 有司述公言, 膚菴曰: “我獨不能爲飽翁耶!” 亦辭謝. 其菴竟存. 嘉靖初, 又有詔毀. 知府伍疇中納金承佃, 都御史毛貞甫亦納金佃, 竟成訟奪. 時人謠曰: 昔日吳與施, 官送猶遜辭. 今日毛與伍, 訐告到官府.’ in Zhang, *Jade Light Sword Qi Collection*, 1049.

40 Fan and Xia, *Social and Economical History of Suzhou Region, Ming and Qing Dynasties*, 187–192. Also mentioned by Chow, *The Rise of Confucian Ritualism in Late Imperial China*, 16–18.

41 Some farmers were even willing to register their lands under the gentry families' title because this could reduce the tax they paid. See ibid.

42 Huang Xingzeng, 'Appeal to Magistrates Nie and Cai, Ministers Huang and Zhu to Stop Seizing Temples 與轟蔡二郡公黃朱二令止奪寺觀書 (Yu Nie Cai Er Jungong Huang Zhu Er Ling Zhi Duo Si Guan Shu),' in *Collection of the Five Sacred Mountains Hermit*, vol. 31, 12–14.

43 Wang, 'Destruction of Illegal Temples and Buddhism Temples in Local Society in Early Period of Emperor Jiajing's Rule'.

44 Schneewind, *Community Schools and the State in Ming China*, 78–91.

45 Cai Yu 蔡羽 (1470–1541), 'Record of Stone Lake Thatched Hall 石湖草堂記 (Shihu caotang ji),' in *WDWCXJ*, vol. 31, 37–38.

46 Wu Kuan, 'Records of Giant Stone in Clouds Spring Cloister in Sun Mountain 陽山大石雲泉菴記,' in *WDWCXJ*, vol. 33, 20–33. Also see Liaoning sheng bowuguan, *Catalogue of Chinese Calligraphy and Paintings Collected Through the Dynasties, Calligraphy Volume*, 210–221.

47 During the 1500s, when monk Baian Jin 白菴金 approached the retired Wu Kuan for a record of the reconstruction of Ten Thousand Longevity Temple. Though not familiar with the temple, Wu accepted the invitation. See Wu, 'Record of Ten Thousand Longevity Temple.'

48 '吳中多佛老之區, 雖更洪武歸併, 而其廬故在也, 既不可廢. 則列其叢林而諸歸併者各附見焉.' in *GSZ*, vol. 29, 1.

49 '至者然即其毀建, 足以见释氏之张弛与世治之污隆也. 近歲, 精蓝古刹, 往往弊于赋役, 夺于豪借. 溷琳宫为贾区, 摧琦木而薪樀. 至有削金于满月之容, 凿寶于不壞之体者. 既而人随渐灭, 居成. 僧舍所得, 亦几何哉. 呜呼.' Peng, 'Record of Reconstructing Auspicious Light Chan Temple.'

50 '今百神之典祀, 儒宫之建修. 敛其财, 役其力. 民以势从之尔. 恶知所谓法施定国之五者哉? 故不知而倡之, 虽从犹勉耳.' Zhu, 'Record of Imperial Endowed Repay the Nation Chan Temple in Suzhou.'

51 Huang, 'Appeal to Magistrates Nie and Cai, Ministers Huang and Zhu to Stop Seizing Temples.'

52 Huang, 'Stele Record of Shrine of the King of Qi.'

53 Wu Kuan, 'Beautiful Prefect Cao Destroys Profane Shrines 美曹太守毀淫祠 (Mei Cao Taishou hui yinci),' in *WDWCXJ*, vol. 51, 49–50.

54 This shrine for Wang Ao was still extant today in front of Bright Virtue Temple. See Tao Wangling 陶望齡 (1542–1609), 'Record of Shrine of Wang Wenke Gong 王文恪公祠記,' in Niu and Wang, *Gazetteer of Wu County in the Reign of Chongzhen*, vol. 19, 62–64.

55 Qian, 'Imperial Censor Wang Destroyed the Temple.'

56 Ibid.

57 Jiang Yingke 江盈科 (1553–1605), 'Record of the Reconstruction of the Buddha Hall and Five Sage Shrine at Forever Pacification Temple 重復永定寺建佛殿五賢祠記,' in *Anthology of Jiang Yingke*, 246–248.

58 Welch, *Buddhism Under Mao*, 42–44.

59 Bernard, 'The Dissolution of the Monasteries'; Hoyle, 'The Origins of the Dissolution of the Monasteries.'

60 Williamson, *Polite Landscapes*, 11.

61 Spence, *Dispossessing the Wilderness*, 1–8.

Conclusion

A New Garden History: Hydro-Social Transformation Illuminated by 3D Modelling and GIS Mapping

The current study's theory, which triangulates garden practice, hydraulic contention, and social transformation, offers architectural insights into the landscape transformation of Suzhou from the 13th to 16th centuries. Is this theoretical framework applicable to understand the global garden practices, and thus shed light on the reciprocal influence between landscape and the society? Furthermore, considering the distinct nature of evidence in Eastern Asian and European studies, could 3D modelling and GIS mapping be applied in both fields? The conclusion addresses these questions, investigating how spatial methods facilitate hydro-social inquiries, and further extending the application of the current methodology to a wider global context.

Spatial Configurations and Garden Style in Conversion

During the hydraulic and estate contention in the 13th to 16th centuries, the adaptation of landscapes by successive landowners led to the emergence of three layouts in garden design. The first involved creating scenes organised by touring paths along mounds and rivers. The second configuration was the linear development of waterfront façades, where scenes were choreographed to provide visual experience for boating or meandering along the shore. The third configuration was the axial courtyard complex orienting towards urban canal with gardens at the front or rear. The three configurations were both enlivened by movement within the landscape, and all of them were developed from previous temple estates. The first pattern is predominantly seen in private gardens constructed by the literati, forming an enclosed pleasure garden. The second and the third were observed in the public Confucian gardens as crafted by the government to visualise Confucian values.

The first spatial pattern is exemplified by Lion Grove Garden as converted from Shizi Lin Temple, and Unsuccessful Politician's Garden, converted from Great Propagation Temple. Inheriting the topography, rivers,

DOI: 10.4324/9781003387169-7

and plantation laid out by monks in the 13th–14th centuries, the gentry re-read these Buddhist scenes, inscribing aesthetic implications, and established literati ways of seeing by album leaf paintings and poems. Their strolling experience identified paths connecting individual scenes, as usually narrated in the garden records (see Appendix, Translations 1, 2). The construct of literati scenes particularly erased previous Buddhist symbology. We also observe a shift from the use of a bodily footpath in the 16th century towards an emphasis on visual plots in the experience of touring gardens from the 17th century onwards. The touring experience in the Middle and High Ming periods created a sense of being lost, during which visual connections between scenes were implicit. However, in the late Ming and Qing periods, touring became visually driven, embodying picturesque features, with vistas established between scenes.

The second configuration, characterised by building façades choreographed towards the public canal, is exemplified by the transformation of the Surging Wave Pavilion into a public Confucian garden. The value of the waterfront façades was particularly recognised by the government, as seen in their consistent efforts to construct structures along the shore. The development of the waterfront along the Surging Wave pond began in the High Ming period, with monks initiating the construction of a waterfront gazebo to cater to gentry activities. Subsequently, from the early Qing period onwards, the waterfront saw ongoing enhancements, as illustrated in Wang Hui's handscroll painting. These waterfront façades, composed of imperial steles, symbols of literati culture, and archways of Confucian academies and schools, conveyed Confucian values promoting movement along the urban canal. This aligns with the government's centralised management of public canals to improve navigability and foster auspicious geomancy for Confucian institutions. The handscroll format of Chinese painting, with a linear progression as it unfolds during viewing, effectively captures the spatial narrative of the waterfront.

The last configuration is characterised by an axial arrangement of buildings orienting towards urban canals, with scenic gardens at the front or rear, which is exemplified by the transformation of five temples into Confucian academies and schools. Such a transformation underscores the versatility of Chinese architecture in accommodating diverse religious practices. The original monastic design, with the Buddha hall and Dharma hall aligned along the central axis, was readily adaptable to Confucian shrines and lecture halls. The axial courtyards, featuring a scenic garden at the front, were represented in short hanging scroll paintings, as exemplified by the two woodblock paintings of the Surging Wave Pavilion (Fig. 4.7a, Fig. 4.12a).

During the Mid-Ming to Qing dynasties, which were often regarded as the zenith of Chinese garden practices, the three spatial configurations emerged as archetypes of Chinese garden design. This research demonstrates that these garden styles evolved when temples were transformed into the estates of the gentry and government. As such, the research challenges the conventional notion that garden styles are static and ascribed to specific periods or social groups. Instead, it shows that garden styles developed dynamically as successive landowners modified previous owner's landscaping. Garden styles both emerged from and facilitated social transformation, offering insights into social changes through spatial evidence.

This observation from the gardens of Suzhou invites reflection on the transformation of gardens in other regions. In picturesque gardens, for instance, lakes were often reconfigured from former fishing ponds.[1] Churches and castles, typically relics of medieval estates, were recontextualised with new meanings.[2] Veteran trees in the picturesque estates were usually planted by previous owners. The use of ha-ha strategies was a deliberate choice by gentry owners, aiming to hide the old boundaries of land parcels, as their vast estates were created by amalgamating these piecemeal lands under the former parish system.[3] The dimension of garden renovation, intertwined with social transformations, provides a spatial perspective that bridges Eastern and Western Garden studies.

Hydraulic Gardens Contested

The contention over hydraulically established estates, the central theme of this book, is exemplified by the nine case studies. Temples such as Shizi Lin, Great Propagation Temple, and the three temples around Surging Wave—Grand Clouds Cloister, Witty Retreat Cloister, and Gathering Clouds Temple, all featured in the mound-field-river topography in the Yuan. This topography, praised by Shen Zhou for its geomantic supremacy, was the coveted aspect of an estate. Similarly, Dragon Revival Temple, Forever Pacification Temple, Bright Virtue Temple, and Ten Thousand Longevity Chan Temple, situated along major canals, enjoyed hydraulic advantages. The transformation of these temples in the Ming era provides valuable insights into the hydraulic status of other converted temples in the Ming. The current research's hydraulic perspective enhances the understanding of Sinologists' observation of land transformation in China during the 16th century: it is the good geomantic estates, notably those featuring well-established hydraulic infrastructure, that were highly contended. The hydraulic contention underpinned a social transformation from the Buddhist to the gentry society.

The hydro-topography, as argued in Chapter 1, was advantageous to agricultural gardens, fostering diverse habitats that enhanced both productivity and sustainability. It was also conducive to the construction of gardens that embody enduring Chinese aesthetics—resembling natural mountains, streams, and rivers, featuring ancient trees, and adorned with sparse artefacts. This landscape aesthetics in literati culture was epitomised by the 12th-century Song dynasty paintings.[4] Acknowledging the Buddhists' foundational role in shaping the site's topography and plantations, it could be concluded that the literati gardens in China were formed out of the social contention between the gentry and Buddhists.

Suzhou was the cultural and economic centre of China during the Ming dynasty.[5] The garden estates played a pivotal role in elevating the social and cultural status of the gentry, transitioning them from affluent landlords to literati elites. Garden estates generated both financial, social, and cultural capital for the gentry. First, gardens were lucrative investments for the gentry, serving agricultural and even industrial purposes. Second, gardens became social places for gatherings and networking, as exemplified by Zhang Shijun's Lion Grove Garden. The tradition of inscribing and sealing garden paintings chronicled the network surrounding the garden, solidifying its status as a nexus of cultural and intellectual exchange.

Third, the gardens, as exclusive realms for the literati, foster an art genre that is only capable of being crafted by the literati. The gardens, together with their artistic representations, composed a distinctive language and corpus in the literati's art criticism. Garden aesthetics largely defined the taste of Ming literati and became their symbol of social status, as advocated by the Late Ming gentry Wen Zhenheng in his *Treatise on Superfluous Things (Zhangwu zhi)*, later by Ji Cheng's *Craft of Gardens (Yuan ye)*.[6] The cultural significance of gardens was established by their artistic portrayals, including paintings, poems, and records. These representations inscribed the current gardens with classical meanings, rendering them as poetic and sentimental. The tranquillity and seclusion of the gardens were associated with the refined characteristics of their owners, thereby positioning them as cultural elites. The garden became a laudatory portrait of its owner.

The triangulated theoretical model of garden practice, hydraulic contention, and social transformation informs the global garden studies. It suggests that specific garden styles, traditionally ascribed to distinct groups or eras, were formed during the rise of new social classes, through their contention with established traditions and former classes. This dynamic could be investigated by examining the way the emergent groups inherit, modify, develop, or completely erase previous landscape features to create new sceneries. Such a perspective underscores the unique contribution of landscape history to the history of the Anthropocene, as defined by social transformations within environmental contexts.

The transformation of gardens from the geometric style of the Renaissance and Baroque periods to the picturesque gardens during 18th-century Britain served as a salient example of how landscape changes reflect social shifts.[7] This shift in landscape preference from the geometric to the irregular pinpointed a historical process whereby the British gentry acquired lands and ascended to the land-holding class. Particularly during the enclosure movement of the mid-18th century, the gentry seized expansive lands from parish churches and the lower social strata in the countryside. Based on these incorporated lands, picturesque gardens were constructed, modifying the topography while retaining the veteran trees and abandoned church relics. Thus, the picturesque style could be interpreted as manifesting from the contention between capitalism and feudalism, between the modern and the imperial.

The formation of gardens within wetland regions is closely tied to the historical reclamation of land from marshes and seas, with notable examples in Venice from the 10th to the 16th centuries, followed by the Netherlands during the 17th century. In Venice, prior to the Grand Canal's construction, religious institutions spearheaded land reclamation, inscribing the changing landscapes with religious ideology.[8] In the 15th century, Jacobo di Babari's map of Venice illustrated gardens as situated on the lowlands at the periphery of Venice's central islands, which were newly dredged from the sea.[9] Archives recorded that owners in this period actively reclaimed lands to establish agricultural estates. Later maps showed that these agricultural gardens diminished, with lots of artefacts newly constructed.[10] In light of the hydraulic gardening in Suzhou, it prompts further examination on how Venetian estate owners during the Medieval and Renaissance periods may similarly use gardens as infrastructure to enhance cultivability and habitability.

In the Netherlands, the 17th-century land reclamation created a distinctive dune-scape that came to be celebrated in Dutch landscape paintings.[11] Ann Jensen Adams has shown that during the late 16th and 17th centuries, the Dutch reclaimed extensive lands from the sea.[12] The reclamation activities shaped the Netherlands' landscape as featured in topography, preserved old trees, and intricate waterways. The constructed polders were valuable assets for both communal and private ownership. The sandy dunes piled up by soils dredged from wetlands, similar to the formation of mounds in Suzhou during the 13th–14th centuries, frequently featured in the works of painters such as Pieter van Sabrvoort. The dredged canal and navigations of boats and ships also became iconic scenes of Dutch landscape painting.

History Enabled by 3D Modelling and GIS Mapping

The 3D modelling used in this research has provided a historical arena for integrating and verifying evidence, providing a visual and spatial reconstruction of the past. This approach is informed by the trend in architectural history in which 3D models are increasingly used to enhance spatial analysis of historical buildings. Frommel's 3D reconstruction of Santa Maria delle Carceri in Prato offered insights into Giuliano de Sangallo's original design intentions.[13] The Visualising Venice Project rendered the urban-scape of Venice in the 16th century into vision.[14] Dany Sandron and Andrew Tallon's book *Notre-Dame Cathedral: Nine Centuries of History* vividly illustrate the cathedral's construction stages: from the choir and ambulatory in 1160–1182, to the nave and aisles in 1180–1200, and finally, the entrance and great organ space in 1200–1210.[15] 3D models reconstructed in these researches referenced measured architectural drawings to visualise the past.

But could historical Chinese gardens be reconstructed by modelling? Chinese architecture, due to the use of timber material, is vulnerable to decay and fire and usually did not leave architectural relics in the same sense as the stone relics in Pompeii or Rome. Second, Chinese garden paintings often lacked precise measurements and were not created with the intent of accurate depiction in scale. Especially after the Song dynasty, the emerging literati paintings emphasised visualising subjective perception of the scene and its spirit rather than recording the exact physical layout.[16] This expressive orientation often resulted in the omission of unwanted site features or an exaggeration of certain elements to convey seeing instead of sight. Even for gazetteer maps prescribed with a documentary nature, such as *Pingjiang Map*, the distortion in scale is still witnessed.

The current research demonstrates that Chinese gardens, particularly those dating from the 16th century onwards, are indeed feasible to be modelled. The viability of modelling hinges on a combination of factors: a specified site with archaeological evidence, rich textual and pictorial documentation, and continuous building records. Together, they formed a closed ring of evidence, a corroborative dataset which supports accurate modelling. The evidence of the three gardens modelled in this research exemplifies these characteristics. Notably, the hydro-topographical features, such as mounds, canals, bridges, ancient trees, and the stony basis of the non-existent timber structures, served as archaeological evidence that corresponds artefacts in the painting to the current site.

As the writings craft historical narrative and discourse, it is questioned what is the arena introduced by 3D modelling to historical investigation?[17] Observed in this research is that 3D models present a field to verify and integrate evidence in diverse formats. The 3D models pinpoint archaeological evidence and, when corroborated with textual records—which are more reliable for location and construction details—effectively verify the pictorial representation in paintings. 3D models also serve as arenas where disparate pieces of evidence converge, as the final model should be the one that best synthesises fragmented pieces of evidence, bridges gaps between different types of data, and aligns with the established patterns and experiences derived from other historical cases. The models' credibility hinges on the historian's process of reasoning and deducing spaces from historical evidence.

The 3D models empower spatial reasoning in historical enquiries. They provide a tangible space for historians to analyse the views and spatial concepts articulated by historical figures. For instance, the isometric drawing of Surging Wave Pavilion illustrates the Surging Wave pond as an end water surrounded by three temples in the Yuan and early Ming (Fig. 4.2), which informs the understanding of monk Baotao's metaphor 'three temples sharing the same root.' Rotating the models in software such as Rhinoceros 3D helps historians to visually engage with scenes depicted in Shen Zhou's poem and further deduce the poet's viewpoint atop the mound. The model of Surging Wave Pavilion in the mid-Qing period visualises the configuration of the whole site, which is key to understand the spatial metaphor of 'myriad stars surrounding the polar star' as highlighted in Wu Cunli's stele memorial. I have discussed these aspects in my presentation *Translating Chinese Painting into Architectural Drawings* in 2021.[18]

The use of chronological 3D models to track the evolution of a garden is a notable feature in the current book's 3D approach. Amongst architectural scholarships in the 3D modelling of historical sites, Sandron and Tallon's modelling of Notre Dame exemplifies modelling a site's development in a chronological way.[19] The application of 3D modelling in garden history, particularly considering aspects such as topography, plantation, and hydrology, was less common and awaits further inquiry.[20]

Successive modelling of a site offers nuances to understand its spatial evolution, an aspect less investigated in architectural history. Comparing models from different time periods shed light on the most contended aspect of a site's history and most revealed the intention of an estate owner: how the owner inherited, modified, adapted, and developed the site. The model drawings serve as illustrated histories, discoursing the evolution of a site comprehensively.[21] The current research particularly used isometric drawings, which can convey the site's evolution without omission.

In short, 3D modelling brings dimensions to historical study which writing does not adequately cover and is insufficient to address, such as the physical consistency of a site, its ecological practicality, and the basic scales and relationships with human activities. While textual history excels in logical reasoning, chronological narrative, and classification, the 3D model as a historical construct efficiently addresses issues such as indexing artefacts and analysing topographical features, situating narratives in spatial contexts. The 3D historical modelling consistently poses questions to a historian such as where did the water resource originate from? Is it reasonable to locate the artefacts and plants there? Did this construct suit the load bearing? Thus 3D models formed an arena to provide architectural verification of the evidence.

A GIS map situates individual sites within the broader urban context. Combining 3D modelling with GIS mapping allows a seamless transition from micro-history, focusing on individual estates, to macro analysis that considers the urban and regional landscape. The individual 3D models in this book could be further integrated into a GIS map by referencing the locations of bridges, streets, and canals. The uniform coordinate locations in GIS mapping facilitate historians to further compare landscapes in different regions and tackle global issues.

As in-depth case studies are contextualised within the broader geography by GIS mapping, historians could zoom out to advance urban inquiries. GIS mapping enables the deduction of specific prototypes towards broader historical patterns. The GIS maps in Figures 1.1 and 1.2 illustrate water flow at regional, urban, and estate scales. A city's elevation model helps validate the selection of cases for investigating how an estate's hydraulic practices influence the overall urban hydrology. For instance, the GIS map in Figure 1.4 reveals that Yuan-era temples were mainly built in low-lying areas. This finding justifies the current approach, which posits that examining the hydraulic strategies of these monastic estates can further illuminate how the wetlands of Suzhou were transformed. GIS mapping also sheds light on the government's scheme of urban control through renovating the waterfront. The lineage of Suzhou's Confucianism, as framed by Hu Zuanzong, was not only inscribed into stele texts for reading, but visually promoted by the scenes organized along waterways. This scenographic urban strategy was similarly observed in the promotion of Christianity in medieval Venice.[22]

3D models and GIS maps could provide arenas for spatial analysis and historical reasoning, far beyond merely serving as illustrations and visual aids. These tools reveal new findings, and support argumentation in writing, providing insights that are overlooked or downplayed in the text-based historical narratives. Furthermore, 3D models and GIS maps encourage scholars to address challenges such as tracing the spatial evolution of architecture and landscape.

The current research advocates a new history highlighting gardens' agency in urban hydraulics and social transformation. It conceives gardens as a city's hydraulic components, participating in the urbanisation process through water circulation. Challenging the traditional notion of garden styles as merely reflective of social tastes, the research proposes that the economy of gardens contributed to establishing the new social structure. The social contentions could be best observed in the evolution of gardens, as their owners adapted and renovated the pre-existing landscapes. These spatial narratives are realised by the modelling of gardens' successive historical phases, with GIS mapping providing a comprehensive view from estates to urban historical trends.

Notes

1 Bishop, *Ornamental Lakes*, 17–61.
2 Williamson and Bellamy, *Property and Landscape*, 54.
3 Williamson, *Polite Landscapes*, 35–46.
4 Murashige, 'Rhythm, Order, Change, and Nature in Guo Xi's Early Spring'.
5 See Marmé, *Suzhou: Where the Goods of All the Provinces Converge*, 154–186.
6 Blishen and Wen, *An Elegant Life of Chinese Literati*. Also see Clunas, *Superfluous Things*; Chen and Ji, *Annotation on the Craft of Gardens*.
7 Williamson, *Polite Landscapes*, 9–18.
8 Ammerman, 'Venice before the Grand Canal', 141–158.
9 Schulz, 'Jacopo de' Barbari's View of Venice'.
10 Hunt, 'The Garden in the City of Venice'; Housley, Ammerman, and McClennen, 'That Sinking Feeling.' I would like to thank Heiner Krellig for his discussion with me on the location of gardens in Jacobo di Babari's map.
11 For examples of these dune-scapes, refer to Stechow, *Dutch Landscape Painting of the Seventeenth Century*.
12 Ann Jensen Adams, 'Competing Communities in the "Great Bog of Europe."'
13 Frommel, Gaiani, and Garagnani, '3-D Digital Modeling and Giuliano Da Sangallo's Designs for Santa Maria Delle Carceri in Prato.'
14 Huffman, Giordano, and Bruzelius, *Visualizing Venice*.
15 See the analysis of historical narrative in White, *Metahistory*, 2–7.
16 See Su Shi's advocacy of expressive painting during the Song dynasty and its profound impact on the later trend of Chinese painting in Bush, *The Chinese Literati on Painting*, 7–13.
17 White, 'The Value of Narrativity in the Representation of Reality'. Also see White's discussion on historical writing as the field of historians' intellectual activity in White, *Metahistory*, 351–353.
18 See https://ea-aaa.eu/online-talk-suzhou-surging-wave-pavilion-translating-landscape-paintings-into-architectural-drawings-by-dr-pania-mu/.
19 Sandron and Tallon, *Notre-Dame Cathedral*.
20 Gleason, 'Porticus Pompeiana.'
21 Bruzelius, 'Digital Technologies and New Evidence in Architectural History.'
22 Savoy, 'Palladio and the Water-Oriented Scenography of Venice.'

Appendix

Translation 1. 游師子林記 **Record of Touring Shizi Lin**

蘇城之東北區, 有林若干畝. 佛者居之, 曰師子. 師子者, 林之一峯如其形, 故名. 而其地特隆然, 以起為丘焉. 雜植竹樹. 丘之北窪然, 以下為谷焉. 皆植竹, 多至數十萬本.

In the northeast area of Suzhou, there is a grove covering several *mu*. It was inhabited by Buddhists who called it 'Lion Grove.' The name comes from one of the peaks in the forest that resembles the form of a lion. The ground there rises to a hill, densely covered with bamboo and trees. North of the hill is a depression, below which lies a valley, also filled with bamboo, amounting to hundreds of thousands of stalks.

始升其邱之南麓, 便仰見師子峯. 高僅若干尺, 如舞且踞. 兩傍復各有峯, 亞匹之東曰含暉, 作人立. 左腋下有穴一, 腹枵然, 有四穴. 日始出, 則其暉映曖相射. 西曰吐月, 頗峭且鋭. 稍夕, 月即見其上.

Climbing the southern slope of the hill, one can look up to Lion Peak (Fig. 2.5(15)), only several *chi* high, appearing as if dancing. On both sides, there are two more peaks with lower heights; to the east of the main peak is Teaming with Rays Peak (Fig. 2.5(16)), resembling a standing person. Under its left armpit is a hole, and the belly is hollow with four more holes. When the sun rises, its light shines through, creating a brilliant display. To the west is Holding Moon Peak (Fig. 2.5(13)), rather steep and sharp, where the moon can be seen on its top in the early evening.

師子之北有茇一, 曰禪窩. 含暉之東有隙地踰尋, 氄以石子為環坐者之所藉, 曰繙經臺. 傍有峯特出, 曰立玉. 然其狀嵌若刀劍, 劃作四五葉者, 或曰以地肺名之為宜. 吐月之西有澗, 自竹谷中來. 因架石為梁, 曰飛虹.

North of Lion Peak is a room called Chan Hut (Fig. 1.5d(8), Fig. 2.5(8)). To the east of Teaming with Rays Peak there is an open area several feet wide, paved with stones and used as a seating area. The place is called 'Reading Sutra Platform' (Fig. 2.5(9)). Beside it, there is a prominent peak named Erected Jade Peak (Fig. 2.5(17)). However, its shape looks like a knife or sword with four or five leaves. Some say it should be named as 'Earth Lung' due to its shape. West of Holding Moon Peak (Fig. 2.5(13)) is a stream coming from the Bamboo Valley (Fig. 2.5(18)). A stone was arched to make a bridge, called 'Flying Rainbow' (Fig. 2.5(5)).

蹦飛虹以西而下其西麓, 乃北入竹谷中, 委蛇東來. 折以南, 出立玉後, 而上其東麓. 復折而南, 且西出師子前, 而下其南麓. 凡丘之巔踵, 自三西峯外, 諸小峯又十數計. 且叢列怪石, 什伯為羣, 而所取道往往經緯其間.

West of Flying Rainbow bridge and down its western slope, the stream enters the bamboo valley, winding eastwards like a snake. Turning south, it passes behind Erected Jade Peak (Fig. 2.5(17)) and climbs its eastern slope, then turns again south and west, emerging in front of Lion Peak and descending its southern slope. On the summit of the hill, besides the three main peaks, there are several smaller peaks, numbering over a dozen. Moreover, the strange rocks are arranged in clusters and groups, thus the paths often crisscross them.

既下南麓, 有二道. 其循麓而東者, 至立雪堂, 方堂. 之南為臥雲室, 又南為指柏軒. 其循麓而西者, 至問梅閣. 問梅與指柏相直. 梅與柏各一, 皆相結為蛟虬, 其壽幾二百年. 柏之南有池, 曰玉鑑. 若鑑影以自媚者. 梅之西有井, 曰冰壺. 初鑿井時得古壺鑄地下, 而其泉冽且甘, 以瀹茗味尤勝云.

Descending the southern slope, there are two paths. The one going east along the slope leads to Stand in Snow Hall (Fig. 2.5(7.2)), a square hall (Fig. 2.5(7.1)), south of which is the Lying Clouds Chamber (Fig. 2.5(12)), and further south is Point to Cypress Hall (Fig. 2.5(11)). Ask Plum Pavilion (Fig. 2.5(6)) and Point to Cypress Hall are directly aligned, with a flowering plum (Fig. 2.5(19)) and cypress tree respectively (Fig. 2.5(20)), both intertwined like dragons, nearly 200 years old. South of the cypress is a pond called 'Jade Mirror Pond' (Fig. 2.5(21)), as if admiring its reflection. West of the plum is a well named 'Icy Pottery' (Fig. 2.5(10)). When the well was first dug, an ancient pot was found underground, and its spring water is cold and sweet, especially good for brewing tea.

余在昔於斯遊也, 蓋屢焉而不厭. 今年秋復與茅宅民, 陳彥廉, 張曼端來游. 而因師者, 予故人也. 止予宿問梅閣, 得咏歌其丘與谷者累日. 師曰: '是果可以咏歌歟, 願有記也.' 故書之石, 而使刊之. 先是有十二咏書石上, 其倡者高太史季廸, 和者張水部子宜, 王文學止仲, 謝翰林玄懿. 今亦為同遊者.

In the past, I often toured here, never tiring of it. This autumn, I came to visit again with Mao Zhaimin, Chen Yanlian, and Zhang Manduan. Master Yin (Ruhai Deyin) is an old friend of mine. I stayed at Ask Plum Pavilion. I enjoyed eulogising and singing about the hills and valleys for several days. The teacher said, 'Can these indeed be sung and composed about? I wish there would be a record.' So, I wrote this as a stele inscription and had it carved. Before this, there were 12 poems inscribed on the stone, initiated by the Grand Historian Gao Jidi (Gao Qi). The persons echoed are Zhang Ziyi (Zhang Shi), the director of the Hydraulic Bureau; Wang Zhi Zhong (Wang Xing), the literati; and Xie Xuanyi (Xie Hui), the Hanlin academician, who were also among the wanderers here.

洪武五年秋七月稽岳王彝記

Recorded by Wang Yi from Jiyue, in the seventh month of autumn, the fifth year of Hongwu. (Note: the Chinese text is transcribed from *JSJ*, juan xia, 10–12, with reference to the text in *WDWCXJ*, vol. 30, 16–18. The indexing numbers of artefacts in Figure 2.5 correspond with Figure 1.5d.)

Translation 2. 王氏拙政園記 Record of Unsuccessful Politician's Garden of Mr. Wang (an Excerpt)

槐雨先生王君敬止所居, 在郡城東北界婁齊門之間. 居多隙地, 有積水亙其中. 稍加濬治, 環以林木. 爲重屋其陽, 曰夢隱樓. 爲堂其陰, 曰若墅堂.

堂之前爲繁香塢, 其後爲倚玉軒. 軒北直夢隱, 絕水爲梁, 曰小飛虹. 踰小飛虹而北, 循水西行, 岸多木芙蓉, 曰芙蓉隈. 又西, 中流爲榭, 曰小滄浪亭. 亭之南, 翳以脩竹. 經竹而西, 出於水滋, 有石可坐, 可俯而濯, 曰志清處.

Master Huaiyu, Mr. Wang Jingzhi, resides to the northeast of the county city, between the boundaries of Lou and Qi gates. His dwelling includes a lot of interstitial land, with accumulated water extending through it. With a bit of dredging and management, the area is encircled with woods. On the sunny side, he built a double-storied building, called Dream of Recluse Tower (no. 1). On the shady side, he constructed a hall, named Hall Like a Villa (no. 2).

In front of the Hall is located the Many Fragrances Valley (no. 3), to its back is Leaning Jade Veranda (no. 4). The veranda faces straight northward to Dream of Recluse Tower. A bridge crosses the water, named Little Flying Rainbow (no. 5). Go across Little Flying Rainbow bridge and head north, then follow the water and go west. Along the bank grow many hibiscus, hence the name Hibiscus Bend (no. 6). Further to the west and in the middle of the stream is a gazebo named Lesser Surging Wave (no. 7). The south of the pavilion is concealed by tall bamboos. Walk through the bamboos and turn westward, and then come out at the waterside. There is a rock for sitting. Here one can overlook the water and wash one's feet, and it is named 'Purifying Will Place' (no. 8).

至是水折而北, 滉漾渺瀰, 望若湖泊. 夾岸皆佳木. 其西多柳, 曰柳隩. 東岸積土爲臺, 曰意遠臺. 臺之下植石爲磯, 可坐而漁, 曰釣䂔.

遵釣䂔而北, 地益迴, 林木益深, 水益清駛. 水盡, 別疏小沼, 植蓮其中, 曰水花池. 池上美竹千挺, 可以追涼, 中爲亭, 曰深净. 循深净而東, 柑橘數十本, 亭曰待霜.

又東, 出夢隱樓之後, 長松數植. 風至泠然有聲, 曰聽松風處. 自此繞出夢隱之前, 古木疎篁, 可以憩息, 曰怡顏處.

From here, the water turns northward. This vast expanse of rolling water looks like a lake. Fine trees grow along its banks. On the west there are many willow trees, hence the name Willow Bend (no. 9). On the east, earth has been piled up to form a terrace, named the Thoughts Afar

Terrace (no. 10). Below the terrace, a rock is set to form a crag where one can sit and fish, and it is named the Angling Rock (no. 11).

Following the Angling Rock and going north, the area is more remote, the groves deeper, and the water more limpid and speedier. At the end of the stream, another little pond has been dredged, with lotus planted inside, and named Water Flower Pond (no. 12). Above the pond are thousands of fine bamboo poles, where one can withdraw to enjoy the cool. A pavilion has been built in the middle and named 'Deep Tranquil (no. 13).' Along the Deep Tranquil and going east, there are several tens of orange trees and a pavilion named 'Waiting for Frost (no. 14).'

Going further east and emerging at the rear of Dream of Recluse Tower (no. 1), there are several tall pine trees. When the wind blows, they rustle and make clear sounds, hence the name Listening to Windblown Pines Place (no. 15). From here, go around [the hill] and emerge at the front of the Tower. There are ancient trees and sparse bamboos where one can have rest, and hence the name 'Flushed with Pleasure Place' (no. 16).

又前, 循水而東. 果林彌望, 曰來禽囿. 囿縛盡, 四檜爲幄, 曰得真亭. 亭之後爲珍李坂. 其前爲玫瑰柴, 又前爲薔薇徑.

至是水折而南, 夾岸植桃, 曰桃花汧. 汧之南爲湘筠塢. 又南, 古槐一株, 敷蔭數弓, 曰槐幄.

其下跨水爲杠. 踰杠而東, 篁竹陰翳, 榆槐蔽虧, 有亭翼然, 西臨水上者, 槐雨亭也. 亭之後爲爾耳軒, 左爲芭蕉檻.

Go forward, follow the water, and head eastward. The view is filled with a fruit orchard named Arriving Birds Park (no. 17). At the end of the park, four juniper trees are bound together to form a tent, which is named Obtaining Essence Pavilion (no. 18). The Precious Plum Slope (no. 19) is located behind the pavilion, and Rose Fence (no. 20) to its front, and further forward is Rosebush Footpath (no. 21).

From here, the water turns towards the south. Peach trees are planted along both its banks, hence the name Peach Blossom Bank (no. 22). To the south of the Bank is located Xiang Bamboo Valley (no. 23). Further south, there is an ancient sophora, casting its shade measuring several Gongs [bows], and it is named Sophora Tent (no. 24).

Below Sophora Tent, a bridge spans the stream. Cross the bridge and go east. Under the shades of bamboos and partially concealed by the elm tree and sophora tree, a pavilion stands like a bird spreading its wings above the water on its west. This is Sophora Rain Pavilion (no. 25). Behind the pavilion located the So-so Gallery (no. 26), on its left Plantain Balustrade (no. 27).

凡諸亭檻臺榭, 皆因水爲面勢. 自桃花汧而南, 水流漸細, 至是伏流而南踰百武, 出於別圃藂竹之間, 是爲竹澗.

　竹澗之東江梅百株，花時香雪爛然，望如瑤林玉樹，曰瑤圃．圃中有亭，曰嘉實亭．泉曰玉泉．凡爲堂一樓一，爲亭六，軒檻池臺塢榭之屬二十有三，總三十有一，名曰拙政園．

All the above pavilions, balustrade, terrace, and gazebo have the form of their front side composed in relation to the water. From Peach Blossom Bank and southward, the stream has been gradually narrowed. From here, it prowls southward for more than a hundred *wu*, and comes out in another orchard and among a bamboo grove; this is the Bamboo Gully (no. 28).

To the east of the Bamboo Gully, there are a hundred flowering plum trees. In the blossom season, these fragrant snows shine and look like an orchard of jade trees, and are thus named Jade Orchard (no. 29). There is a pavilion in the orchard named Fine Fruit Pavilion (no. 30), and a spring named Jade Spring (no. 31). In all there are one hall, one tower, six pavilions, and 23 in the categories of gallery, balustrade, pond, terrace, valley, and gully, making a total of 31, by name the garden of the Unsuccessful Politician.

Note: the thirty-one scenes are indexed from no. 1 to no. 31. Translated by Lu Andong in 'Deciphering the Reclusive Landscape,' with author's modifications.

Table A.1 Temples Inside Suzhou City during the Yuan Dynasty

ID	Pinyin	Chinese Character	English Name	Location
a1	Kaiyuan chan si	開元禪寺	Opening Origin Chan Temple	盤門内
a1-1	Baoguo chansi	報國院, alt. 報國禪寺	Repay the Nation Temple, alt. Repay the Nation Chan Temple	長洲縣一都
a1-2	Huacheng an	化城菴	Transformation City Cloister	城西南隅
a2	Bao'en si	報恩講寺	Repay Kindness Teaching Temple; Alt. Repay Kindness Temple	在城北陲
a2-1	Dong Huayuan si	東華嚴寺	East Avatamsaka Temple	府治北
a2-2	Xi huayan si	西華嚴寺	Western Avatamsaka Temple	城西北隅
a2-3	Jiqing an	集慶菴	Gathering Joy Cloister	城西北隅
a2-4	Xizi an	西資菴	Western Resource Cloister	吳縣一都吳山南麓
a3	Ruiguang chan si	瑞光禪寺	Auspicious Light Chan Temple	開元寺南
a3-1	Shunxin an	順心菴	Follow the Mind Cloister	吳山下
a4	Wanshou chan si	萬壽禪寺	Ten Thousand Longevity Chan Temple	府治東北
a4-1	Pumen si	普門寺	Universal Gate Temple	城東北隅
a5	Zhengjue chansi	正覺禪寺	Right Enlightenment Chan Temple	城東南隅
a5-1	Baoshan chanyuan	包山禪院	Chan Temple of Mount Bao	洞庭西山
a5-2	Mingyin chanyuan	明因禪院	Mingyin Chan Monastery	橫山下
a5-3	Chongqing an	崇慶菴	Cloister of Celebrated Joy	城東北隅
a5-4	Shanhui an	善慧菴	Kind Wisdom Cloister (*Shanhui an*)	長洲縣九都
a5-5	Shanxian an	善現菴	Temple of Benevolent Presence	城東北隅
a5-6	Donglin an	東林菴	Eastern Grove Cloister	城東北隅
a5-7	Jishan an	積善菴	Cloister of Accumulated Virtue	城東南隅
a5-8	Fuhui an	福慧菴	Temple of Blessed Wisdom	長洲縣二十四都
a5-9	Xingfu an	興福菴	Revive Blessing Cloister	長洲縣三十一都
a6	Chengtian Nengren si	承天能仁寺	Heavenly Bestowed Benevolence Temple	府治北,甘節坊
a6-1	Fuchang si	福昌寺	Fortune and Prosperity Temple	寺内, 寺之子院
a6-2	Yuantong si	圓通寺	Universal Access Temple	寺内,寺之子院

(*Continued*)

Table A.1 (Continued)

ID	Pinyin	Chinese Character	English Name	Location
a6-3	Chongyi yuan	崇義院	Venerating Righteousness Temple	城西北隅
a6-4	Jifu an	集福菴	Assemble Benevolence Cloister	城西北隅
a6-5	Dacheng an	大乘菴	Great Vehicle Cloister	城東北隅
a6-6	Qiyuan an	祇園菴	Cloister of Jetevana-vihara[1]	許蘊子巷內
a6-7	Shizi Lin an	獅子林菴	Shizi Lin Cloister (Shizi Lin Temple)	城東北隅
a6-8	Xiuxiu an	休休菴	Restful Cloister	吳縣治西北
a6-9	Dinghui chan si	定慧禪寺	Determined Wisdom Chan Temple	城西北隅
a6-10	Nengren an	能仁菴	Capable Kindness Cloister	城西北隅
a6-11	Shanhu an	善護菴	Cloister of Virtuous Protection	城西北隅
a6-12	Xinxin an	信心菴	Sincere Faith Cloister	閶門外南濠
a6-13	Guanyin an	觀音菴	Avalokitesvara Cloister	城西北隅
a7	Lingjiu jiaosi	靈鷲教寺	Vulture Teaching Temple	城東北隅
a7-1	Lingjiu an	靈鷲菴	Vulture Cloister	城東南隅
a7-2	Qingyun an	慶雲菴	Celebratory Cloud Cloister	報恩寺西
a7-3	Baozhi an	寶志菴	Cloister of Precious Aspiration	城西北隅
a7-4	Tianfu an	天福菴	Heavenly Blessing Temple	城东北隅
a7-5	Jiqing an	集慶菴	Gathering Joy Cloister	長洲縣十九都
a7-6	Baoqin an	報親菴	Repay Parents Cloister	長洲縣二十九都
a8	Dongchan jiaosi	東禪教寺	Eastern Chan Teaching Temple	萬壽寺東
a8-1	Chongsheng yuan	重昇院	Double Ascending Yard	城東南隅,即南園地
a8-2	Ningshou an	寧壽菴	Peaceful Longevity Cloister	城東南隅
a8-3	Donglin an	東林菴	Eastern Grove Cloister	城東南隅
a8-4	Ciji an	慈濟菴	Compassionate Aid Cloister	城東南隅
a8-5	Tianning an	天寧菴	Heavenly Peace Cloister	城東南隅
a8-6	Tongxuan an	通玄菴	Penetrating Mystery Cloister	城東南隅
a8-7	Guanyin an	觀音菴	Avalokitesvara Cloister	城東南隅

(Continued)

Table A.1   (Continued)

ID	Pinyin	Chinese Character	English Name	Location
a8-8	Huacheng an	化城菴	Transforming City Cloister	長洲縣二十九都
a8-9	Puli an	普利菴	Universal Benefit Cloister	葑門外
a8-10	Fushou an	福壽菴	Blessing and Longevity Cloister	城東南隅
a8-11	Guangming an	光明菴	Cloister of Brightness and Illumination	城東南隅
a9	Yongding jiangsi	永定講寺 alt. 永定普慈天台講寺	Forever Pacification Teaching Temple, alt. Forever Pacification Universal Compassion Tiantai Teaching Temple, Forever Pacification Teaching Temple	吳縣東南
a9-1	Jingde si	景德寺, alt. 景德教寺	Bright Virtue Temple, alt. Bright Virtue Teaching Temple	黃牛坊橋東
a9-2	Xichan si (Beiguanyin an)	西禪寺, alt. 北觀音菴	Western Chan Temple, alt. North Avalokitesvara Cloister	吳縣學前
a9-3	Tianwang si	天王寺	Heavenly King Temple	城東南隅
a9-4	Dahong si	大弘寺	Great Propagation Temple	城東北隅
a9-5	Hongfan jinghui yuan	洪範淨慧院	Hongfan Pure Wisdom Institute	城西北隅
a10	Beichan jiangsi	北禪講寺	Northern Chan Teaching Temple	齊門內
a10-1	Zhaoqing si	昭慶寺	Illuminous Celebration Temple	城東北隅
a10-2	Wuliangshou si	無量壽寺	Infinite Longevity Temple, alt. Longer Sukhavativyuha Temple	長洲縣十五都
a10-3	Pufu yuan	普福院	Universal Blessing Temple	城東北隅
a10-4	Yuanming an	圓明菴	Bright Roundness Cloister	城東北隅
a10-5	Fohui an	佛慧菴	Buddha's Wisdom Retreat	城東北隅
a10-6	Anyin an	安隱菴	Peaceful Seclusion Retreat	城東北隅
a10-7	Tianlong an	天龍菴	Heavenly Dragon Retreat	城東北隅
a10-8	Jiantai an	揀汰菴	Selective Elimination Cloister	長洲縣十五都
a10-9	Deqing an	德慶菴	Virtue Celebration Cloister	長洲縣十五都

(*Continued*)

Table A.1 (Continued)

ID	Pinyin	Chinese Character	English Name	Location
a11	Nanchan chansi, alt. Jiyun si	南禪禪寺, alt. 集雲寺	Southern Chan Temple, alt. Gathering Clouds Temple	城東南隅
a11-1	Miaoyin an	妙隱菴	Witty Retreat Cloister	城東南隅
a11-2	Dayun an	大雲菴	Grand Clouds Cloister	城東南隅
a12	Shuangta chansi	雙塔禪寺	Twin Pagoda Chan Temple	城東南隅
a12-1	Yongshou an	永壽菴	Eternal Longevity Cloister	城東南隅
a12-2	Yanqing an	衍慶菴	Extending Celebration Cloister	城東南隅
a12-3	Yuantong an	圓通菴	Universal Access Cloister	城東南隅
a13	Dinghui chansi	定慧禪寺, alt. 定慧寺	Determined Wisdom Chan Temple, alt. Determined Wisdom Temple	雙塔寺西
a13-1	Huacheng an	化城菴	Transformation City Cloister	長洲縣二十九都
a13-2	Jingzhu an	淨住菴	Pure Residence Cloister	城東北隅
a14	Baoguang jiangsi	寶光講寺	Precious Light Teaching Temple	城東北隅
a14-1	Fayun tayuan	法雲塔院	Dharma Cloud Pagoda Yard	城西北隅
a14-2	Yongqing an	永慶菴	Eternal Celebration Cloister	長洲縣二十二都
a14-3	Chongfu an	崇福菴	Revered Blessing Cloister	長洲縣二十三都
a14-4	Dingming an	定銘菴	Fixed Inscription Cloister	城東北隅
a14-5	Guanyin an	觀音菴	Avalokitesvara Cloister	長洲縣二十九都
a14-6	Xianqing an	顯慶菴	Manifesting Celebration Cloister	長洲縣六都
a14-7	Fuxing an	福興菴	Blessing Revival Cloister	長洲縣二十九都
a15	Guanghua jiaosi	廣化教寺	Broad Transformation Teaching Temple	城東北隅
a15-1	Longxing si	龍興寺	Dragon Revival Temple	吳縣治西
a15-2	Tiangong si	天宮寺	Heavenly Palace Temple	城東北隅
a15-3	Baolin si	寶林寺	Precious Grove Temple	城西北隅
a15-4	Dingguang an	定光菴	Fixed Light Cloister	城西北隅
a15-5	Jiqing an	積慶菴	Accumulated Celebration Cloister	城西北隅

(Continued)

Table A.1   (Continued)

ID	Pinyin	Chinese Character	English Name	Location
a15-6	Mingxin an	銘心菴	Inscribed Heart Cloister	城東北隅
a15-7	Jishan an	即山菴	Immediate Mountain Cloister	not recorded
a15-8	Huazang an	華藏菴	Flower Treasure Cloister	not recorded
a16	Jiqing chansi	積慶禪寺	Accumulated Celebration Chan Temple	善教橋北
a16-1	Chanxing si	禪興寺	Chan Revival Temple	乘鯉坊
a16-2	Shengan an	聖安菴	Sacred Peace Cloister	長洲縣二十九都
a16-3	Bao'en an	報恩菴	Repaying Kindness Cloister	not recorded
a16-4	Baoqin an	報親菴	Repaying Parents Cloister	not recorded
a16-05	Guanyin an	觀音菴	Avalokitesvara Cloister	not recorded
a16-6	Fuqing an	福慶菴	Blessing and Celebration Temple	not recorded
a17	Baochuang jiangsi	寶幢講寺	Precious Sutra Pillar Teaching Temple	承天寺內東偏
a17-1	Shangshan an	上善菴	Supreme Kindness Cloister	城西南隅
a17-2	Jixiang an	吉祥菴	Auspicious Cloister	中街路
a17-3	Shengfu an	聖福菴	Sacred Blessing Cloister	穹窿山之南
a18	Sizhou si	泗洲寺	Four Continents Temple	城西南隅
a18-1	Xiying an	西隱菴	Western Seclusion Cloister	城西南隅
a18-2	Fayuan an	法圓菴	Dharma Completion Cloister	城西南隅
a18-3	Lingyin an	靈隱菴	Spiritual Seclusion Cloister	盤門內
a18-4	Dasheng an	大聖菴	Great Sage Cloister	城西南隅陸家巷
a18-5	Qingyin an	清隱菴	Pure Seclusion Cloister	盤門外舊堰橋
a19	Zhuming si	朱明寺	Zhuming Temple	城隍廟西
a20	Xizhu si	西竺寺	West India Temple	平權坊內
a21	Miaozhan si	妙湛寺	Splendid Profound Temple	長洲縣東
a22	Zishou nisi	資壽尼寺	Supporting Longevity Nun Temple	長洲縣治東北

Note: Temples in grey cells are identified as abbeys under the Early Ming amalgamation law and are designated with IDs such as a1, a2, and a3. Cloisters under the supervision of these abbeys are indexed with IDs in the format a1-1, a2-1, a2-2, a2-3, etc.

Table A.2 Yuan Gardens

ID	Pinyin	Chinese Character	English Name	Owner
c1	Pan Shiyong ju	潘時用居	Pan Shiyong's Dwelling	Pan Shiyong 潘時用
c2	Xiao danqiu	小丹丘	Small Vermilion Hill	Chen Ji
c3	Lepu linguan	樂圃林館	Pleasure Patch Grove Residence	Zhang Shi 張適
c4	Lüshui yuan	綠水園	Green Water Garden	Chen Weiyin 陳惟寅, Chen Weiyuan 陳惟元
c5	Lu Zhining yuguan	陸志寧寓館	Lu Zhining's Residence	Lu Zhining 陸志寧

Table A.3 Yuan Academies

ID	Pinyin	Chinese Character	English Name	Location
d1	Fuli shuyuan	甫里书院	Fuli Academy	長洲縣治東
d2	Wenzheng shuyuan	文正书院	Wenzheng Academy	吴县治东北
d3	Heshan shuyuan	鶴山書院	Heshan Academy	南宮坊
d4	Fuxue	府學	Prefectural School	

Table A.4 Ming Gardens

ID	Pinyin	Chinese Character	English Name	Owner
c1	Xiao taoyuan	小桃園	Little Peach Garden	
c2	Caixiang an	采香庵	Fragrance Gathering Hermitage	
c3	Taohua an [1]	桃花庵	Peach Blossom Hermitage	Tang Yin
c4	Yuan jia zhai	袁家宅	Yuan Family Residence	
c5	Taohua an [2]	桃花庵	Peach Blossom Hermitage	
c6	Julin	橘林	Orange Grove	
c7	Tang jia yuan	唐家園	Tang Family Garden	Tang Yin
c8	Qiuzhi yuan	求志園	Aspiration Garden	
c9	Fangcao yuan	芳草園	Fragrant Grass Garden	
c10	Zhangshi meiyuan	張氏梅園	Zhang Family Plum Garden	Yang Weizhen 楊維楨
c11	Gui tianyuan ju	歸田園居	Returning to Farm Dwelling	Wang Xinyi
c12	Pan Shiyong ju	潘時用居	Pan Shiyong Residence	Pan Shiyong 潘時用
c13	Shizi Lin	獅子林	Lion Grove Garden	Lu family and etc.
c14	Wufeng yuan	五峰園	Five Peaks Garden	Yang Chen 楊成
c15	Xiaoqi yuan	小漆園	Little Lacquer Garden	
c16	Yu qing shanfang	玉磬山房	Jade Chime Mountain House	Wen Zhengming
c17	Xiangcao cha	香草垞	Fragrant Herb Hut	Wen Zhenheng 文震亨
c18	Yipu	藝圃	Garden of Cultivation	Wen Zhenmeng
c19	Dong yuan	東園	Eastern Garden	Du Qiong
c20	Qu yuan	蘧園	Qu Garden	Shen Shixing 申时行
c21	Xu Youzhen zhai	徐有貞宅	Xu Youzhen Residence	Xu Youzhen 徐有貞
c22	Zhu Yunming zhai	祝允明宅	Zhu Yunming Residence	Zhu Yunming
c23	Tingyun guan	停雲館	Clouds Resting Pavilion	Wen Zhengming
c24	Dongxi shushe	東溪書舍	Eastern Creek Book Chamber	Tang Yi'an 唐以安
c25	Zhuozheng yuan	拙政園	Unsuccessful Politician's Garden	Wang Xianchen and etc.
c26	Guanjia yuan	管家園	Guan Family Garden	Guan Zhengxin 管正心
c27	Guishi yuan	歸氏園	Gui Family Garden	Gui Zhanchu 歸湛初

(Continued)

Table A.4 (Continued)

ID	Pinyin	Chinese Character	English Name	Owner
c28	Qiayin shanfang	恰隱山房	Hidden Secluded Mountain Villa	Hu Ruchun 胡汝淳
c29	Qiayin yuan	洽隱園	Hidden Secluded Garden	
c30	Wang Ao bieye	王鏊別業	Wang Ao's Eastate	Wang Ao
c31	Lüshui yuan	綠水園	Green Water Garden	Zhu Xiang 朱襄
c32	Wu Wending Gong zhai	吳文定公宅	Master Wu Wending's Residence	Wu Kuan
c33	Gu Qiguo zhaiyuan	顧其國宅園	Gu Qiguo Residence Garden	Gu Qiguo 顧其國
c34	Shuizhu zhuang	水竹莊	Water Bamboo Villa	Gu Rongfu 顧榮夫
c35	Wan pu	晚圃	Evening Garden	Qian Menghu 錢孟虎
c36	Huanghuang zhai	荒荒宅	Wilderness Chamber	Tang Chuanying 湯傳楹
c37	Pijiang yuan	辟疆園	Pijiang Garden	Gu Kuang 顧況
c38	Mi yuan	泌園	Mi Garden	Zhang Shiwei 張世偉
c39	Zhou Tianqiu yuan	周天球園	Zhou Tianqiu Garden	Zhou Tianqiu 周天球
c40	Dong zhuang	東莊	Eastern Estate	Wu Mengrong 吳孟融
c41	Huangfu Lu zhaiyuan	黃甫錄宅園	Huangfu Lu Residence Garden	Huangfu Lu 黃甫錄
c42	Mochi yuan	墨池園	Ink Pond Garden	Kong Yong 孔鏞
c43	Fengxi caotang	葑溪草堂	Feng River Thatched Hall	Han Yong 韓雍
c44	Zhou Jiading zhai	周嘉定宅	Zhou Jiading Residence	Zhou Jiading 周嘉定

Table A.5 Ming Academies

Id	Name	Chinese Character	English Name
d1	Xuedao shuyuan	學道書院	Learn the Way Academy
d2	Jiu Changzhou xianxue	舊長洲縣學	(old) Changzhou County School
d3	Jinxiang shuyuan	金鄉書院	Golden Country Academy
d4	Hejing shuyuan	和靖書院	Hejing Academy
d5	Changzhou xianxue	長洲縣學	Changzhou County School
d6	Fuxue	府學	Prefectural school
d7	Qi Wang miao	蘄王廟	Shrine of the King of Qi

Table A.6 Temples in Suzhou's Extra-mural Area During the Yuan Dynasty

ID	Pinyin	Name In Chinese Character	English Name	Location
b1	Huqiu chansi	虎丘禪寺	Tiger Hill Chan Temple	虎丘山
b1-1	Guiyuan si, alt. Guiyuan xingguo chan si	歸源寺, alt. 歸元興國禪寺	Returning to Origin Temple, alt. Returning to Origin and Reviving the Nation Chan Temple	長蕩東
b1-2	Huanzhu an	幻住菴	Illusory Abiding Cloister	閶門外雁蕩村
b1-3	Yuanzhao an	圓昭菴	Perfect Illumination Cloister	閶門外
b1-4	Zhengxin an	正信菴	True Faith Cloister	陽山
b1-5	Huiyun an	會雲菴	Meeting Clouds Cloister	中峯嶺
b1-6	Yunyin an	雲隐菴	Cloud Retreat Cloister	閶門外
b1-7	Yongfu an	永福菴	Eternal Benevolence Cloister	長洲縣九都
b1-8	Yuanjue an	圓覺菴	Complete Awakening Cloister	長洲縣九都
b1-9	Shanfu an	善福菴	Good Fortune Cloister	長洲縣九都
b2	Bantang shousheng jiaosi	半塘壽聖教寺, alt. 半塘壽聖寺	Sacred Longevity Teaching Chan Temple at Bantang (Sacred Longevity Teaching Temple at Bantang)	九都綵雲里

(Continued)

Table A.6 (Continued)

ID	Pinyin	Name In Chinese Character	English Name	Location
b2-1	Zhenfu an	臻福菴	Ultimate Benevolence Cloister	長洲縣十都
b2-2	Zifu an	資福菴	Accumulated Benevolence Cloister	長洲縣九都
b2-3	Yuanming an	原明菴	Pristine Brightness Cloister	長洲縣九都
b2-4	Cixiao an	慈孝菴	Kindness and Filial Piety Cloister	長洲縣九都
b2-5	Fohui an	佛慧菴	Buddha Wisdom Cloister	虎丘寺東
b2-6	Decheng an	得成菴	Attained Achievement Cloister	長洲縣八都
b2-7	Puguang an	普光菴	Universal Light Cloister	長洲縣一都
b2-8	Pufu an	普福菴	Universal Benevolence Cloister	虎丘山東
b3	Yinghu chansi	迎湖禪寺	Welcoming Lake Chan Temple	長洲縣六都
b3-1	Guangji an	廣濟菴	Extensive Aid Cloister	長洲縣四都
b3-2	Chongfu an	崇福菴	Esteemed Benevolence Cloister	長洲縣四都
b4	Bailian jiaosi	白蓮教寺	White Lotus Teaching Temple	長洲縣十五都
b4-1	Hujing shanqing an	湖徑善慶菴	Lake Path Kindness Celebration Cloister	長洲縣十六都
b4-2	Purun shoushan an	普潤壽山菴	Universal Moistening Longevity Mountain Cloister	長洲縣十六都
b4-3	Puqing shenqi an	普慶深棲菴	Universal Celebrating Deep Abode Cloister	長洲縣十六都
b4-4	Baode an	報德菴	Repaying Virtue Cloister	長洲縣九都
b4-5	Meilin an	梅林菴	Plum Grove Cloister	長洲縣九都
b4-6	Xizi an	西資菴	Western Resource Cloister	長洲縣十五都
b4-7	Guanyin an	觀音菴	Avalokitesvara Cloister	長洲縣十五都
b5	Lingyuan jiaosi	靈源教寺	Spiritual Source Teaching Temple	太湖洞庭東山碧螺峯下
b5-1	Nengren si	能仁寺	Capable Benevolence Temple	長圻嶺

(Continued)

Table A.6 (Continued)

ID	Pinyin	Name In Chinese Character	English Name	Location
b5-2	Xingfu si	興福寺	Prosperous Blessing Temple	洞庭東山
b5-3	Zhongfeng si	中峯寺	Central Peak Temple	洞庭東山
b5-4	Sanfeng si	三峯寺	Three Peaks Temple	洞庭東山
b5-5	Mile si	弥勒寺	Maitreya Temple	洞庭東山
b5-6	Yuanzhao an	圓照菴	Complete Illumination Cloister	洞庭東山
b6	Zhiping jiaosi, alt. Lengqie si	治平教寺, alt.楞伽寺	Peaceful Governance Teaching Temple (Lankavatara Temple)	上方山下
b6-1	Baoji si	寶積寺	Treasure Accumulation Temple	橫山下
b6-2	Zhenru an	真如菴	True Suchness Cloister	縣西南三十里
b6-3	Shusheng an	殊勝菴	Exceptional Victory Cloister	縣西南三十里
b6-4	Huayan an	華嚴菴	Avatamsaka Cloister	縣西南二十里
b6-5	Yuanjue an	圓覺菴	Complete Awakening Cloister	縣南十里
b7	Lengjia jiangsi (Shangfang si)	楞伽講寺, alt.上方寺	Lankavatara Lecture Temple, alt. Supreme Direction Temple	楞伽山上
b7-1	Bailong an	白龍菴	White Dragon Cloister	長洲縣二十二都
b7-2	Miaoming an	妙明菴	Witty Brightening Cloister	長洲縣二十二都
b8	Qionglong chansi	穹窿禪寺	Qinglong Chan Temple	吳縣西南四十五里穹窿山
b8-1	Baohua si	寶華寺	Precious Flower Temple	吳縣西南三十里寶華山
b8-2	Jianfu chansi	薦福禪寺	Recommended Blessing Chan Temple	吳縣西南薦福山
b8-3	Longchi an	龍池菴	Dragon Pond Cloister	吳縣西二十五里
b8-4	Zhenru an	真如菴	True Suchness Cloister	吳縣西四十五里
b9	Baima chansi	白馬禪寺	White Horse Chan Temple	穹窿山西址
b9-1	Shengen an	聖恩菴	Sacred Grace Cloister	鄧尉山之南岡

(Continued)

Table A.6 (Continued)

ID	Pinyin	Name In Chinese Character	English Name	Location
b9-2	Wenshu an, alt. Jinshan si	文殊菴, alt. 金山寺	Manjusri Cloister, alt. Golden Mountain Temple	金山東嶺
b9-3	Yuquan an	玉泉菴	Jade Spring Cloister	吳縣西十八里
b10	Lingyan chansi	靈岩禪寺	Spiritual Rock Chan Temple	靈巖山上
b10-1	Daolin an	道林菴	Path Grove Cloister	吳縣西四十五里
b10-2	Yuanming an	圓明菴	Perfect Brightness Cloister	吳縣西二十七里
b10-4	Shou'an an	壽安菴	Longevity Peace Cloister	吳縣西十八里
b10-5	Shusheng an	殊勝菴	Exceptional Victory Cloister	吳縣西十九里
b11	Chongfu jiaosi (Yinshan si)	崇福教寺, alt. 尹山寺	Esteemed Blessing Teaching Temple, alt. Yinshan Temple	長洲縣三十一都 尹山
b11-1	Jishan an	集善菴	Gather Kindness Cloister	長洲縣三十一都
b11-2	Fuyuan an	福源菴	Blessing Origin Cloister	長洲縣二十九都
b11-3	Mingyuan an	明遠菴	Bright Distance Cloister	長洲縣三十一都
b11-4	Nengren qingfu an	能仁慶福菴	Capable Benevolence Celebrating Blessing Cloister	長洲縣二十九都
b12	Lingdian jiaosi	靈澱教寺	Spiritual Sediment Teaching Temple	長洲縣十四都
b12-1	Fahua yuan	法華院	Lotus Sutra Yard	長洲縣十四都
b12-2	Jingfu an	景福菴	Scenic Blessing Cloister	長洲縣十九都
b12-3	Ersheng an	二聖菴	Two Saints Cloister	長洲縣十七都
b12-4	Fahui an	法惠菴	Dharma Benevolence Cloister	長洲縣十四都
b12-5	Huayan an	華嚴菴	Avatamsaka Cloister	長洲縣二十二都
b12-6	Zhijue an	至覺菴	Ultimate Awareness Cloister	長洲縣十四都
b12-7	Guangming an	光明菴	Bright Light Cloister	長洲縣一都
b12-8	Baihe an	白鶴菴	White Crane Cloister	長洲縣十七都
b12-9	Fahua an	法華菴	Lotus Sutra Cloister	長洲縣十四都
b12-10	Xifeng an	息峯菴	Resting Peak Cloister	長洲縣十四都
b12-11	Maota an	毛塔菴	Fur Tower Cloister	長洲縣二十二都
b12-12	Chunshen an	春申菴	Chunshen Cloister	長洲縣十七都

(*Continued*)

Table A.6 (Continued)

ID	Pinyin	Name In Chinese Character	English Name	Location
b12-13	Jishan an	積善菴	Accumulate Virtue Cloister	長洲縣二十一都
b13	Dajue jiaosi, alt.	大覺教寺, alt. 大覺寺	Great Awakening Temple	長洲縣二十七都大姚山
b14	Bangjian si	邦建寺	State Constructed Temple	not recorded
b14-1	Songyin an	松隱菴	Pine Hidden Cloister	長洲縣二十六都
b14-2	Wenshu an	文殊菴	Manjushri Cloister	not recorded
b15	Tiangong jiaosi	天宮教寺	Heavenly Palace Teaching Temple	胥山南
b15-1	Shousheng guangfu chansi	壽聖廣福禪寺	Longevity Saint Extensive Blessing Chan Temple	吳山嶺上
b15-2	Shouning an	壽寧菴	Longevity Serenity Cloister	吳縣西十三里
b15-3	Zhongfeng an	中峯菴	Central Peak Cloister	吳縣西十三里
b15-4	Jishan an	積善菴	Accumulate Virtue Cloister	吳縣西十三里
b15-5	Jingju an	淨居菴	Pure Dwelling Cloister	吳縣西十三里
b15-6	Baode an	報德菴	Repaying Virtue Cloister	吳縣西十三里
b15-7	Fuhai an	福海菴	Blessing Sea Cloister	吳縣西十三里
b16	Zhaoming jiaosi	昭明教寺	Illuminating Teaching Temple	吳縣錦峯山
b16-1	Longhua an	龍華菴	Dragon Transformation Cloister	吳縣西六十里
b16-2	Fengxiang an	鳳祥菴	Phoenix Auspicious Cloister	吳縣西六十里
b17	Guangfu jiangsi, alt. Guangfu si	光福講寺	Light Benevolence Temple	吳縣鄧蔚山峯上
b17-1	Daci si	大慈寺	Great Compassion Temple	陽山之右
b17-2	Zunxiang chansi	尊相禪寺	Honored Appearance Zen Temple	長洲縣三都
b17-3	Chongfu an	崇福菴	Esteemed Blessing Cloister	吳縣西七十里
b17-4	Fengci an	奉慈菴	Offer Compassion Cloister	吳縣西七十里
b17-5	Shuangshou an	雙壽菴	Double Longevity Cloister	吳縣西七十里

(Continued)

Table A.6 (Continued)

ID	Pinyin	Name In Chinese Character	English Name	Location
b17-6	Hefeng an	和豐菴	Harmonious Abundance Cloister	吳縣西七十里
b17-7	Yuwen an	宇文菴	Literary Grace Cloister	吳縣西七十里
b17-8	Zhongfu an	種福菴	Plant Benevolence Cloister	吳縣西七十里
b17-9	Yuanjue an	圓覺菴	Complete Awakening Cloister	吳縣西七十里
b18	Changshan jiaosi	長山教寺	Long Mountain Teaching Temple	吳縣二十三都
b18-1	Zhengjue an	證覺菴	Proven Awakening Cloister	吳縣西六十里
b18-2	Shanqing an	善慶菴	Good Celebration Cloister	吳縣西六十里
b18-3	Jizhao an	寂照菴	Silent Illumination Cloister	吳縣西六十里
b18-4	Chongfu an	崇福菴	Esteemed Blessing Cloister	吳縣西六十里
b18-5	Pusa an	菩薩菴	Bodhisattva Cloister	吳縣西六十里
b19	Shixiang jiaosi	實相教寺	True Appearance Teaching Temple	吳縣香山, 西太湖之濱
b19-1	Shensha an	深沙菴	Deep Sand Cloister	吳縣西南五十里
b19-2	Guanyin an	觀音菴	Avalokitesvara Cloister	吳縣西南五十里
b19-3	Shufu an	殊福菴	Special Blessing Cloister	吳縣西南七十里
b19-4	Guangfu an	廣福菴	Wide Blessing Cloister	吳縣西北六十里
b19-5	Anyin an	安隱菴	Peaceful Concealment Cloister	吳縣西六十四里
b19-6	Xingfu an	興福菴	Prosperous Blessing Temple	吳縣西六十五里
b19-7	Chongfu an	崇福菴	Esteemed Blessing Cloister	吳縣西北六十五里
b20	Fahai si	法海寺	Dharma Sea Temple	洞庭東山
b20-1	Huayan si	華嚴寺	Avatamsaka Temple	翠峯之楊家灣
b20-2	Jishan an	積善菴	Accumulate Virtue Cloister	洞庭東山
b20-3	Beiqi an	北奇菴	North Marvel Cloister	白沙之南
b20-4	Shouning an	壽寧菴	Longevity Serenity Cloister	武山

(*Continued*)

Table A.6 (Continued)

ID	Pinyin	Name In Chinese Character	English Name	Location
b21	Hanshan chansi	寒山禪寺	Cold Mountain Zen Temple	閶門西十里楓橋下
b21-1	Xiufeng chansi	秀峯禪寺	Show Peak Chan Temple	吳縣西華鄉
b21-2	Huiqing chansi	慧慶禪寺	Wisdom Celebration Chan Temple	去郡城西七里
b21-3	Nanfeng si, alt. Guzhixing si	南峰寺, alt. 古支硎寺	South Peak Temple, alt. Ancient Support Temple	支硎之南峯
b21-4	Wenshu an	文殊菴	Manjushri cloister	吳縣十一都
b21-5	Yunquan an	雲泉菴	Clouds Spring cloister	吳縣十一都
b21-6	Shedu an	射瀆菴	Shooting Gorge cloister	長洲縣一都
b22	Cuifeng chansi	翠峯禪寺	Emerald Peak Chan Temple	莫釐山之陰
b22-1	Guangji an	廣濟菴	Broad Relief cloister	莫釐山渡口
b23	Shangfang jiaosi	上方教寺	Upper Direction Teaching Temple	洞庭西山
b23-1	Qiyuan si	祇園寺	Jetavana Temple	五峯嶺下
b23-2	Shuiyue si	水月寺	Water Moon Temple	縹緲峯下
b23-3	Xi xiaohu tiantai jiaosi	西小湖天台教寺	West Small Lake Tiantai Teaching Temple	縹緲峯北
b23-4	Fuyuan si	福源寺	Benevolence Origin Temple	not recorded
b23-5	Guanyin yuan, alt. Huashan si	觀音院, alt. 華山寺	Avalokitesvara Yard, alt. Huashan Temple	龍頭山之西
b23-6	Luohan si	羅漢寺	Arhat Temple	洞庭西山
b23-7	Ziqing yuan	資慶院	Resource Celebration Yard	洞庭西山
b23-8	Kanjing yuan	看經院	Scripture Viewing Yard	洞庭西山
b23-9	Changshou yuan	長壽院	Longevity Yard	洞庭西山
b23-10	Shiji yuan	實際院	Reality Yard	洞庭西山
b23-11	Houhuang yuan	后黃院	Empress Huang Yard	洞庭西山
b23-12	Donghu an	東湖菴	East Lake cloister	洞庭西山
b23-13	Puji an	普濟菴	Universal Relief Cloister	洞庭西山

(*Continued*)

Table A.6 (Continued)

ID	Pinyin	Name In Chinese Character	English Name	Location
b24	Baiyun chansi/ Tianping si	白雲禪寺, alt. 天平寺	White Cloud Zen Temple, alt. Tianping Temple	天平山
b24-1	Guanyin si, alt. (Gu Bao'en si)	觀音寺, alt.古報恩寺	Avalokitesvara Temple, alt. Ancient Repaying Kindness Temple	支硎山東麓
b24-2	Yanqing si	延慶寺	Prolonged Celebration Temple	吳縣西南四十里, 地名橫金
b24-3	Haiyun an	海雲菴	Sea Cloud Cloister	吳縣穹窿山之北麓
b24-4	Shenyin an	深隱菴	Deeply Hidden Cloister	吳縣西四十里
b24-5	Haihui an	海慧菴	Sea Wisdom Cloister	吳縣西三十里
b24-6	Juelin an	覺林菴	Awakening Forest Cloister	吳縣西四十五里
b25	Lianhua jiaosi	蓮花教寺	Lotus Flower Teaching Temple	陽山西太湖之濱
b25-1	Luxiang si	陸香寺	Land Fragrance Temple	長洲縣西北三都
b25-2	Qingci an	慶慈菴	Celebratory Mercy Cloister	長洲縣七都
b25-3	Qibaoquan an	七寶泉菴	Seven Treasures Spring Cloister	吳縣十九都
b25-4	Shenju an	深居菴	Deep Residence Cloister	長洲縣三都
b26	Baoshou jiaosi	寶壽教寺	Treasure Longevity Teaching Temple	吳縣八都,地名黃蘆
b26-1	Yongxi si	雍熙寺	Harmonious Splendor Temple	城武狀元坊內
b26-2	Gusu si	姑蘇寺	Gusu Temple	吳縣橫山之錢莊
b26-3	Yuqing yuan	餘慶院	Remaining Joy Yard	吳縣南五十里, 地名殷涇
b26-4	Puzhao an	普照菴	Universal Illumination Cloister	吳縣西南五十里
b26-5	Jianming an	建名菴	Establishing Fame Cloister	吳縣西南四十里
b26-6	Yongshou an	永壽菴	Eternal Longevity Cloister	吳縣西南四十里
b26-7	Guanyin an	觀音菴	Avalokitesvara Cloister	吳縣西南四十里

(Continued)

Table A.6 (Continued)

ID	Pinyin	Name In Chinese Character	English Name	Location
b26-8	Liuyun an , alt. Liuyun jiedai yuan)	留雲菴, alt. 留雲接待院	Staying Cloud Cloister, alt. Staying Cloud Reception Yard	吳縣西南五十里
b27	Xingguo jiaosi	興國教寺	Prosperous Country Teaching Temple	長洲縣十一都黃埭
b27-1	Zhenru an	真如菴	Suchness Cloister	長洲縣十一都
b27-2	Puji an	普濟菴	Universal Relief Cloister	長洲縣十都
b28	Baosheng jiaosi	保聖教寺	Protecting the Sacred Teaching Temple	長洲縣二十都甫里
b28-1	Dingshan an	定善菴	Establishing Goodness Cloister	長洲縣二十一都
b28-2	Fuyuan an	福源菴	Blessing Source Cloister	長洲縣十九都
b28-3	Zifu an	資福菴	Resource Blessing Cloister	長洲縣二十都
b28-4	Wufu an	五福菴	Five Blessings Cloister	長洲縣二十都
b28-5	Liutong an	流通菴	Circulating Cloister	長洲縣十九都
b29	Yongshou jiaosi	永壽教寺	Eternal Longevity Teaching Temple	長洲縣二十七都籍墟陳湖之上
b29-1	Shouning an	壽寧菴	Longevity Peace Cloister	長洲縣二十都
b29-2	Chongyuan an	崇遠菴	Esteemed Faraway Cloister	長洲縣二十六都
b29-3	Baofu an	報福菴	Repaying Blessing Cloister	長洲縣二十六都
b29-4	Chongxian an	崇先菴	Esteemed Ancestor Cloister	長洲縣二十七都
b29-5	Guanyin an	觀音菴	Avalokiteshvara Cloister	長洲縣十九都
b29-6	Xigui an	西歸菴	West Return Cloister	長洲縣十九都
b29-7	Zhifu chantang	植福懺堂	Planting Blessings Repentance Hall	長洲縣十九都
b30	Juelin jiaosi	覺林教寺	Awakening Forest Teaching Temple	長洲縣十三都冶長涇
b30-1	Miaozhi an	妙智菴	Witty Wisdom Cloister	長洲縣十八都
b30-2	Putuo an	濮陀菴	Putuo Cloister	長洲縣十八都
b30-3	Chongfu an	崇福菴	Esteemed Blessing Cloister	長洲縣十八都

(Continued)

Table A.6 (Continued)

ID	Pinyin	Name In Chinese Character	English Name	Location
b30-4	Juedi an	覺地菴	Awakening Ground Cloister	長洲縣十二都
b31	Chengzhao jiaosi	澄照教寺	Clear Illumination Teaching Temple	陽山下
b31-1	Jingfu an	景福菴	Scenic Blessing Cloister	長洲縣二都
b31-2	Wenshu an	文殊菴	Manjushri Cloister	長洲縣二都
b32	Zengshan jiaosi	甑山教寺	Steaming Mountain Teaching Temple	陽山北竹青塘
b32-1	Fengxian an	奉先菴	Offering to Ancestors Cloister	長洲縣二都
b32-2	Guangfu an	廣福菴	Broad Blessing Cloister	長洲縣八都
b32-3	Guanyin an [1]	觀音菴 (其一)	Avalokiteshvara Cloister (First)	長洲縣六都
b32-4	Guanyin an [2]	觀音菴 (其二)	Avalokiteshvara Cloister (Second)	長洲縣八都
b33	Fahua jiaosi	法華教寺	Lotus Sutra Teaching Temple	葑門外斜塘
b33-1	Yanqing an	衍慶菴	Spreading Joy Cloister	長洲縣二十九都
b33-2	Miaoming an	妙名菴	Wonderful Name Cloister	長洲縣二十九都
b33-3	Xingfu an	興福菴	Prosperous Blessing Cloister	長洲縣二十九都
b33-4	Ouye an	歐冶菴	Eulogising Metalworking Cloister	長洲縣二十九都
b34	Bailian jiangsi	白蓮講寺	White Lotus Lecture Temple	長洲縣二十都 甫里
b34-1	Lutang tayuan	陸塘塔院	Land Pond Pagoda Yard	長洲縣十五都
b34-2	Liutong an	流通菴	Circulating Cloister	長洲縣十九都
b34-3	Jinming an	晉明菴	Advance Brightness Cloister	長洲縣十九都
b34-4	Ciyun an	慈雲菴	Merciful Cloud Cloister	長洲縣二十一都
b35	Liji jiaosi	利濟教寺	Benefit Salvation Teaching Temple	閶門外南濠
b35-1	Shifo jingtusi	石佛净土寺	Stone Buddha Pure Land Temple	閶門外南濠
b35-2	Queyun yuan	闕雲院	Missing Cloud Yard	城西北隅

(Continued)

Table A.6 (Continued)

ID	Pinyin	Name In Chinese Character	English Name	Location
b36	Dajue jiaosi	大覺教寺	Great Awakening Teaching Temple, Great Awakening Temple	陽山南
b36-1	Beifeng si	北峯寺	North Peak Temple	吳縣支硎山
b36-2	Huashan si	華山寺	Huashan Temple	華山蓮花峯下
b36-3	Juelin yuan	覺林院	Awakening Forest Yard	長洲縣一都
b36-4	Shouci an	壽慈菴	Longevity Mercy Cloister	吳縣西北四十里
b36-5	Juehai an	覺海菴	Awakening Sea Cloister	長洲縣一都
b36-6	Xizi an	西資菴	West Resource Cloister	長洲縣六都
b36-7	Fuyan an	福嚴菴	Blessing Strictness Cloister	長洲縣一都
b37	Qisha jiaosi	磧砂教寺	Gravel Sand Teaching Temple	長洲縣二十六都陳湖之北
b37-1	Jifu an	集福菴	Assembling Benevolence Cloister	長洲縣十九都
b37-2	Jiqing an	集慶菴	Gathering Celebration Cloister	長洲縣十九都
b37-3	Yingfu an	迎福菴	Welcoming Blessing Cloister	長洲縣二十都
b38	Chongfu jiaosi	崇福教寺	Esteemed Blessing Teaching Temple	長洲縣二十八都章練塘
b38-1	Mingyin si	名因寺	Name Cause Temple	not recorded
b38-2	Fushou an	福壽菴	Blessing Longevity Cloister	not recorded
b38-3	Yuantong an	圓通菴	Universal Access Cloister	not recorded
b38-4	Fuji an	福濟菴	Blessing Relief Cloister	not recorded
b38-5	Dingxiang an	定祥菴	Establishing Auspiciousness Cloister	not recorded
b39	Jiedai jiaosi	接待教寺	Reception Teaching Temple	婁門外
b39-1	Puming an	普明菴	Universal Brightness Cloister	長洲縣二十三
b39-2	Chongzhen an	崇真菴	Esteemed Truth Cloister	長洲縣二十三

(*Continued*)

Table A.6 (Continued)

ID	Pinyin	Name In Chinese Character	English Name	Location
b39-3	Yuantong an	圓通菴	Universal Access Cloister	長洲縣二十三
b39-4	Xiaoyou an	孝友菴	Filial Friendship Cloister	長洲縣二十三
b39-5	Bore an	般若菴	Prajna Cloister	長洲縣二十三
b39-6	Yanfu an	延福菴	Prolonging Blessing Cloister	長洲縣十九都
b39-7	Daowang an	道王菴	Way King Cloister	長洲縣十九都
b39-8	Yiwang an	夷王菴	Yi King Cloister	長洲縣十九都
b39-9	Yuanming an	圓明菴	Complete Brightness Cloister	長洲縣二十四都
b39-10	Chongshou an	崇壽菴	Esteemed Longevity Cloister	長洲縣二十四都
b39-11	Bailong an	白龍菴	White Dragon Cloister	長洲縣二十二都
b40	Puxian jiaosi	普賢教寺	Universal Virtue Teaching Temple	天平山之東, 牛頭峯塢內
b40-1	Mingyue si	明月寺	Bright Moon Temple	吳縣木瀆鎮
b40-2	Jiqing an	積慶菴	Accumulating Celebration Cloister	吳縣西三十里
b40-3	Yingfu an	迎福菴	Welcoming Blessing Cloister	吳縣西三十里
b40-4	Jishan an	積善菴	Accumulating Virtue Cloister	吳縣西三十二里
b40-5	Baojue an	寶覺菴	Precious Enlightenment Cloister	吳縣西十五里
b40-6	Huaijing an	懷敬菴	Embracing Reverence Cloister	吳縣西十五里
b40-7	Zifu an	資福菴	Supporting Blessings Cloister	吳縣西十五里
b41	Quanfu jiangsi	全福講寺	Complete Blessing Lecture Temple	長洲縣二十六都周莊
b41-1	Shanzhu an	善住菴	Good Dwelling Cloister	長洲縣二十七都
b41-2	Guanmiao an	觀妙菴	Observing Wonders Cloister	長洲縣二十六都
b41-3	Yongqing an	永慶菴	Eternal Celebration Cloister	長洲縣二十六都
b41-4	Qingyuan an	清遠菴	Clear Distant Cloister	長洲縣二十六都

Note: Temples in grey cells were identified as abbeys under the Early Ming amalgamation law and are designated with IDs such as b1, b2, and b3. Cloisters under the supervision of these abbeys are indexed with IDs in the format b1-1, b1-2, b2-1, etc.

Bibliography

Abbreviations

CLTXZ Jiang Jinghuan 蔣鏡寰. *Canglang Ting xinzhi* 滄浪亭新志 (New Gazetteer of Surging Wave Pavilion). Suzhou: Suzhou meishu guan shengli tushu guan, 1929.

CLXZ Song Luo, ed. *Canglang xiaozhi* 滄浪小志 (A Small Gazetteer of Surging Wave Pavilion), Sq. ed., printed in the 1700s.

CLTZ Suzhou Shi Yuanlin lühua guanli ju 蘇州市園林綠化管理局. *Canglang Ting zhi* 滄浪亭志 (Gazetteer of Surging Wave Pavilion). Wenhui chubanshe, 2016.

CXTMSZ Xia Hongbing, ed. *Chongxiu Xi Tianmu shan zhi* 重修西天目山志 (Re-editing Gazetteer of Western Mount Heaven Eye). Beijing: Fangshi chubanshe, 2009.

GSZ Wang Ao, ed. *Gusu zhi* 姑蘇志 (Gazetteer of Gusu [Suzhou]), Sq. ed. Taiwan: Taiwan shangwu yinshuguan, printed in the Ming.

JSJ Daoxun 道恂 (fl. the Ming dynasty). *Shizi Lin Jisheng Ji* 師子林紀勝集 (Anthology of the Recorded Scenery of Shizi Lin). Yangzhou: Guanglin Shushe, 2007.

JSJXJ Xu Lifang 徐立方 (fl. Qing dynasty). *Shizi Lin Jisheng Xuji* 師子林紀勝續集 (Addition to Anthology of the Recorded Scenery of Shizi Lin). Yangzhou: Guanglin Shushe, 2007.

SZLZ Suzhou shi Yuanlin Lühua guanliju 蘇州市園林綠化管理局. *Shizi Lin Zhi* 獅子林志 (Gazetteer of Shizi Lin). Shanghai: Wenhui chubanshe, 2015.

TMSZ Xi *Tianmu zushan zhi* 西天目祖山志 (Gazetteer of Western Mount Heaven Eye). Edited by Guangbin 廣賓 in *Zhongguo fosi zhi huikan* 中國佛寺史志彙刊 (The Collected Annals of Chinese Buddhist Temples), vol. 33. Taipei: Zongqing tushu chuban gongsi, 1994.

WDFC Zhou Yongnian 周永年 (1730–1791). *Wudu facheng* 吳都法乘 (Dharma Vehicle of Wu Capital). Taipei: Xin wenfeng chuban gongsi, 1987.

WDWCXJ Qian Gu 錢穀 (b. 1508). *Wudu wencui Xuji* 吳都文粹續集 (Supplementary Anthology of Literary Essence from the Capital of Wu [Suzhou]), Ming. 1565. Sq. ed. Shanghai: Shanghai guji chubanshe, 1989.

YL *Tianru Weize chanshi yulu* 天如惟則禪師語錄 (Recorded Sayings of Master Tianru Weize). Compiled by Shanyu 善遇 (fl. 14th c.). Z 70, no. 1403, in CBETA.

ZZYSBY Suzhou Zhuozhengyuan guanli chu 蘇州拙政園管理處. *Ming Wen Zhengming Zhuozheng Yuan Shi Bianyi* 明 文徵明拙政園詩編譯 (Anthology and Translation of Wen Zhengming's Poems on Unsuccessful

Politician's Garden of the Ming Dynasty). Hong Kong: Hong Kong Tianma chuban youxian gongsi, 2011.

ZZYZ *Suzhou shi Yuanlin Lühua guanliju, Zhuozheng Yuan Zhi* 拙政园志 (Gazetteer of Unsuccessful Politician's Garden). Shanghai: Wenwu chubanshe, 2012.

ZZYZG Suzhou Shi difang zhi bianzuan weiyuan hui bangong shi 蘇州市地方志編纂委員會辦公室 and Suzhou shi yuanlin guanli ju. *Zhuozheng Yuan Zhigao* 拙政園誌稿 (Gazetteer of Zhuozheng yuan). Suzhou: Neibu faxing, 1986.

Primary Texts

Cai, Xingyi. *Yun Shouping* 惲壽平. Shijia zhuang: Hebei jiaoyu chubanshe, 2006.

Canglang qu zhi bianzuan weiyuan hui 滄浪區志編撰委員會. *Canglang Qu Zhi* 滄浪區志 (*Gazetteer of Canglang District*). Shanghai: Shanghai shehui kexueyuan chubanshe, 2006.

Cao, Yunyuan 曹允源 (1856–1927), and Li Gengyuan 李根源 (1879–1965), eds. *Wuxian Zhi* 吳縣志 (Gazetteer of Wu County). Suzhou: Suzhou wenxin gongsi, 1933.

Chen, Ji 陳基 (1314–1370). *Yibai Zhai Gao* 夷白齋稿. Sq. eds., fl. 1352.

Chen, lü 陳旅 (1288–1343). *Anya Tang Ji* 安雅堂集 (Anthology of Anya Hall), Sq. ed., 1335.

Chen, Pan-Show 陳彭壽. *Tianmu Shengjing* 天目勝景 (Excellent Scenery of Mount Heaven Eye). Shanghai: Wen Hwa Fine Arts Press, 1933.

Dōgen, Kigen 道元希玄 (1200–1253). *The True Dharma Eye, Zen Master Dōgen's Three Hundred Kōans*. Translated by John Daido Loori and Kazuaki Tanahashi. Boulder: Shambhala Publications, 2009.

Fan, Chengda 范成大 (1126–1193). *Wujun Zhi* 吳郡志 (Gazetteer of Wu Prefecture). Nanjing: Jiangsu guji Chubanshe, 1986.

Fu, Chun 傅椿 (fl. the 1700s), and Jueluo Ya'er Hashan 覺羅雅爾哈善 (d. 1759). *Suzhou Fu Zhi* 蘇州府志 (Gazetteer of Suzhou Prefecture), Sq. ed., 1748.

Gao, Jin 高晋 (1707–1778). *Nanxun Shengdian Mingsheng Tulu* 南巡盛典名勝圖錄 (Pictorial Anthology of the Famous Scenic Sites from the Grand Ceremony of the Southern Inspections). Suzhou: Guwuxuan Chubanshe, 1999.

Ge, Yinliang 葛寅亮 (1570–1646). *Jinling fancha zhi* 金陵梵刹志 (Gazetteer of Temples in Jinling). Edited by He Xiaorong and Pu Xiaonan. Nanjing: Nanjing chubanshe, 2011.

Gu, Zhentao 顾震濤 (fl. 1790). *Wumen Biaoyin* 吳門表隐 (Wumen's Manifest and Hidden). Nanjing: Jiangsu guji chubanshe, 1999.

Guo, Zhongheng 郭衷恆 (ca. 18th c.). *Suzhou Mingsheng Tu Yong* 蘇州名勝圖詠 (Famous Places of Suzhou, with Pictures and Odes). Suzhou: Yuanfentang block print, 1759.

Hu, Zuanzong 胡纘宗 (1480–1560). *Niaoshu Shanren Xiao Ji* 鳥鼠山人小集 (Small Collection of the Bird and Mouse Hermit), Vol. 11, printed in Jiajing's reign.

Huang, Jin 黃溍 (1277–1357). *Huang Jin Quanji* 黃溍全集 (Comprehensive Anthology of Huang Jin). Edited by Wang Ting. Tianjin: Tianjin guji chubanshe, 2008.

Huang, Xingzeng 黃省曾 (1496–1546). *Wuyue Shanren Ji* 五嶽山人集 (Collection of the Five Sacred Mountains Hermit), Sq., printed in the Ming.

Jiang, Yingke 江盈科 (1553–1605). *Jiang Yingke Ji Shang Ce* 江盈科集 上冊 (Anthology of Jiang Yingke, vol. 1). Edited by Huang Rensheng. Changsha: Yuelu shushe, 2008.

Li, Guangzuo 李光祚 (fl. 18th c.), and Gu Yilü 顧詒祿 (fl. 18th c.). *Changzhou Xian Zhi* 長洲縣志 (Gazetteer of Changzhou County), Sq. ed., 1753.

Li, Mingwan 李銘皖 (circa. 19th c.), and Feng Guifen. *Suzhou Fu Zhi* 蘇州府志 (Gazetteer of Suzhou Prefecture), printed in the 19th c. Taipei: Chengwen Chubanshe, 1970.

Li, Xiusheng 李修生 ed. *Quan Yuan Wen* 全元文 (Complete Anthology of Yuan Literature). Nanjing: Fenghuang chubanshe, 1998.

Liang, Zhangju 梁章鉅 (1775–1849). *Canglang Ting Zhi* 滄浪亭志 (Gazetteer of Surging Wave Pavilion), beiji, 1–2, printed in the 1800s.

Lu, Xiaojun. *Qingdai Shuyuan Keyi Zonglu Xulu (Shang)* 清代書院課藝總集敘錄 (上) (Comprehensive Collection of Academy Curriculum and Arts in the Qing Dynasty (vol. 1)). Wuhan: Wuhan University Press, 2015.

Lu, Xiong 盧熊 (1331–1380). *Hongwu Suzhou fu zhi (Also Called 'Wujun Guangji')* 洪武蘇州府志 (Gazetteer of Suzhou in the Reign of Hongwu). Edited by Suzhou shi Difangzhi ban'gongshi 蘇州地方志辦公室. Yangzhou: guanglin shushe, 2020.

Miaosheng. *Donggao lu* 東皋錄 (Anthology of Donggao), Sq. ed. Taipei: Taipei Commercial Press, 1986.

Minxi 敏曦 (fl. 17–19th c,). *Suzhou Fu Bao'en Si Zhi* 蘇州府報恩寺志 (Gazetteer of Repay Kindness Temple in Suzhou Prefecture). Yangzhou: Guanglin shushe, 2005.

Musō, Soseki 夢窓疎石. *Dialogues in a Dream: The Life and Zen Teachings of Musō Soseki*. Translated by Thomas Yūhō Kirchner. Somerville: Wisdom Publication, 2015.

Niu, Ruolin 牛若麟 (16th–17th c.), and Wang Huanru 王焕如 (16th–17th c.). *Chongzhen Wuxian Zhi* 崇禎吳縣志 (Gazetteer of Wu County in the Reign of Chongzhen). 1642 Block Printed. Shanghai: Shanghai shudian, 1989.

Pingjiang qu zhi bianzuan weiyuan hui 平江區志編撰委員會. *Pingjiang Qu Zhi* 平江區志 *(Gazetteer of Pingjiang District)*. Shanghai: Shanghai shehui kexueyuan chubanshe, 2006.

Qingquan 清泉 (fl. 19th C.). *Zhutang Si Zhi* 竹堂寺志 (Gazetteer of Bamboo Hall Temple), in vol. 43. Zhongguo Fosi Zhi Cong Kan 中國佛寺志叢刊. Yangzhou: Guanglin shushe, 2006.

Shen, Deqian 沈德潛, and Gu Yilu 顧詒祿, eds. *Qianlong Yuanhe xianzhi* 乾隆元和縣志 (Gazetteer of Yuanhe County in Qianlong's reign). Nanjing: Jiangsu guji chubanshe, 1991.

Shen, Zhou 沈周 (1427–1509). *Shen Zhou Ji (Xia)* 沈周集 下 (Anthology of Shen Zhou, Xia). Edited by Zhang Xiuling and Han Xingyin. Shanghai: Shanhai guji chubanshe, 2013.

Shenzhou guoguang she 神州國光社. *Shenzhou Guoguang Ji*, vol. 10 神州國光集 第十集. Shanghai: Shenzhou guoguang she, 1909.

Song, Lian. *Song Lian Quanji* 宋濂全集 (Complete Works of Song Lian). Hangzhou: Zhejiang guji chubanshe, 2014.

Suzhou Shi difang zhi bianzuan weiyuan hui bangong shi 蘇州市地方志編纂委員會辦公室, and Suzhou shi yuanlin guanli ju 蘇州市園林管理局. *Zhuozheng Yuan Zhigao* 拙政園誌稿 (Gazetteer of Zhuozheng yuan). Suzhou: Neibu faxing, 1986.

Wang, Shizhen 王世貞. *Yanzhou xugao* 弇州續稿 (Continued Manuscript of Yanzhou), Sq. ed., printed in the Ming.

Wang, Xing. *Banxuan Ji Fangwai Buyi* 半軒集方外補遺 (Supplementary Additions to the Banxuan Collection, Buddhism and Daoism), Sq. ed., printed in the 14th c.

Wang, Zhen. *Wang Zhen nongshu* 王禎農書 (Wang Zhen's Agricultural Manual). Edited by Sun Xianbin 孫顯斌 and You Xingchao 攸興超. Changsha: Hu'nan kexue jishu chubanshe, 2014.

Wen, Zhengming 文徵明 (1470–1559). *Wen Zhengming ji* 文徵明集 (Anthology of Wen Zhengming). Edited by Zhou Daozhen. Shanghai: Shanghai guji chubanshe, 2014.

Wu, Kuan 吳寬. *Jiacang ji* 家藏集 (Anthology of Family Treasures). Shanghai: Shanghai guji chubanshe, 1991.

Xu, Fang 徐枋 (1622–1694). *Yiju Tang Ji* 居易堂集 (Anthology of Yiju Hall), Sq. eds., printed in the Ming.

Xu, Jie 徐階 (1503–1583). *Shi Jing Tang Ji* 世經堂集 (Collection of the Shijing Hall), printed in the reign of Wanli.

Xu, Shupi (fl. 17th c.) 徐樹丕. *Shi Xiao Lu* 識小錄 (Small Record of Observations), printed in the 17th c.

Xu, Song 徐崧 (1617–1690), and Zhang Dachun 張大純 (1637–1720). *Baicheng Yanshui* 百城煙水 (Smoke and Water of a Hundred Cities). Nanjing: Jiangsu guji Chubanshe, 1999.

Yang, Er'zeng 楊爾曾 (fl. 1575–1609). *Xinjuan Hainei qiguan* 新鐫海內奇觀 (Marvelous Sights Within Seas). 1609. Print from Yibai tang. In the collection of Berlin State Library.

Yang, Xunji 楊循吉 (1458–1546). *Wuyi Zhi, Changzhou xian zhi* 吳邑志 長洲縣志 (Gazetteer of Wu County and Changzhou County). Edited by Chen Qidi. Yangzhou: Guanglin shushe, 2006.

Yang, Yu 楊瑀, and Kong Qi 孔齊. *Shanju Xinyu Zhizheng Zhiji* 山居新語 至正直記 (New Sayings on Mountain Living, Straight Records of Zhizheng). Edited by Li Mengsheng and Zhuang Wei. Shanghai: Guji chubanshe, 2012.

Yao, Guangxiao. *Yao Guangxiao Ji* 姚廣孝集 (Anthology of Yao Guangxiao). Edited by Le Guiming, 293–294. Beijing: The Commercial Press, 2016.

Yuan, Hongdao 袁宏道 (1568–1610). *Yuan Hongdao Ji Jianjiao* 袁宏道集箋校 (Collated Anthology of Yuan Hongdao). Edited by Qian Bocheng, 180–182. Shanghai: Shanghai guji chubanshe, 1981.

Yun, Shouping 惲壽平 (1633–1690). *Ouxiang Guan Ji* 甌香館集 (A Compendium of Fragrant Cup Studio). Edited by Lü Fengshang. Hangzhou: Xiling yinshe chubanshe, 2012.

Zhang, Guowei 張國維 (1595–1646), and Cai Yiping, eds. *Wuzhong Shuili Quanshu* 吳中水利全書 (Complete Record of Hydraulic Management in the Region of Wu). Hangzhou: Zhejiang guji chubanshe, 2014.

Zhang, Yi 張怡 (1608–1695). *Yuguang jianqi ji* 玉光劍氣集 (Jade Light Sword Qi Collection). Edited by Wei Lianke. Beijing: Zhonghua shuju, 2006.

Zhang, Yinglin 張英霖. *Suzhou Gucheng Ditu Ji* 蘇州古城地圖集 (The Atlas of Ancient Suzhou). Suzhou: Guwuxuan Chubanshe, 2004.

Zhongfeng Mingben. *Zhongfeng Mingben quanji* 中峰明本全集 (Complete Anthology of Zhongfeng Mingben). Edited by Yu Delong. Beijing: Jiuzhou chubanshe, 2020.

Zhengguo 正果. *Chanzong Dayi* 禪宗大意 (Principles of Chan Buddhism). Beijing: Shangding shuwei chuban, 1990.

Zhongfeng Mingben. *Zhongfeng Mingben quanji* 中峰明本全集 (Complete Anthology of Zhongfeng Mingben). Edited by Yu Delong. Beijing: Jiuzhou chubanshe, 2020.

Zhongshan daxue lishi xi, ed. *Lin Zexu Ji Riji* 林則徐集 日記 (Anthology of Li Zexu's Diaries). Beijing: Zhonghua shuju, 1962.

Zhu, Changwen 朱長文(1041–1098). *Wujun Tujing Xuji* 吳郡圖經續記 (Supplementary Records to the 'Illustrated Guide to Wu Commandery'). Nanjing: Jiangsu guji chubanshe, 1999.

Zhu, Yunming. *Zhu Yunming Ji* 祝允明集 (Anthology of Zhu Yunming). Edited by Xue Weiyuan. Shanghai: Shanghai guji chubanshe, 2016.

Zongze 宗賾. *The Origins of Buddhist Monastic Codes in China: An Annotated Translation and Study of the Chanyuan Qinggui.* Translated by. Yifa. Honolulu: University of Hawaii Press, 2002.

Secondary Texts

Ackerman, James S. 'The Photographic Picturesque'. *Artibus et Historiae* 24, no. 48 (2003): 73–94.

Adams, Ann Jensen. 'Competing Communities in the "Great Bog of Europe": Identity and Seventeenth-Century Dutch Landscape Painting'. In *Landscape and Power*, edited by W. J. T. Mitchell, 2nd ed., 35–76. Chicago and London: University of Chicago Press, 2002.

Amherst, Alicia. *A History of Gardening in England.* Cambridge Library Collection–Botany and Horticulture. Cambridge: Cambridge University Press, 2013.

Ammerman, Albert J. 'Venice before the Grand Canal'. *Memoirs of the American Academy in Rome* 48 (2003): 141–158.

Archaeology Team of Suzhou Museum 蘇州博物館考古組. 'Suzhou Faxian Qimen Gu Shuimen Nei Jichu 蘇州發現齊門古水門基礎 (Reports on Excavation of the Foundation of Ancient Water Gates in Suzhou)'. *Wenwu* 文物, no. 05 (1983): 55–59.

Bassnett, Susan. *Constructing Cultures: Essays on Literary Translation.* Clevedon: Multilingual Matters, 1998.

Baudrillard, Jean. *Simulacra and Simulation.* Translated by Sheila Faria Glaser. Ann Arbor: University of Michigan Press, 1994.

Bernard, George W. 'The Dissolution of the Monasteries'. *History* 96, no. 324 (2011): 390–409.

Bishop, Wendy. *Ornamental Lakes: Their Origins and Evolution in English Landscapes.* London and New York: Routledge, 2021.

Blishen, Tony, and Wen Zhenheng 文震亨 (1585–1645). *An Elegant Life of Chinese Literati: From the Chinese Classic, 'Treatise on Superfluous Things', Finding Harmony and Joy in Everyday Objects.* New York: Better Link Press, 2019.

Bourdieu, Pierre. *Distinction: A Social Critique of the Judgement of Taste.* Translated by Nice Richard. Cambridge: Harvard University Press, 1996.

Brook, Timothy. *The Chinese State in Ming Society.* New York: RoutledgeCurzon, 2005.

Brook, Timothy. *Praying for Power: Buddhism and the Formation of Gentry Society in Late-Ming China.* Cambridge: Harvard University Press, 1993.

Brown, David, and Williamson Tom. *Lancelot Brown and the Capability Men: Landscape Revolution in Eighteenth-Century England.* London: Reakton Books, 2016.

Bruzelius, Caroline. 'Digital Technologies and New Evidence in Architectural History'. *Journal of the Society of Architectural Historians* 76, no. 4 (2017): 436–439.

Bush, Susan. *The Chinese Literati on Painting: Su Shih (1037–1101) to Tung Ch'i-Ch'ang (1555–1636)*. Hong Kong: Hong Kong University Press, 2012.

Buswell Jr., Robert E., and Donald S. Lopez Jr. *The Princeton Dictionary of Buddhism*. Princeton: Princeton University Press, 2013.

Cahill, James. *Parting at the Shore: Chinese Painting of the Early and Middle Ming Dynasty, 1368–1580*. New York: Weatherhill, 1978.

Cahill, James, Huang Xiao, and Liu Shanshan. *Buxiu de Linquan: Zhongguo Gudai Yuanlin Huihua* 不朽的林泉: 中國古代園林繪畫 (Imperishable Groves and Spring: Garden Paintings in Old China). Beijing: Joint Publishing, 2016.

Cai, Rixin. *Linji Xia Huqiu Chanxi Gai Su* 臨濟下虎丘禪系概述 (An Outline of the Lineage of Tiger Hill in Linji Chan Buddhism). Nanzhou: Gansu People's Press, 2008.

Campbell, Aurelia. *What the Emperor Built: Architecture and Empire in the Early Ming*. Seattle: University of Washington Press, 2020.

Chambers, William. *Designs of Chinese Buildings, Furniture, Dresses, Machines and Utensils*. London: Published for the author, 1757.

Chang, Kang-I Sun. 'Literature of the Early Ming to Mid-Ming (1375–1572)'. In *The Cambridge History of Chinese Literature*, edited by Chang Kang-i Sun and Stephen Owen, 1st ed., 1–62. Cambridge: Cambridge University Press, 2010.

Charlton, Rosemary. *Fundamentals of Fluvial Geomorphology*. London: Routledge, 2008.

Chen, Congzhou. *Chinese Gardens: Theory and Practice*. Translated by Ling Yuan. Singapore: Gale Asia, 2016.

Chen, Congzhou. *Shuoyuan* 说园 (On gardens). Shanghai: Tongji University Press, 1984.

Chen, Congzhou. *Shuoyuan Zhongguo Mingyuan* 說園 中國名園 (On Gardens: Famous Chinese Gardens). Nanjing: Jiangsu wenyi chubanshe, 2013.

Chen, Congzhou. *Suzhou Yuanlin* 蘇州園林 (Suzhou Gardens). Shanghai: Tongji daxue Jiaocaike, 1956.

Chen, Congzhou. *Zishi Yu Mo* 梓室餘墨 (Remaining Ink in the Carpenter's Studio). Shanghai: Shanghai shudian chubanshe, 2019.

Chen, Gaohua. 'Yuandai Fojiao Siyuan Foyi de Yanbian 元代佛教寺院賦役的演變 (The Development of Taxes and Corvee of Buddhist Temples in the Yuan Dynasty)'. *Journal of Beijing Union University (Humanities and Social Sciences)* 北京聯合大學學報 (人文社會科學版) 11, no. 03 (2013): 5–15.

Chen, Gaohua. 'Yuandai Fojiao Yu Yuandai Shehui 元代佛教與元代社會 (Buddhism and Society in the Yuan Dynasty)'. In *Commemorative Collection of the Inaugural Meeting of the Chinese Mongolian Historical Association* 中國蒙古史學會成立大會紀念集刊, edited by Chinese Mongolian Historical Association 中國蒙古史學會, 312–326. Huhe haote: Zhongguo menggushi xuehui, 1979.

Chen, Gaohua. 'Yuandai neiqian Weiwu'er ren yu fojiao 元代内遷畏兀兒人与佛教 (The Yuan Dynasty's Internal Migration of the Uyghur People and Buddhism)'. 中國史研究 *Studies in Chinese History*, no. 1 (2011): 33–52.

Chen, Jinhua. *Philosopher, Practitioner, Politician: The Many Lives of Fazang (643–712)*. Leiden: Brill, 2007.

Chen, Qidi. 'Hongying Ganshan *Yu Su Bei Yiming* "紅蠅趕散"與蘇北移民 (The Scattering of the 'Red Flies' and the Northern Jiangsu Migrants)'. *Xun Gen* 尋根, no. 01 (2012): 130–135.

Chen, Wangheng. *Chinese Environmental Aesthetics*. Edited by Cipriani Gerald. Translated by Su Feng. Routledge Contemporary China Series. London and New York: Taylor & Francis, 2015.

Chen, Zhi, and Ji Cheng (1582– circa.1642). *Yuanye Zhushi* 園治注釋 (Annotation on the Craft of Gardens). Beijing: Zhongguo jianzhu gongye chubanshe, 1988.

Cheng, Liao. *Ru xinzhong you yiwu: chejie Chan zong gong'an* 汝心中有一物: 徹解禪宗公案 (There is One Thing in Your Heart: Thoroughly Understanding of the Gong'ans of Chan School). Guangdong: Shantou University Press, 2014.

Chiang, Nicole T. C. *Emperor Qianlong's Hidden Treasures—Reconsidering the Collection of the Qing Imperial Household*. Emperor Qianlong's Hidden Treasures. Hong Kong: Hong Kong University Press, 2019.

Chow, Kai-Wing. *The Rise of Confucian Ritualism in Late Imperial China: Ethics, Classics, and Lineage Discourse*. Stanford: Stanford University Press, 1994.

Clunas, Craig. *Art in China*. Oxford and New York: Oxford University Press, 1997.

Clunas, Craig. *Elegant Debts: The Social Art of Wen Zhengming, 1470–1559*. Honolulu: University of Hawai'i Press, 2004.

Clunas, Craig. *Fruitful Sites: Garden Culture in Ming Dynasty China*. London: Reaktion Books, 1996.

Clunas, Craig. 'Nature and Ideology in Western Descriptions of the Chinese Gardens'. *Extrême-Orient, Extrême-Occident* 22 (2000): 153–166.

Clunas, Craig. 'Social History of Art'. In *Critical Terms for Art History*, edited by Robert S. Nelson and Richard Shiff, 2nd ed., 465–477. Chicago and London: The University of Chicago Press, 2003.

Clunas, Craig. *Superfluous Things: Material Culture and Social Status in Early Modern China*. Cambridge: Polity Press, 1991.

Cosgrove, Denis. *The Palladian Landscape: Geographical Change and Its Cultural Representations in Sixteenth Century Italy*. University Park: Penn State University Press, 1993.

Cosgrove, Denis. *The Palladian Landscape: Geographical Change and Its Cultural Representations in Sixteenth Century Italy*. University Park: Penn State University Press, 1993.

Cosgrove, Denis. 'Prospect, Perspective and the Evolution of the Landscape Idea'. *Transactions of the Institute of British Geographers* 10, no. 1 (1985): 45–62.

Cosgrove, Denis. *Social Formation and Symbolic Landscape*. Croom Helm Historical Geography Series. London and Sydney: Croom Helm, 1984.

Cosgrove, Denis, and Stephen Daniels, eds. *The Iconography of Landscape: Essays on the Symbolic Representation, Design and Use of Past Environments*. Cambridge: Cambridge University Press, 1988.

Damian, Michelle. 'As Estates Faded: Late Medieval Maritime Shipping in the Seto Inland Sea'. In *Land, Power, and the Sacred*, edited by Janet R. Goodwin and Joan R. Piggott, 351–374. Honolulu: University of Hawai'i Press, 2018.

Dardess, John W. *Confucianism and Autocracy: Professional Elites in the Founding of the Ming Dynasty*. Berkeley: University of California Press, 1984.

Dong, Shouqi. *Suzhou Yuanlin Shanshui Hua Xuan* 蘇州園林山水畫選 (Selected Landscape Paintings of Suzhou Gardens). Shanghai: Joint Publishing, 2007.

Eyres, Patrick, and Karen Lynch. *On The Spot: The Yorkshire Red Books of Humphry Repton, Landscape Gardener*. Huddersfield: New Arcadian Press, 2018.

Fan, Jinmin, and Xia Weizhong. *Suzhou Diqu Shehui Jingji Shi* 蘇州地區社會經濟史(明清卷) Social and Economical History of Suzhou Region, Ming and Qing Dynasties. Edited by Lun Luo. Nanjing: Nanjing University Press, 1993.

Feng, Chengjun, trans. *Travels of Marco Polo* 馬可波羅行紀. Beijing: The Commercial Press, 2017.

Feng, Jin. 'Jing, the Concept of Scenery in Texts on the Traditional Chinese Garden: An Initial Exploration'. *Studies in the History of Gardens & Designed Landscapes* 18, no. 4 (December 1998): 339–365.

Finnane, Antonia. *Speaking of Yangzhou: A Chinese City, 1550–1850*. Cambridge: Harvard University Asia Centre, 2004.

Fong, Wen. *Beyond Representation: Chinese Painting and Calligraphy, 8th–14th Century*. New Haven: Yale University Press, 1992.

Frommel, Sabine, Marco Gaiani, and Simone Garagnani. '3-D Digital Modeling and Giuliano Da Sangallo's Designs for Santa Maria Delle Carceri in Prato'. *Journal of the Society of Architectural Historians* 80, no. 1 (1 March 2021): 30–47.

Fu, Huayang. 'Shijing shanshui yu Xufang *Jianshang Caotang tu* xinkao 實景山水與徐枋《澗上草堂圖》新考 (New Examination on Real Landscape and Xu Fang's Painting of Thatched Cottage on the Gully)'. *Shoucang* 收藏, no. 23 (2015): 56–63.

Fu, Li-Tsui. *Mingdai Suzhou Huihua Yu Wenren de Seng Si You* 明代蘇州繪畫與文人的僧寺遊 (Ming Dynasty Suzhou Paintings and the Literati's Monastery Tours). Edited by Uematsu Mizuki, 153–166. Nara: Yamato Bunkakan, 2016.

Fueoka, Jishō. *Eiheiji*. Saitama: Kokeisō, 1966.

Fung, Stanislaus. 'Non-Perspectival Effects in the Liu Yuan'. Paper presented at the conference in EAAC Gwangju, Korea, 2015.

Gao, Jun. 'Yuanmo Ming Chu Shiseng Shi Miaosheng Kaolun 元末明初詩僧釋妙聲考論 (Study and Discussion on the Poet Monk Miaosheng from the Late Yuan to Early Ming Dynasty)'. *Wenjiao Ziliao* 文教資料, no. 33–34 (2017): 123–126.

Gilchrist, Roberta. *Norwich Cathedral Close: The Evolution of the English Cathedral Landscape*. Woodbridge: The Boydell Press, 2005.

Glahn, Richard von. *The Economic History of China: From Antiquity to the Nineteenth Century*. Cambridge: Cambridge University Press, 2016.

Gleason, Kathryn L. 'Porticus Pompeiana: A New Perspective on the First Public Park of Ancient Rome'. *The Journal of Garden History* 14, no. 1 (March 1994): 13–27.

Goodwin, Janet R. 'Claiming the Land: Chōgen and the Development of Ōbe Estate'. In *Land, Power, and the Sacred*, edited by Janet R. Goodwin and Joan R. Piggott, 231–252. Honolulu: University of Hawai'i Press, 2018.

Gu, Kai. *Mingdai Jiangnan yuanlin yanjiu* 明代江南園林研究 (Gardens of the Jiangnan Area in the Ming Period). Nanjing: Southeast University Press, 2010.

Gu, Kai. 'Zhongguo Chuantong Yuanlin Zhong "Ting Ju Shandian" de Zai Renshi: Zuoyong, Wenhua, Yu Guannian Bianqian 中國傳統園林中'亭踞山巔'的再認識：作用，文化與觀念變遷 (Rethinking on Rockery-Top Pavilion in Traditional Chinese Gardens: Function, Culture, and Idea Change)'. *Zhongguo yuanlin* 中國園林, no. 07 (2016): 78–83.

Guo, Daiheng. *Nansong jianzhushi* 南宋建築史 (History of Architecture in Southern Song Dynasty). Shanghai: Guji chubanshe, 2014.

Guo, Lu, and Rong Xiao. 'Taking the High Ground: Construction of the Regional Spatial Order of Chang'an Area in Tang Dynasty'. *Planning Perspectives* 35, no. 1 (2 January 2020): 115–141.

Guo, Mingyou. *Mingdai Suzhou Yuanlin Shi* 明代蘇州園林史 (The Garden History of Suzhou in the Ming Dynasty). Beijing: Zhongguo jianzhu gongye chubanshe, 2013.

Hamar, Imre, ed. *Reflecting Mirrors: Perspectives on Huayan Buddhism*. Wiesbaden: Harrassowitz, 2007.

Han, Seunghyun. *After the Prosperous Age: State and Elites in Early Nineteenth-Century Suzhou*. Cambridge: Harvard University Asia Centre, 2016.

Hardie, Alison, and Duncan M. Campbell, eds. *The Dumbarton Oaks Anthology of Chinese Garden Literature*. Washington: Dumbarton Oaks, 2020.

Heine, Steven, and Dale S. Wright. *From Chinese Chan to Japanese Zen: A Remarkable Century of Transmission and Transformation*. New York: Oxford University Press, 2017.

Heine, Steven, and Dale S. Wright. *The Kōan: Texts and Contexts in Zen Buddhism*. New York: Oxford University Press, 2000.

Heine, Steven, and Dale S. Wright. *Opening a Mountain: Kōans of the Zen Masters*. Oxford: Oxford University Press, 2002.

Heine, Steven, and Dale S. Wright. *Zen Ritual: Studies of Zen Buddhist Theory in Practice*. New York: Oxford University Press, 2007.

Heller, Natasha. *Illusory Abiding: The Cultural Construction of the Chan Monk Zhongfeng Mingben*. Cambridge: Harvard University Asia Center, 2014.

Hon, Tze-ki. *The Yijing and Chinese Politics: Classical Commentary and Literati Activism in the Northern Song Period, 960–1127*. New York: State University of New York Press, 2005.

Housley, R. A., A. J. Ammerman, and C. E. McClennen. 'That Sinking Feeling: Wetland Investigations of the Origins of Venice'. *Journal of Wetland Archaeology* 4, no. 1 (June 2004): 139–153.

Hoyle, R. W. 'The Origins of the Dissolution of the Monasteries'. *The Historical Journal* 38, no. 2 (1995): 275–305.

Huang, Chenying. 'A Building as a Biography: I.M. Pei and His Suzhou Museum'. Charlottesville: University of Virginia, 2014.

Huang, Xiao, and Liu Shanshan. 'Yuanlinhua, cong xingle tu dao shijing tu 園林畫: 從行樂圖到實景圖 (Garden Paintings: From Depicting Pleasure to Actual Scene)'. *Zhongguo shuhua* 中國書畫 9 (2015): 6.

Huffman, Kristin L., Andrea Giordano, and Caroline Bruzelius. *Visualizing Venice: Mapping and Modeling Time and Change in a City*. Abingdon: Routledge, 2017.

Hunt, John Dixon. 'The Garden in the City of Venice: Epitome of State and Site'. *Studies in the History of Gardens & Designed Landscapes* 19, no. 1 (March 1999): 46–61.

Hunt, John Dixon. *The Picturesque Garden in Europe*. London: Thames & Hudson, 2002.

Inn, Henry. *Yuanting huacui* 園庭畫萃 (Chinese Houses and Gardens). Edited by Lee Shao Chang. New York: Bonanza Books, 1940.

Isozaki, Arata 磯崎新. *Japan-Ness in Architecture*. Cambridge: MIT Press, 2006.

Jia, Jun. 'Qianlong *di xuejing xingle tu* yu Changchun yuan Shizilin xukao《乾隆帝雪景行樂圖》與長春園獅子林續考 (Further Exploration into Shizilin Garden in Changchunyuan and the Picture of Emperor Qianlong in Snowscape)'. *Zhuangshi* 裝飾 03, no. 239 (2013): 52–53.

Jiang, Zhaoshen. *Gugong Canghua Daxi, Ba* 故宮藏畫大系 八 (A Panorama of Paintings in the Collection of The National Palace Museum, vol. VII). Taipei: Guoli gugong bowuyuan, 1994.

Jiangsu wenwu zonglü bianji weiyuanhui 江蘇文物綜錄編輯委員會, and Nanjing bowuyuan 南京博物院. *Jiangsu Wenwu Zonglu* 江蘇文物綜錄 (Comprehensive Catalogue of Cultural Relics in Jiangsu). Jiangsu Wuxian: Wuxian wenyi yinshuachang, 1988.

Johnston, R. Stewart. *Scholar Gardens of China: A Study and Analysis of the Spatial Design of the Chinese Private Garden* 中國文人園. Cambridge: Cambridge University Press, 1992.

Keirstead, Thomas. 'Gardens and Estates: Medievality and Space'. *Positions: Asia Critique* 1, no. 2 (May 1993): 289–320.

Kerby, Kate. *An Old Chinese Garden: A Three-Fold Masterpiece of Poetry, Calligraphy and Painting* 拙政園圖. Shanghai: Chung Hwa Book Company, 1923.

Keswick, Maggie. *The Chinese Garden: History, Art & Architecture*. New York: Rizzoli International Publications, 1978.

Kopytoff, Igor. 'The Cultural Biography of Things: Commoditization as Process'. In *The Social Life of Things: Commodities in Cultural Perspective*, edited by Arjun Appadurai, 64–94. Cambridge: Cambridge University Press, 1988.

Li, Cho-Ying. '地方性與跨地方性–從子游傳統之論述與實踐看蘇州在地文化與理學之競合 (Locality and Trans-locality: Understanding the Negotiation and Competition between Suzhou Local Cultural and Neo-Confucianism Based on the Discourse and Practice of the "Ziyou Heritage")'. *Bulletin of the Institute of History and Philology* 中央研究院歷史語言研究所集刊 82, no. 2 (2011): 325–398.

Li, Yifan. 'Making the Qianlong Emperor's Private Garden: Imperialization of the Lion Grove in Eighteenth-Century China'. Master Thesis, University of Alberta, 2017.

Liang, Sicheng, *Tuxiang zhongguo jianzhushi* 圖像中國建築史 (A Pictorial History of Chinese Architecture). Beijing: Joint Publishing, 2011.

Liaoning sheng bowuguan 遼寧省博物館. *Guancang Zhongguo Lidai Shuhua Zhulu Shufa Juan* 館藏中國歷代書畫著錄 書法卷 (Catalogue of Chinese Calligraphy and Paintings Collected Through the Dynasties, Calligraphy Volume). Shenyang: Liaoning meishu chubanshe, 2015.

Lippit, Yukio. *Japanese Zen Buddhism and the Impossible Painting*. Los Angeles: Getty Research Institute, 2017.

Liu, Dunzhen. *Suzhou de Yuanlin* 苏州的園林 (Gardens of Suzhou), 1957. Unpublished archive.

Liu, Dunzhen. *Suzhou Gudian Yuanlin* 苏州古典園林 (Classical Gardens of Suzhou). Beijing: Zhongguo jianzhu gongye chubanshe, 1979.

Liu, Hui. '*Qianlong Guzhuang Xuejing Xingle Tu Kaozheng*'. 乾隆古裝雪景行樂圖考證 (Research on "Qianlong Ancient Costume Snow Scene Pleasure Trip Picture")'. *Wenwu* 文物, no. 08 (2013): 88–94.

Liu, Hui. 'Cong gongting xieshi huihua dao Yangshi Lei tuyang: dui Changchunyuan Shizilin de zhenshi miaohui 從宮廷寫實繪畫到樣式雷圖樣: 對長春園獅子林的真實描繪 (From Imperial Realistic Painting to Yangshi Lei Pattern: The True Portray on Lion Grove of the Garden of Eternal Spring (Changchun Yuan))'. *Zijin Cheng* 紫禁城, no. 02 (2019): 98–111.

Liu, Jianguo, Peng Xiaojun, Tao Yang, and Xiang Qifang. *Jianghan Pingyuan Shiqian Zhishui Wenming* 江漢平原史前治水文明 (Prehistoric Water Management Civilization in the Jianghan Plain). Beijing: Zhongguo shehui kexue chubanshe, 2023.

Long, Bo. 'Tianru Weize Chanshi Jiqi Chanfa Yanjiu 天如惟則禪師及其禪法研究 (Study on Chan Master Tianru Weize and His Chan Buddhism)'. Master Thesis, Xi'nan University, 2017.

Loori, John Daido. *The Eight Gates of Zen: A Program of Zen Training.* Boulder: Shambhala Publications, 2002.

Lu, Andong. 'Deciphering the Reclusive Landscape: A Study of Wen Zheng-Ming's 1533 Album of the Garden of the Unsuccessful Politician'. *Studies in the History of Gardens & Designed Landscapes* 31, no. 1 (25 March 2011): 40–59.

Lu, Andong. 'Lost in Translation: Modernist Interpretation of the Chinese Garden as Experiential Space and Its Assumptions'. *The Journal of Architecture* 16, no. 4 (2011): 499–527.

Ma, Caixia. 'Yuandai Sengren Shisu Hua Tanxi 元代僧人世俗化探析 (Study on the Secularization of Monks in the Yuan Dynasty)'. Master Thesis, Hebei Shifan University, 2005.

Macdougall, Elisabeth Blair, ed. *Medieval Gardens.* Washington: Dumbarton Oaks, 1986.

Marmé, Michael. *Suzhou: Where the Goods of All the Provinces Converge.* Stanford: Stanford University Press, 2005.

Mclean, Teresa. *Medieval English Gardens.* Mineola: Dover Publications, 2014.

Mei, Jing. 'Mingqing Suzhou Yuanlin Jizhi Guimo Bianhua Jiqi Yu Chengshi Bianqian Zhi Guanxi Yanjiu 明清蘇州園林基址規模變化及其與城市變遷之關係研究 (Study on the Changes in the Scale of Ming and Qing Dynasty Suzhou Gardens and Their Relationship with Urban Transformation)'. Master Thesis, Tsinghua University, 2009.

Milburn, Olivia. 'From Hero to Ancestor, God, and Ghost: The Posthumous Career of Han Shizhong'. *Archiv orientální* 84, no. 1 (2016): 189–229.

Mitchell, W. J. T., ed. *Landscape and Power.* 2nd ed. Chicago and London: University of Chicago Press, 2002.

Morris, Edwin T. *The Gardens of China: History, Art, and Meanings.* New York: Charles Scribner's Sons, 1983.

Morrison, Elizabeth. *The Power of Patriarchs: Qisong and Lineage in Chinese Buddhism.* Leiden: Brill, 2010.

Mote, Frederick W. *The Poet Kao Ch'i, 1336–1374.* Princeton: Princeton University Press, 1962.

Murashige, Stanley. 'Rhythm, Order, Change, and Nature in Guo Xi's Early Spring'. *Monumenta Serica* 43 (1995): 337–364.

Murck, Alfreda, and Fong Wen. *A Chinese Garden Court: The Astor Court at the Metropolitan Museum of Art.* New York: The Metropolitan Museum of Art, 2012.

Nolf, Christian, Xie Yuting, Florence Vannoorbeeck, and Chen Bing. 'Delta Management in Evolution: A Comparative Review of the Yangtze river basin and Rhine-Meuse-Scheldt Delta'. *Asia-Pacific Journal of Regional Science* 5, no. 2 (June 2021): 597–624.

Pan, Guxi. *Jiangnan Lijing Yishu* 江南理景藝術 (The Art of Crafting Scenes in Jiangnan). Nanjing: Dongnan daxue chubanshe, 2001.

Pan, Guxi, ed. *Zhongguo Gudai Jianzhu Shi, Di Si Juan, Yuan Ming Jianzhu* 中國古代建築史 第4卷 元明建築 (History of Ancient Chinese Architecture: vol. 4, Architecture in the Yuan and Ming Dynasties). Beijing: China Architecture & Building Press, 2001.

Pan, Junting, and Zhao Bing. 'Qing chu Suzhou yuanlin Jian shang caotang gouchen 清初蘇州園林澗上草堂鉤沉 (Study of Xu Fang's Jian Shang Garden in Early Qing Dynasty)'. *Landscape Architecture* 風景園林 26, no. 03 (2019): 116–120.

Pellatt, Valerie, and Eric T. Liu. *Thinking Chinese Translation, a Course in Translation Method: Chinese to English*. London and New York: Routledge, 2010.

Pelowski, Matthew. 'Satori, Koan and Aesthetic Experience: Exploring the "Realization of Emptiness" in Buddhist Enlightenment via an Empirical Study of Modern Art'. *Psyke & Logos* 33, no. 2 (2012): 236–268.

Piggott, Joan R. 'Estates: Their History and Historiography'. In *Land, Power, and the Sacred*, edited by Joan R. Piggott and Janet R. Goodwin, 3–36. Honolulu: University of Hawai'i Press, 2018.

Piper, Fredrik Magnus. *Ängelsk Lustpark, English Park*. Stockholm: The Royal Swedish Academy of Fine Arts Publication, 2004.

Poceski, Mario. *The Records of Mazu and the Making of Classical Chan Literature*. Oxford: Oxford University Press, 2015.

Powell, Florence Lee. *In the Chinese Garden: A Photographic Tour of the Complete Chinese Garden, with Text Explaining Its Symbolism, As Seen In The Liu Yuan (The Liu Garden) And The Shih Tzu Lin (The Forest of Lions), Two Famous Chinese Gardens in the City of Soochow, Kiangsu Province, China*. New York: The John Day Company, 1943.

Prip-Møller, Johannes. *Chinese Buddhist Monasteries: Their Plan and Its Function as a Setting for Buddhist Monastic Life*. Hong Kong: Hong Kong University Press, 1967.

Qi, Gong. 'Huang Zijiu Qiushan Tu Zhi Zhenwei 黄子久《秋山圖》之真偽 (The Authenticity of Huang Zijiu's 'Autumn Mountain Painting)'. In *Qi Gong lunyi* 启功论艺 (Qi Gong on Art), edited by Shen Peifang, 123–125. Shanghai: Shanghai shuhua chubanshe, 2010.

Qin, Boqian, ed. *Lake Taihu, China*. Dordrecht: Springer, 2008.

Qu, Weizu, ed. *Suzhou Hedao Zhi* 蘇州河道志 (Gazetteer of Rivers and Canals of Suzhou). Changchun: Jilin renmin chubanshe, 2007.

Rinaldi, Bianca Maria, ed. *The Chinese Garden: Garden Types for Contemporary Landscape Architecture*. Basel: Birkhäuser, 2011.

Rinaldi, Bianca Maria, ed. *Ideas of Chinese Gardens*. Pennsylvania: University of Pennsylvania Press, 2016.

Rotherham, Ian D., and Christine Handley, eds. *What Did Capability Brown Do for Ecology? The Legacy for Biodiversity, Landscapes, and Nature Conservation*. Sheffield: Wildtrack Publishing, 2017.

Russell-Smith, Lilla. *Uygur Patronage in Dunhuang: Regional Art Centres on the Northern Silk Road in the Tenth and Eleventh Centuries*. Leiden: Brill, 2005.

Sandron, Dany, and Andrew Tallon. *Notre-Dame Cathedral: Nine Centuries of History*. Translated by Lindsay Cook. University Park: Pennsylvania State University Press, 2020.

Savoy, Daniel. 'Palladio and the Water-Oriented Scenography of Venice'. *Journal of the Society of Architectural Historians* 71, no. 2 (1 June 2012): 204–225.

Schneewind, Sarah. *Community Schools and the State in Ming China*. Stanford: Stanford University Press, 2006.

Schopen, Gregory. 'The Buddhist "Monastery" and the Indian Garden: Aesthetics, Assimilations, and the Siting of Monastic Establishments'. *Journal of the American Oriental Society* 126, no. 4 (2006): 487–505.

Schulz, Juergen. 'Jacopo de' Barbari's View of Venice: Map Making, City Views, and Moralized Geography before the Year 1500'. *The Art Bulletin* 60, no. 3 (1978): 425–474.

Sekiguchi, Kinya 関口欣也. *Chūgoku Kōnan no Daizen'in to Nansō Gozan* 中国江南の大禅院と南宋五山 (The Great Zen Temples of Jiangnan, China, and the Five Mountains of the Southern Song Dynasty). Tokyo: Shogakukan, 1984.

Sekiguchi, Kinya 関口欣也. Kōnan Zen'in no Genryū, Kōrai no Hatten 江南禅院の源流、高麗の発展 (The Origins of Chan Temples in Jiangnan and the Development of Goryeo). Tokyo: Chūō Kōron Bijutsu Shuppan, 2012.

Sekiguchi, Kinya. *Gozan to Zen'in* 五山と禅院 (Monasteries of the Five Mountains). Kyoto: Shōgakkan, 1990.

Sensabaugh, David. 'Life at Jade Mountain: Notes on the Life of Man of Letters in Fourteenth Century Wu Society'. In *Chūgoku kaigashi ronshū: Suzuki Kei sensei kanreki kinen* 中國繪畫史論集:鈴木敬先生還曆記念 (Collected Essays on the History of Chinese Painting: Commemorating the 60th Birthday of Professor Suzuki Kei), edited by Suzuki Kei sensei kanreki kinen kai 鈴木敬先生還曆記念会, 43–69. Tokyo: Yoshikawa Kōbunkan, 1981.

Sensabaugh, David. 'The Lion Grove in Space and Time'. In *Bridges to Heaven: Essays on East Asian Art in Honor of Professor Wen C. Fong (Volumn II)*, edited by Jerome Silbergeld, 2:638–639. Princeton: Princeton University Press, 2011.

Shen, Fu. *Six Records of a Floating Life*. Translated by Chiang Su-Hui and Leonad Pratt. London: Penguin Books Limited, 2004.

Shigemori, Mirei 重森三玲. *Nihong Teienshi Zukan* 日本庭園史図鑑 (Japanese Garden History with Drawings). Vol. 1. Tokyo: Yūkōsha, 1936.

Shinohara, Hisao 篠原寿雄, and Satō Tatsugen 佐藤達玄. *Zusetsu Zen No Sube Te Ikiteiru Zen* 圖說禅のすべて-- 生きている禅 (Illustriation on Zen Life: the Living Zen Buddhism). Tokyo: Mokujisha, 1989.

Sirén, Osvald. *Gardens of China*. New York: Ronald, 1949.

Siu, Victoria M. *Gardens of a Chinese Emperor: Imperial Creations of the Qianlong Era, 1736–1796*. Bethlehem: Lehigh University Press, 2013.

Smith, Joanna F. Handlin. 'Gardens in Ch'i Piao-Chia's Social World: Wealth and Values in Late-Ming Kiangnan'. *The Journal of Asian Studies* 51, no. 1 (1992): 55–81.

Spence, Mark David. *Dispossessing the Wilderness: Indian Removal and the Making of the National Parks*. Oxford: Oxford University Press, 2000.

Stackelberg, Katharine T. von. *The Roman Garden: Space, Sense, and Society*. London and New York: Routledge, 2009.

State Hermitage Museum. *Eluosi Guoli Aiermitashi Bowuguan Cang Dunhuang Yishupin Ecang Dunhuang Yishupin* (Dunhuang Artistic Relics in the Collection of the State Hermitage Museum) 俄羅斯國立艾爾米塔什博物館 俄羅斯國立艾爾米塔什博物館藏敦煌藝術品. 5 vols. Shanghai: Shanghai guji chuban she, 1997.

Stechow, Wolfgang. *Dutch Landscape Painting of the Seventeenth Century*. London: Phaidon Press, 1966.

Stein, Rolf Alfred. *The World in Miniature: Container Gardens and Dwellings in Far Eastern Religious Thought*. Translated by Phyllis Brooks. Stanford: Stanford University Press, 1990.

Steinhardt, Nancy. *Chinese Architecture: The Culture and Civilization of China*. New Bhaven: Yale University Press, 2003.

Steinhardt, Nancy. *Liao Architecture*. Honolulu: University of Hawaii Press, 1997.

Suzhou difangzhi bangongshi 蘇州地方志辦公室. *Suzhou Laoqiao Zhi* 蘇州老橋志 (Gazetteer of Old Bridges in Suzhou). Yangzhou: Guangling shushe, 2013.

Suzhou shi Yuanlin Lühua guanliju 蘇州市園林和綠化管理局. *Shizi Lin Zhi* 獅子林志 (Gazetteer of Shizi Lin). Shanghai: Wenhui chubanshe, 2015.

Suzhou Shi Zhuozhengyuan Guanlichu 蘇州市拙政園管理處, ed. *Ninggu de Yuezhang, Suzhou Zhuozheng Yuan Jianzhu* 凝固的樂章 蘇州拙政園建築 (Frozen Symphony, the Architecture of Unsuccessful Politician's Garden in Suzhou). Beijing: China Architecture & Building Press, 2018.

Tian, Ye, and Fang Hai. 'Research on the Historic Appearance of the Lion Grove from the Yuan Dynasty to the Republic of China'. *Studies in the History of Gardens & Designed Landscapes* 37, no. 1 (21 April 2016): 1–14.

Tong, Jun. 'Chinese Gardens, Especially in Kiangsu and Chekiang'. *T'ien Hsia Monthly* III, no. 3 (1936): 220–244.

Tong, Jun. *Jiangnan Yuanlin Zhi* 江南園林志 (Gazetteer of Jiangnan Gardens). Beijing: Zhongguo jianzhu gongye chubanshe, 1984.

Tongji Daxue Jianzhu Gongcheng Xi Jianzhu Yanjiu Shi 同濟大學建築工程系建築研究室. *Suzhou Jiu Zhuzhai Cankao Tu Lu* 蘇州舊住宅參考圖錄 (Suzhou Old Residences Reference Drawing Catalogue). Shanghai: Tongji University, 1958.

Twitchett, Denis. 'The Fan Clan Charitable Estate, 1050–1760'. In *Confucianism in Action*, edited by David S. Nivison and Arthur F. Wright, 97–133. Stanford: Stanford University Press, 1959.

Valleriani, Matteo, and Dagmar Schäfer, eds. 'Knowledge by Design: Architectural and Jade Models During the Qianlong Reign (1735–1796)'. In *The Structures of Practical Knowledge*, edited by Matteo Valleriani, 271–286. Switzerland: Springer International Publishing, 2017.

Walton, Linda A. *Academies and Society in Southern Sung China*. Honolulu: University of Hawai'i Press, 1999.

Wang, Eugene. *Shaping the Lotus Sutra: Buddhist Visual Culture in Medieval China*. Washington: University of Washington Press, 2005.

Wang, Fushan. 'Wenyuan de zaoyuan yishu 文園的造園藝術 (The Garden-Making Art of Wen Garden)'. *Gujian yuanlin jishu* 古建園林技術, no. 3 (1987): 32–36.

Wang, Guixiang. *Zhongguo Hanchuan Fojiao Jianzhu Shi* 中國漢傳佛教建築史: 佛寺的建造,分布與寺院格局,建築類型及其變遷 (History of Buddhist Architecture in China: Buddhist Temple' Construction, Distribution, Layout, Type, and Evolution). Beijing: Tsinghua University Press, 2016.

Wang, Guoping, Wu Enpei, Zhu Xiaotian, and Li Feng, eds. *Suzhou Shigang* 蘇州史綱 (Outlined History of Suzhou). Suzhou: Gu wuxuan chuban she, 2009.

Wang, Jian. '嘉靖初期毀淫祠與廢佛寺政策的地方實踐——以江南,福建為重點 (Destruction of Illegal Temples and Buddhism Temples in Local Society in Early Period of Emperor Jiajing's Rule: Focus on Jiangnan and Fujian Area)'. *Shi Lin* 史林, no. 3 (2016): 94–102.

Wang, Jin. 'Qi Zhuo Si Zu Qing Si Ying, Canglang Ting Zhi Ming Shi Bianqian Kao 豈濁斯足清斯纓——滄浪亭之名實變遷考 (How Could This Turbidity Clarify the Tassel: A Study on the Name and Reality Changes of the Surging Wave Pavilion)'. *Jianzhushi* 建築師, no. 5 (2011): 66–72.

Wang, Jingya. 'Qianlong Huangdi yu Shizilin 乾隆皇帝與獅子林 (Emperor Qianlong and the Lion Grove Garden)'. *Studies in Qing History* 清史論叢, no. 1 (2018): 124–141.

Wang, Jinping. *In the Wake of the Mongols: The Making of a New Social Order in North China, 1200–1600*. Cambridge and London: The Harvard University Asia Center, 2018.

Weber, Susan. *William Kent: Designing Georgian Britain*. New Haven: Yale University Press, 2013.

Wei, Jiazan. *Suzhou gudian yuanlinshi* 蘇州古典園林史 (History of Classical Gardens of Suzhou). Shanghai: Joint Publishing, 2005.

Weiss, Allen S. *Zen Landscapes: Perspectives on Japanese Gardens and Ceramics*. London: Reaktion Books, 2013.

Welch, Holmes. *Buddhism Under Mao*. Cambridge: Harvard University Press, 1972.

White, Hayden. *Metahistory: The Historical Imagination in Nineteenth-Century Europe*. Baltilnore and London: The John Hopkins University Press, 1973.

White, Hayden. 'The Value of Narrativity in the Representation of Reality'. *Critical Inquiry* 7, no. 1 (October 1980): 5–27.

Williamson, Tom. *Environment, Society and Landscape in Early Medieval England: Time and Topography*. Woodbridge: The Boydell Press, 2012.

Williamson, Tom. 'Estate Landscapes in England: Interpretive Archaeologies'. In *Interpreting the Early Modern World: Transatlantic Perspectives*, edited by Mary C. Beaudry and James Symonds, 25–42. Boston: Springer US, 2011.

Williamson, Tom. *Polite Landscapes: Gardens and Society in Eighteenth-Century England*. Baltimore: Johns Hopkins University Press, 1995.

Williamson, Tom. 'Understanding Enclosure'. *Landscapes* 1, no. 1 (April 2000): 56–79.

Williamson, Tom, and Liz Bellamy. *Property and Landscape: A Social History of Land Ownership and the English Countryside*. London: George Philip, 1987.

Wright, Mary Clabaugh. *The Last Stand of Chinese Conservatism: The T'ung-Chih Restoration, 1862–1874*. Stanford: Stanford University Press, 1957.

Wu, Hung. *The Double Screen: Medium and Representation in Chinese Painting*. London: Reaktion Books, 1996.

Wu, Hung. 'Reborn in Paradise: A Case Study of Dunhuang Sutra Painting and Its Religious, Ritual and Artistic Context'. *Orientations* 23 (1992): 52–60.

Wu, Hung. *Spatial Dunhuang: Experiencing the Mogao Caves*. Seattle: University of Washington Press, 2023.

Wu, Hung. *The Wu Liang Shrine: The Ideology of Early Chinese Pictorial Art*. Stanford: Stanford University Press, 1989.

Xie, Chongguang, and Bai Wengu. *Zhongguo Sengguan Zhidu Shi* 中國僧官制度史 (History of Policies of Buddhist Officials in China). Qinghai: Qinghai renmin chuban she, 1990.

Xie, Hongquan. *Tiantai zong fojiao jianzhu yanjiu* 天台宗佛教建築研究 Research on Tiantai Buddhism Architecture. Beijing: Jianzhu gongye chubanshe, 2018.

Xu, Bangda. *Gushu hua wei e Kaobian Vol. 3* 古書畫偽訛考辨 叁, 徐邦達集十二 (Study on the Authentication of Ancient Books and Paintings Vol. 3, Collected Works of Xu Bangda, No. 12). Edited by Gugong bowuyuan 故宮博物院. Beijing: Gugong chubanshe, 2015.

Xu, Yinong. *The Chinese City in Space and Time: The Development of Urban Form in Suzhou*. Honolulu: University of Hawaii Press, 2000.

Xu, Yinong. 'Interplay of Image and Fact: The Pavilion of Surging Waves, Suzhou'. *Studies in the History of Gardens & Designed Landscapes* 19, no. 3–4 (September 1999): 288–301.

Yang, Hongxun. *Jiangnan Yuanlin lun: Zhongguo gudian zaoyuan yishu yanjiu* 江南園林論: 中國古典造園藝術研究 (On Gardens in Jiangnan: A Study of the Art of Crafting Chinese Classical Gardens). Shanghai: Renmin chubanshe, 1994.

Yang, Zhaohui, Shi Xun, and Su Qun. 'Knowledge-Based Raster Mapping Approach to Wetland Assessment: A Case Study in Suzhou, China'. *Wetlands* 36, no. 1 (February 2016): 143–158.

Ye, Ruei-chen. 'The Construction of the Confucian Garden: The Reconstruction of Chang-Lang Ting and Interpretation of Its Spatial Meaning in the Qing Dynasty 儒家園林的建構 清代滄浪亭的重修與空間意義詮釋'. Master Thesis, National Taiwan University, 2014.

Ye, Xianyun. 'Yuandai zhuming gaoseng Zhongfeng Mingben yu Suzhou de fomen yuanyuan 元代著名高僧中峰明本與蘇州的佛門淵緣 (Famous Yuan Dynasty Monk Zhongfeng Mingben and His Deep Connection with Buddhism in Suzhou)'. In *Proceedings of the Sixth Hanshan Temple Cultural Forum* 第六屆寒山寺文化論壇論文集, edited by Qiu Shuang and Yao Yanxiang, 251–257. Shanghai: Joint Publishing, 2012.

Yu, Jimmy. *Reimagining Chan Buddhism: Sheng Yen and the Creation of the Dharma Drum Lineage of Chan*. London and New York: Routledge, 2022.

Yue, Isaac. '蛇妖與英雄: 淺論宋代俗文學與民間傳説中的韓世忠形象 (The Serpent and the Hero: Folk Literature and the Public Imagination of Han Shizhong during the Song Dynasty)'. *Journal of Oriental Studies* 47, no. 1 (2014): 17–28.

Zeitlyn, David. 'Anthropology in and of the Archives: Possible Futures and Contingent Pasts. Archives as Anthropological Surrogates'. *Annual Review of Anthropology* 41 (2012): 461–480.

Zhang, Guang Wei. 'The Documentation of Historic Maps of World Heritage Site City Suzhou'. *ISPRS-International Archives of the Photogrammetry* (19 July 2013): 295–307.

Zhang, Hui, and Bai Qianshen. 'Qing Chu Guiqi Shoucang Jia Wang Yongning (Shang) 清初貴戚收藏家王永寧 (上) (Early Qing Noble Collector Wang Yongning (Part One))'. 新美術 *New Art*, no. 06 (2009): 36–41.

Zhang, Hui, and Qianshen Bai. 'Qing Chu Guiqi Shoucang Jia Wang Yongning (Xia) 清初貴戚收藏家王永寧 (下) (Early Qing Noble Collector Wang Yongning (Part One))'. 新美術 *New Art*, no. 02 (2010): 11–20.

Zhang, Jiaji. *Zhongguo* zaoyuanlun 中國造園論 (Treatise on Crafting Chinese Garden). Taiyuan: Shanxi renming chubanshe, 1991.

Zhang, Jie. 'Shizi lin: Yuandai linji shanlinchan de jueyi 獅子林: 元江南臨濟山林禪的孑遺 The Lion Grove Garden: The Only Yuan Dynasty Garden of Rinzai Zen in Regions South of the Yangtze River'. *Zhongguo yuanlin* 中國園林, no. 7 (2016): 97–100.

Zhang, Shiqing. *Wushan Shicha Tu Yu Nansong Jiangnan Chansi* 五山十刹圖與南宋江南禪寺 (The Painting of Five Mountains and Ten Monasteries and Chan Temples in Northern Song Dynasty). Nanjing: Southeast University Press, 2000.

Zhang, Shiqing. *Zhongguo jiangnan chanzong siyuan jianzhu* 中國江南禪宗寺院建築 (Buddhist Temples of Jiangnan in China). Wuhan: Hubei jiaoyu chubanshe, 2002.

Zhao, Yanzhe. *Rugu Hanjin: Qing Qianlong Fanggu Huihua Yanjiu* 茹古涵今: 清乾隆朝仿古繪畫研究 (Absorbing the Ancient, Embracing the Present: Study of Antique-Imitating Paintings during the Qing Dynasty's Qianlong Era). Nanning: Guangxi meishu chubanshe, 2017.

Zhao, Yanzhe. '*Yixun Qingbi, Ni Zan Kuan Shizilin Tu Jiqi Qinggong Fanghua Yanjiu* 藝循清閟—倪瓚(款)《獅子林圖》及其清宮仿畫研究 (Artistic Conformity to the Secluded Path: The Lion Grove Garden Painting attributed to Ni Zan and Its Qing Dynasty Palace Imitation Studies)'. Zhongguo Shuhua 中國書畫 (Chinese Calligraphy and Painting), no. 2 (2019): 37–70.

Zhen, Zhaojing. *Taihu shuili jishu shi* 太湖水利技術史 (History of Hydraulic Technologies in Lake Tai Region). Beijing: Nongye chuban she, 1987.

Zhou, Hongjun. 'Shakkei no Tenkai to Kōsei: Nihon, Chūgoku Zōen ni okeru Hikaku Kenkyū 借景の展開と構成:日本.中国造園における比較研究 (The Development and Composition of Borrowed Scenery: A Comparative Study in Japanese and Chinese Gardening)'. PhD Thesis, Tokyo University, 2012.

Zou, Hui. 'The "True Wonder" in Emperor Qianlong's Garden Labyrinths'. In *Entangled Landscapes, Early Modern China and Europe*, edited by Zhang Yue and Andrea M. Riemenschnitter, 187–209. Singapore: NUS Press, 2017.
Zürcher, Eric. *The Buddhist Conquest of China: The Spread and Adaptation of Buddhism in Early Medieval China*. Leiden: Brill, 2007.

Glossary

People and Title Names

An Ning (fl. 18th c.) 安寧
Azala (fl. 14th c.) 阿扎剌
Bai Juyi (772–846) 白居易
Baojue Jian (fl. 14th c.) 寶覺簡
Baotan Shiying (fl. late 14th c.) 寶曇示應
Beichan Jingfan (d. 1128) 北禪淨梵
Beiguo shiyou 北郭十友 Ten Friends of the North Wall
Biechuan Jiao (fl. 14th c.) 別傳教
Bodhidharma (d. 536) 菩提達摩
Cao Ju (fl. 14th c.) 曹聚
Cao Kai (fl. 17th c.) 曹凱
Cao Ruli (fl. 1281) 曹如理
Chen Hao (1032–1085) 程顥 One of the two Chen Brothers
Chen Ji (1314–1370) 陳基
Chen Ji (1370–1434) 陳継
Chen Lü (1288–1343) 陳旅
Chen Yi (1033–1107) 程頤 One of the two Chen Brothers
Chen Zhilin (1605–1666) 陳之遴
Chunyu Fen (8th c.) 淳于棼
Damei Fachang (752 –839) 大梅法常
Dōgen Kigen (1200–1253) 道元希玄
Dotung (fl. 14th c.) 道童
Du Qiong (1396–1474) 杜瓊
Duanya Liaoyi (d.1293) 斷崖了義
Dushui yingtian shi 都水營田使 Official of Managing Water and Constructing Fields
En Shou (d. 1911) 恩壽
Falei Zuzhen (fl. early 14th c.) 法雷祖震
Fan Zhongyan (989–1052) 范仲淹
Fang Shishu (1693–1751) 方士庶

Fazang (643–712) 法藏
Feng Guifen (1809–1874) 馮桂芬
Fori Qisong (1007–1073) 佛日契嵩
Gao Jian (1634–1707) 高簡
Gao Jin (1706–1779) 高晉
Gao Qi (1336–1373) 高啟
Gaofeng Yuanmiao (1238 –1296) 高峰原妙
Geng Ju (fl. 17th c.) 耿橘
Gu Wenbin (1811–1889) 顧文彬
Guangyun Fazhi (fl. 17th c.) 廣運法智
Guanxiu (832–912) 貫休
Guangxuan Wuyan (fl. 14th c.) 廣宣無言
Gui Youguang (1506–1571) 歸有光
Guting Shanxue (1307–1370) 古庭善學
Han Feng (1783–1860) 韓蒶
Han Shizhong (1089–1151) 韓世忠
He Zhen (fl. late 14th c.) 何禎
Hejing 和靖
Hu Yuan (993–1059) 胡瑗
Hu Zuanzong (1480–1560) 胡纘宗
Huaishou (fl. early 14th c.) 懷壽
Huaiyu xiansheng 槐雨先生
Huang Jin (1277–1357) 黃溍
Huang Xingzeng (1496–1546) 黃省曾
Huang Xuan (fl. 18th c.) 黃軒
Huike (478–593) 慧可
Huishen (fl. 12th c.) 惠深
Huqiu Shaolong (1077–1136) 虎丘紹隆
Jiang Qi (ca. 18th c.) 蔣棨
Jiang Yun (b. 1816) 江鋆
Jihang (fl. late 19th c.) 濟航
Jueyi Jieli (fl. early 15th c.) 覺義戒栗
Kong Qi (d. 1367) 孔齊
Li Bu 禮部 Bureau of Ritual
Li Xiaoguang (1285–1350) 李孝光
Li Xiucheng (1823–1864) 李秀成
Li Yingzhen (1431–1394) 李應禎
Liang Zhangju (1775–1849) 梁章鉅
Lin Zexu (1785–1850) 林則徐
Linwu Qinggong (1272–1352) 石屋清珙
Liu Yu (fl. 18th c.) 柳遇
Lu Zhi (754–805) 陸贄
Maiju (fl. 14th c.) 買住, alt. 拜住

Master Jing (fl. 14th c.) 淨法師
Mazu Daoyi (709–788) 馬祖道一
Meng'an Nan (fl. 1420s–1430s) 夢庵南
Miaosheng Jiugao (1309–1384) 妙聲九皋
monk Master Jing 淨法師
Mus ō Soseki (1275 –1351) 夢窗疎石
Nanxuan Xun (fl. 13th c.) 南軒熏
Ni Zan (1306–1374), alt. Yunlin 倪瓚 alt. 雲林
Ozawa Kei 小澤圭
Pan Yijun (1740–1830) 潘奕雋
Pan Yuanshao (fl. 14th c.) 潘元紹
Panquan Youlan (fl. 1290s) 判荃友蘭
Peng Nian (1505–1566) 彭年
Qi Wuqian (b. 692) 綦毋潜
Qi Zongyou (1334–1407) 啟宗祐
Qian Xiyan (1562–1638) 錢希言
Qianlong (1711–1799) 乾隆
Qing'an Mingxing (fl. 16th c.) 清庵明性
Qingliang Chengguan (738–839) 清涼澄觀
Renzong (1285–1320) 元仁宗
Rinchen Gyaltsen (1256–1305) 輦真監藏
Ruhai Deyin (fl. 14th c.) 如海德因
Rushan (fl. 14th c.) 如山
Sengge Sagi (1283–1331) 祥哥剌吉
Shajian (fl. early 14th c.) 沙監
Shanqing (ca. 14th c.) 善慶
shen 绅 the gentry
Shen Fu (1763–1832) 沈復
shi 士 scholar, official
Shi Yunyu (1755–1837) 石韞玉
Shishi Zuying (1291–1342) 石室祖瑛
Shu Ting (fl. 16th c.) 舒汀
Song Lian (1310–1381) 宋濂
Song Luo (1634–1714) 宋犖
Su Shunqin (1008–1048) 蘇舜欽
Taibuhua (fl. 14th c.) 泰不花
Tang Yin (1470–1524) 唐寅
Tani Sōyō (1526–1563) 谷宗養
Tao Hongjing (456–536) 陶弘景
Tao Shu (1779–1839) 陶澍
Tao Yuanming (365–427) 陶淵明
Tianquan Yuze (fl. 14th c.) 天泉餘澤
Tianru Weize (1286–1354) 天如惟則

Tianyin Yuanzhi (d. 1635) 天隱圓至
Tiyeboo (1752–1824) 鐵保
Tu Zhen (fl. late 14th c.) 塗禎
Tuglug (fl. early 14th. c.) 禿魯
Wang Ao (1450–1524) 王鏊
Wang Ruyu (d. 1415) 王汝玉
Wang Xianchen (fl. 1490s–1540s) 王獻臣
Wang Xing (1331–1395) 王行
Wang Xinyi (1572–1645) 王心一
Wang Yi (d.1374) 王彝
Wang Ying (1651–1726) 王煐
Wang Yongning (d.1672) 王永寧
Wang Zhen (1271–1368) 王禎
Wei Liaoweng (1178–1237) 魏了翁
Wei Yingwu (737–792) 韋應物
Wen Boren (1502–1575) 文伯仁
Wen Zhengming (1470–1559) 文徵明
Wen Zhenmeng (1574–1636) 文震孟
wenren 文人 literati
Wenying (fl. 16th c.) 文瑛
Wo'an Benwu (1286–1343) 我庵本無
Wu Cunli (fl. 18th c.) 吳存禮
Wu Jin (fl. 18th c.) 吳晉
Wu Jun (fl. 19th c.) 吳儁
Wu Kuan (1435–1504) 吳寬
Wu Sangui (1612–1678) 吳三桂
Wu Yun (1811–1883) 吳雲
Wu Zixu (559–484 B.C.) 伍子胥
xianshou xuekui 賢首學魁 leading scholar of Avatamsaka teaching
Xiao Wen (fl. 16th c.) 蕭文
Xie Hui (fl. 14th c.) 謝徽
Xingzhong Ren (fl. 14th c.) 行中仁
Xiuxue (fl. early 15th c.) 修學
Xu Ben (1335–1380) 徐賁
Xu Fang (1622–1694) 徐枋
Xu Jie (1503–1583) 徐階
Xu lüxiang (ca. 16th c.) 徐履祥
Xu Shaoquan, alt. Xu Jia (fl. 16th c.) 徐少泉, alt. 徐佳
Xu Shupi (1596–1683) 徐樹丕
Xu Taishi (1540–1598) 徐泰時
Xuannan Shishe 宣南詩社 Xuannan Poetry Club
Xuechuang Guang (fl. early 14th c.) 雪窗光
Yan Ciping (fl. 1163–1189) 閻次平

Yan Fu (1236–1312) 閻復
Yang Xunji (1458–1546) 楊循吉
Yang Zhu (fl. late 14th c.) 楊鑄
Yao Guangxiao (1335–1418) 姚廣孝
Ye Shikuan (1689–1755) 葉士寬
Yin Jishan (1689–1755) 尹繼善
Yin Tun (1071–1142) 尹焞
Yingzhong Huan (fl. late 14th c.) 瑩中瓛
Yongzhen Fuliang (1317–1371) 用貞輔良
Yu Ji (1272–1348) 虞集
Yuan Hongdao (1568–1610) 袁宏道
Yuan Zhi (1502–1547) 袁袠
Yujia seng 瑜伽僧 yoga monks
Yun Shouping (1633–1690) 惲壽平
Zhang Dun (1035–1106) 章惇
Zhang Guowei (1595–1646) 張國維
Zhang L ü qian (1838–1915) 張履謙
Zhang Shicheng (1321–1367) 張士誠
Zhang Shijun (fl. 18th c.) 張士俊
Zhang Shusheng (1824–1884) 張樹生
Zhang Zhu (1287–1368) 張翥
Zhao Mengfu (1254–1322) 趙孟頫
Zhao Zhixin (1662–1744) 趙執信
Zhaozhou (778–897) 趙州
Zheng Yuanyou (1292–1364) 鄭元祐
Zhenguo (fl. 16th c.) 貞果
Zhengxing (fl. early 14th c.) 正性
Zhongfeng Hai (fl. early 14th. c.) 中峯海
Zhongfeng Mingben (1263–1323) 中峯明本
Zhou Chen (1460–1535) 周臣
Zhou Ji (fl. late 14th c.) 周稷
Zhou Lun (1463–1542) 周倫
Zhu Derun (1294–1365) 朱德潤
Zhu Jian (1769–1850) 朱珔
Zhu Xi (1130–1200) 朱熹
Zhu Xizhou (1473–1557) 朱希周
Zhu Yizun (1629–1709) 朱彝尊
Zhu Yuanzhang (1328–1398) 朱元璋
Zhu Yunming (1460–1526) 祝允明
Zhuofeng (fl. early 14th c.) 卓峰
Ziyou (506–443 B.C.) 子游
Ziyu (b. 512 B.C.) 子羽
Zongjing (fl. 1314–1320) 宗敬

Scenes, Places, and Others

an ji 庵基 the basis of the cloister
Angxiao feng 昂霄峰 Lofty Clouds Peak
Ayu Wang si 阿育王寺 King Ashoka Temple
Ba dong 八洞 Eight Caves
Bajiao jian 芭蕉檻 Plantain Balustrade
Bantang shengshou jiao si 半塘壽聖教寺 Sacred Longevity Teaching
    Temple at Bantang
Bao'en si 報恩寺 Repay Kindness (Teaching) Temple
Baoguo chansi 報國禪寺 Repay the Nation Chan Temple
Baqi fengzhi huiguan 八旗奉直會館 Guild of Eight-Banner Members from
    Fengtian and Metropolitan Provinces
Bashang xiang 壩上巷 Bashang alley
Beichan si 北禪寺 Northern Chan Temple
Beishan ting 北山亭 Northern Mountain Pavilion
ben 本 root
Bieyou dongtian 別有洞天 A Cave of Its Own
Binghu jing 冰壺井 Icy Pottery Well
Bu qi 步碕 Walking Along the Winding Bank
Bu yuan 補園 Additional Garden
cangjing ge 藏經閣 sutra storage pavilion
Canglang guanyu 滄浪觀魚 Watching Fish in Surging Wave
Canglang qingxia 滄浪清夏 Brisk Summer in Surging Wave
Canglang shengji 滄浪勝蹟 Extraordinary Traces of Surging Wave
Canglang ting 滄浪亭 Surging Wave Pavilion
Cao an tu juan 草菴圖卷 *Painting of Thatched Cloister*
ceng ci 層次 layers of scenes
Chan wo 禪窩 Chan Hut
Chan, or Zen in Japanese 禪
Chang men 閶門 Chang gate
Changchun yuan 長春園 Everlasting Spring Garden
Changzhou xianxue 長洲縣學 Changzhou County School
Chengtian nengren si 承天能仁寺 Heavenly Bestowed Benevolence Temple
Chongyun ta 重雲塔 Double Clouds Pagoda
Chunyan yuan 春宴園 Spring Banquet Garden
Cuanyu feng 攢玉峯 Heap Jade Peak
Daci si 大慈寺 Great CompassionTemple
Dahong si 大弘寺 Great Propagation Temple
Daishuang ting 待霜亭 Waiting for Frost Pavilion
Dajue si 大覺寺 Great Awakening Temple
Damo mianbi 達摩面壁 Bodhidharma Meditating Facing a Cliff
Daoshan ting 道山亭 Way Mountain Pavilion
Daoying lou 倒影樓 Reflection Tower

Datong ge 大通閣 Great Transmission Pavilion

Dayun an (alt. Jiecao an) 大雲菴 (alt. 結草菴) Grand Clouds Cloister, alt. Thatched Cloister

Dezhen ting 得真亭 Obtaining Essence Pavilion

Di fei 地肺 Earth Lung

Diao gong 釣碧 Angling Rock

Diezuo shi 跌坐石 Fallen Sitting Rock

Dinghui chan si 定慧禪寺 Determined Wisdom Chan Temple

Dong Yuan 東園 Eastern Garden

Dong yujian 東玉澗 Eastern Jade Ravine

Dong zhai 東斋 Eastern Chamber

Dong Zhuang 東莊 Eastern Estate

Dong Zhuang tuce 東莊圖冊 *Painting Album of Eastern Estate*

Dongnan qulu 東南趨魯 The Southeast Hasten Towards the State of Lu

dui jing 對景 facing a scene, oppose the view, facing the view, facing the scene

Eihei-ji 永平寺 Eihei-ji Temple

Er'er xuan 爾耳軒 So-so Gallery

fang 坊 residential wards

Fangsheng chi 放生池 Release Life Pond

fangzhang shi 方丈室 abbot's quarter

Fanjing tai 翻经臺 Reading Sutra Platform

Fanxiang wu 繁香塢 Many Fragrances Valley

Feihong jian 飛虹澗 Flying Rainbow Ravine. The name of a ravine in Mount Heaven Eye.

Feihong qiao 飛虹橋 Flying Rainbow Bridge

Feng men 葑門 Feng Gate

Feng xi 葑溪 Feng River

fu xue 府學 prefectural school

Fu yuan 復園 Recovery Garden

Fucui ge 浮翠閣 Floating Emerald Pavilion

Fugu 復古 Restoration of the Antiquity

funing si 福寧寺 Blessed Peace Temple

Furong wei 芙蓉隈 Hibiscus Bend

Fuyuan si 福源寺 Benevolence Origin Temple

Gaochang 高昌 Kucha, alt. Khocho

gong'an 公案

Gu Shilin si 古師林寺 The Ancient Shilin Temple

Guangfu si 光福寺 Light Benevolence Temple

Guanyu chu 觀魚處 Place for Watching Fish

Gui tianyuan ju 歸田園居 Returning to Farm Dwelling

Guiyuan chan si, alt. Guiyuan xingguo chan si 歸源禪寺 Return to the Origin Chan Temple

Gusu chengtu 姑蘇城圖 *Map of Gusu City*

Gusu mingyuan Shizi Lin 姑蘇名園獅子林 *The Famous Lion Grove Garden in Suzhou*

Gusu shi jing tu 姑蘇十景圖 *Ten Scenes of Gusu*

Haitang chunwu 海棠春塢 Flowering Crab Apple Spring Dock

Haiying tang 海印堂 Ocean Seal Hall

Hanhui feng 含暉峰 Teeming with Rays Peak

Hanqing Chi 涵青池 Containing Blueness Pond

Hao Pu 濠濮 Hao and Pu rivers

Haoshang guan 濠上觀 Gazing from the Bank of River Hao

Hefeng simian ting 荷風四面亭 Lotus Breeze at Four Sides Pavilion

Hejing shuyuan 和靖書院 Hejing Academy

Hong qiao 紅橋 Red Bridge

Hu qiu 虎丘 Tiger Hill

Huachan si 畫禪寺 Painting Chan Temple

Huai River 淮河 Huai River

Huai wo 槐幄 Sophora Tent

Huaiyu ting 槐雨亭 Sophora Rain Pavilion

Huaiyu xiansheng yuanting tuzhou 槐雨先生園亭圖軸 *hanging scroll of Garden and Pavilions for Mr. Sophora Rain*

Huanxiu shanzhuang 環秀山莊 Mountain Villa of Embracing Beauty

Huanzhu an 幻住庵 Illusory Abiding Cloister

Huomai an 活埋庵 Entombment Cloister

Huqiu chan si 虎丘禪寺 Tiger Hill Chan Temple (alt. Clouds Cliff Chan Temple)

Jianfu Temple 薦福寺 Jianfu Temple

Jianghan plain 江漢平原

Jiangzhe jiyou tu 江浙紀遊圖 Painting of Journeys in Jiangzhe Region

Jianshan lou 見山樓 Seeing Mountain Tower

Jianyan 建炎

jiao hua 教化 transforming through teaching

Jiaoyu ting 教諭廳 Education Office Hall

Jiashi ting 嘉实亭 Fine Fruit Pavilion

Jiayuan fang 解元坊 Second Scholar Arch

jie hua 界畫 ruled-line painting

jie jing 借景 borrowing a scene, borrow the view

Jiechuang Lü Yuan 戒幢律院

Jifu an 集福庵 Assemble Benevolence Cloister

Jin shanlin 近山林 Near Mountain and Grove

Jinfan jing 錦帆涇 Jinfan canal

jing de 敬德 Respect Ritual

jing shen 景深 depth of scenes, spatial depth

Jingde si 景德寺 Bright Virtue Temple

Jingxing weixian 景行維賢 Exalted Go Only the Wise and Good
Jingyi ting 敬一亭 Revere the One Pavilion
Jingzhong you 鏡中遊 Roving in Mirror
Jinwa chi 禁蛙池 Forbidding Frog Pond
Jinxiang shuyuan 金鄉書院 Golden Country Academy
Jiyun si 集雲寺 Gathering Clouds Temple
Juntian zhi 均田製 Equal Field System
kai shan 開山 open the mountain, meaning inaugurating a new temple
Ke Yuan 可園 Garden of Suitability
Kenninji 建仁寺 Kennin-ji Temple
kuang jing 框景 framing a scene, framing the view
Kuayu qiao 跨玉橋 Crossing Jade Bridge
Kunshan 昆山 Kunshan
Laiqin you 來禽囿 Arriving Birds Park
Laixiu qiao 來秀橋 Coming Elegance Bridge
Lake Tai 太湖 Lake Tai
Lanxue tang 蘭雪堂 Orchid Snow Hall
Li men 禮門 Ritual Gate
li she 吏舍 officials' residence
Li ting 笠亭 Hat Pavilion
Lianbi Feng 聯壁峯 Allying Walls Peak
Lianhua feng 蓮花峰 Lotus Peak
Lianhua tai 蓮花臺 Lotus Platform
Lin'an 臨安 (the name of Hangzhou in the Song dynasty)
Linglong guan 玲瓏館 Delicate Pavilion
Lingxing men 欞星門 Lattice Star Gate
Linji 臨濟 Linji
Liu ao 柳隩 Willow Bend
Liu yuan 留園 Lingering Garden
Liuting ge 留聽閣 Lingering Sounds Pavilion
Lixue tang 立雪堂 Stand in Snow Hall
Liyu feng 立玉峯 Erected Jade Peak
Liyu ting 立玉亭 Erected Jade Pavilion
Longmen qiao 龍門橋 Dragon Gate Bridge
Longxing si 龍興寺 Dragon Revival Temple
Lou River 婁江 Lou River
luozuo 落座 landing place
Meigui chai 玫瑰柴 Rose Fence
Meilin 梅林 Plum Grove
Mengyin lou 夢隱樓 Dream of Recluse Tower
Mianshui xuan 面水軒 Facing Water Gazebo
Miaoyin an 妙隱庵 Witty Retreat Cloister
Miaozhi an 妙智庵 Witty Wisdom Cloister
Ming Xuan 明軒

Mingdao Tang 明道堂 Illuminate the Way Hall
Minghuan ci 名宦祠 Shrine of Famous Officials
Minglun tang 明倫堂 Illuminate Ethics Hall
Mountain Estate to Escape the Heat in Chengde 承德避暑山莊 Chengde
    bishu shanzhuang
Mountain Potalaka 普陀山 Mount Potalaka
Nan xuan 南軒 Southern Veranda
nanmu 楠木 Phoebe Zhennan
Nanxun Shengdian 南巡盛典 *Grand Ceremony of the Southern Inspections*
Nengren an 能仁庵 Capable Kindness Cloister
Ōgon-chi 黃金池 Golden Pond
Fangyan ting 放眼亭 Open Vista Pavilion
pan chi 泮池 Pan pond
Pingjiang fu 平江府 Pingjiang prefecture (The name of Suzhou in the Song
    dynasty)
Pingjiang lu 平江路 (The name of Suzhou in the Yuan dynasty)
Pingjiang tu 平江圖 *Pingjiang Map*
Pipa yuan 枇杷園 Loquat Courtyard
pu 浦 longitudinal canal, vertical canal, refers to the north-south canals
Puti lanruo 菩提蘭若 Bodhi Villa
Qi Wang Miao 蘄王廟 Shrine of the King of Qi
Qiangwei jing 薔薇徑 Rosebush Footpath
Qifeng ting 栖鳳亭 Reside Phoenix Pavilion
Qingbi ge 清閟閣 Pure Secluded Pavilion
Qingfu an 慶福菴 Celebrate Benevolence Cloister
Qingshou si 慶壽寺 Celebrate Longevity Temple
Qisheng ci 啟聖祠 Shrine of Giving Birth to the Sages
Qiyou tu 七友圖 *Painting of Seven Friends*
Qiyun feng 棲雲峰 Residing Clouds Peak
Qiyun shi 棲雲室 Residing Clouds Chamber
Ruiguang chan si 瑞光禪寺 Auspicious Light Chan Temple
Ruosu tang 若墅堂 Hall Like a Villa
Ruxue men 儒學門 Confucian Gate
Saiho-ji Temple 西芳寺 Saiho-ji Temple
Sanguan Ting 三關亭 Three Gates Pavilion
Sanshiliu yuanyang guan 三十六鴛鴦館 Thirty-Six Pairs of Mandarin
    Ducks Hall
Seiun-ji 棲雲寺 Residing Clouds Temple
Sengfang 僧房, alt. 僧堂 monks' residence, alt. Sangha hall, monks' quarter
sha tian 沙田 sandy field
shan men 山門 mountain gate, alt. the entrance gate of a temple
Shang Tianzhu Temple 上天竺寺 Shang Tianzhu Temple
Shanhui an 善慧庵 Kind Wisdom Cloister
Shantang district 山塘區 Shantang District

she pu 射圃 archery patch
shen ping 神品 the deified work
Sheng'en si 聖恩寺 Imperial Benevolence Temple
shenglong qiao 昇龍橋 Ascend Dragon Bridge
Shenjing ting 深淨亭 Deep Tranquil Pavilion
Shiba Mantuoluo hua guan 十八曼陀羅花館 Hall of Eighteen Mandala Flowers
Shilin quanjing tu 獅林全景圖 Complete View of the Lion Grove
Shiliu guantang 十六觀堂 Sixteen Contemplation Hall
Shixuan men fu 十玄門賦 Odes on Ten Mysterious Gates
Shizi feng 師子峰 Lion Peak
Shizi kou 獅子口 Lion Mouth
Shizi Lin 獅子林 Lion Grove Garden
Shizi Lin si 獅子林寺, alt. 獅子正宗禪寺 Lion Grove Temple, alt. Lion Grove Authentice Chan Temple (the name of the temple in Mount Heaven Eye)
Shizi lin tu 獅子林圖 *Painting of Shizi Lin*
Shizi Lin xiang 獅林寺巷 Shizi Lin alley
Shizi yan 獅子巖 Lion Cliff
shu yuan 書院 academy
Shu yuan 書園 Book Garden
Shuangqing zhuang 雙清莊 Double Purity Cloister
Shuihua chi 水華池 Water Flower Pond
Shunxin an 順心庵 Follow the Mind Cloister
shupu yuan 蔬圃園 vegetable patch garden
Shushi ting 漱石亭 Rinsing Water Pavilion
Si guan 死關 Death Pass
Song Mountain 嵩山
Songfeng ge 鬆風閣 Pine Breeze Tower
Songfeng ting 鬆風亭 Pine Breeze Pavilion
Southern Chan Temple 南禪禪寺 Southern Chan Temple
Su Shunqin ci 蘇舜欽祠 Shrine of Su Shunqin
Su Song Chang dao xinshu 蘇松常道新署 Bureau of the Circuit of Suzhou, Songjiang, and Changzhou
Suzhou 蘇州
Suzhou fu quanjing shuili tushuo 蘇州府全境水利圖說 *Pictorial Explanation of Water Hydraulics of the Entire Region of Suzhou Prefecture*
Suzhou fucheng nei fenzhi shuidao tu 蘇州府城內分治水道圖 *Map of the Waterways Management within the Suzhou Prefecture City*
Taihu shi 太湖石 Lake Tai rocks
tang 塘 latitudinal canal, horizontal canal, refers to the east-west canals
tang pu wei tian 塘浦圍田 canal and polder planning, canal-and-polder scheme
tang yang 燙樣 ironed paper model

Taohua An ge 桃花菴歌 The Song of Peach Blossom Cloister
Taohua pan 桃花泮 Peach Blossom Bank
Taohua yuan 桃花園 Peach Blossom Garden
Taoli yuan 桃李園 Garden of Peach and Plum
Tengjiao 騰蛟 Soaring Dragon
Thirty-One Scenes of Unsuccessful Politician's Garden 拙政園三十一景圖冊
Tianhui Xianshou jiao si 天惠賢首教寺 Heavenly Benevolence Avatamsaka
    Teaching Temple
Tianmu 天目 Heaven Eye (the name of the pond at the summit of Mount
    Heaven Eye)
Tianmu shan 天目山 Mount Heaven Eye
Tianquan ting 天泉亭 Heaven Spring Pavilion
Tiantai zong 天台宗 Tiantai Sect, Tiantai Buddhism
Tiantong si 天童寺 Tiantong Temple
Tianwang dian 天王殿 hall of the Heavenly King
Tingsongfeng chu 聽松風處 Listening to Windblown Pines Place
Tingyu xuan 聽雨軒 Listening to Rain Pavilion
Tuyue feng 吐月峯 Holding Moon Peak
wandai zongshi 萬代宗師 Ten Thousand Generations' Teacher
Wang Ao ci 王鏊祠 Shrine of Wang Ao
Wangguan qiao 望關橋 Watch Gate Bridge
Wangshi yuan 網師園 Master of Fish-nets Garden
Wanshou chan si 萬壽禪寺 Ten Thousand Longevity Chan Temple
wei tian 圍田 enclosed field
Wen yuan 文園 Garden of Literature
Wenmei ge 問梅閣 Ask Plum Pavilion
Wenmiao men 文廟門 Gate of Confucius Temple
wenren yuan 文人園 literati garden
Wenxue tang 文學堂 Literature Hall
Wolong 臥龍 Lying Dragon
Woyun shi 臥雲室 Lying Clouds Chamber
Wu men 吳門 Wu Gate
Wu yuan 吳園 Wu Garden
Wubai mingxian ci 五百名賢祠 Shrine of Five Hundred Worthies
Wulao tu 五老圖 *Painting of Five Old Men*
Wuliangshou yuan 無量壽院 Infinite Longevity Temple
Wusong River 吳淞江 Wusong River
Wusong yuan 五松園 Garden of Five Pines
Wuzhu youju 梧竹幽居 Secluded Residence of Parasol and Bamboo
Xi Huayan si 西華嚴寺 Western Avatamsaka Temple
Xi Tianmu shan shengjing quantu 西天目山勝境全圖 *Complete Map of
    the Excellent Realm of Western Mount Heaven Eye*
Xi yuan 西園 Western Garden
Xi yujian 西玉澗 Western Jade Ravine

Xian zhai 閒齋 Idling Chamber
Xiange lou 弦歌樓 Tower of Strumming and Singing
Xianglu feng 香爐峯 Censor Peak
Xiangxian ci 鄉賢祠 Shrine of Country Worthies
Xiangyun wu 湘筼塢 Xiang Bamboo Valley
Xiangzhou 香洲 Fragrance Islet
Xianshi dian 先師殿 Primordial Teacher's Hall
Xiao canglang 小滄浪 Lesser Surging Wave
Xiao canglang shuige 小滄浪水閣 Lesser Surging Wave Water Pavilion
Xiao canglang shuiyuan 小滄浪水院 water courtyard of Lesser Surging Wave
Xiao feihong 小飛虹 Little Flying Rainbow
Xiao qiyuan 小淇園 Little Qi Garden
Xiao taoyuan 小桃源 Little Peach Origin
Xima chi 洗馬池 Washing Horse Pond
Xing tan 杏壇 Apricot Altar
Xinghua cun 杏花村 Apricot Flower Cottage
xiu ye 修業 Cultivate Learning
Xiuqi ting 繡綺亭 Embroidered Damask Pavilion
Xiyuan si 西園寺 Western Garden Temple
Xu jiang 胥江 Xu river
Xuanzheng yuan 宣政院 Commission for Buddhist and Tibetan Affairs
Xuedao shuyuan 學道書院 Learn the Way Academy
Xuegu tang 學古堂 Learn the Classics Hall
Xuekong tang 學孔堂 Learn Confucianism Hall
Xundao xie 訓導廨 Instructing and Guiding Office
Yangshan 陽山 Sun Mountain
Yangtze river 長江 Yangtze river
Yanqing Temple 延慶寺 Extended Celebration Temple
Yantie tang 鹽鐵塘 Salt Iron canal
Yanxue ting 艷雪亭 Bright Snow Pavilion
yao men 腰門 waist gate
Yao pu 瑤圃 Jade Orchard
Yiliang ting 宜兩亭 Pavilion of Two Suitabilities
Yingtian shi 營田使 Commissioners of Agriculture
Yiqing chi 挹清池 Ladling Purity Pool
Yiqing tang 挹清堂 Ladling Purity Hall
Yiyan chu 怡顏處 Flushed with Pleasure Place
Yiyu xuan 倚玉軒 Leaning Jade Veranda
Yiyuan tai 意遠臺 Thoughts Afar Terrace
yong qian 詠錢 Odes on Money
Yongding si 永定寺 Forever Pacification Temple
Yu jian 玉澗 Jade Ravine
Yu quan 玉泉 Jade Spring

yuan ji 園記 garden record
Yuanming yuan 圓明園 Garden of Perfect Brightness
Yuanxiang tang 遠香堂 Far Fragrance Hall
yubei ting 御碑亭 imperial stele pavilion
Yudai 玉帶橋 Jade Belt Bridge
Yudai 玉帶河 Jade Belt River
Yugong tang 寓公堂 Accommodating Master's Hall
Yuhua ji 雨花集 Collection of Rain Flowers
Yuhua tang 雨華堂 Rain Flower Hall
Yujian chi 玉鑑池 Jade Mirror Pond
Yunquan an 雲泉庵 Clouds Spring Cloister
Yunxiang xuewei ting 雲香雪蔚亭 Fragrant Snow and Splendid Clouds
    Pavilion
Yushan caotang 玉山草堂 Jade Mountain Thatched Hall
yushu lou 御書樓 imperial book tower
Yushui tongzuo xuan 與誰同坐軒 Who Sits Together with Me pavilion
Zaisheng suo 宰牲所 slaughterhouse
zhai 斎 students' residence
Zhanhua lou 湛華樓 Profound Lustre Loft
Zhangwu zhi 長物志 *Treatis on Superflous Things*
Zhaoqing si 昭慶寺 Illuminous Celebration Temple
Zheng da tang 正大堂 Righteous Grand Hall
Zhengyi shuyuan 正誼書院 Rectify Meanings Academy
Zhenli ban 珍李坂 Precious Plum Slope
Zhibai xuan 指柏軒 Point to Cypress Hall
Zhiping jiao si 治平教寺 Peaceful Governance Teaching Temple
Zhiqing chu 志清處 Purifying Will Place
Zhongfu an 種福庵 Plant Benevolence Cloister
Zhongshan shuyuan 鐘山書院 Zhongshan Academy
Zhongwang fu 忠王府 Mansion of the Loyal Prince
Zhou Li 周禮 *Rites of Zhou*
Zhu gu 竹谷 Bamboo Valley
Zhu jian 竹澗 Bamboo Gully
Zhuangyuan fang 狀元坊 First Scholar Arch
Zhufang jiangjun fu 駐防將軍府 Garrison General's Mansion
Zhuiyun feng 綴雲峯 Linked Clouds Peak
zhuoying chu 濯纓處 Wash Tassel Spot
Zhuozheng yuan 拙政園 Unsuccessful Politician's Garden
Zisheng xuan 自勝軒 Overcome Self Gazebo
Zunjing ge 尊經閣 Revere the Classics Pavilion
Zuo zhi shi 作之師 Teachers of Deeds

Index

For Product Safety Concerns and Information please contact our EU
representative GPSR@taylorandfrancis.com
Taylor & Francis Verlag GmbH, Kaufingerstraße 24, 80331 München, Germany

www.ingramcontent.com/pod-product-compliance
Ingram Content Group UK Ltd.
Pitfield, Milton Keynes, MK11 3LW, UK
UKHW022330100726
473146UK00010B/718